FOOD SCIENCE AND TECHNOLOGY

FOOD SCIENCE AND TECHNOLOGY

— *Editor* —
Bhupendar Singh Khatkar

2026
Daya Publishing House®
A Division of
Astral International Pvt. Ltd.
New Delhi - 110 002

First Published, 2007
Reprinted 2026

ISBN 978-81-7035-422-2 (Hardbound)
ISBN 978-93-5130-644-3 (International Edition)

Publisher's note:

Every possible effort has been made to ensure that the information contained in this book is accurate at the time of going to press, and the publisher and author cannot accept responsibility for any errors or omissions, however caused. No responsibility for loss or damage occasioned to any person acting, or refraining from action, as a result of the material in this publication can be accepted by the editor, the publisher or the author. The Publisher is not associated with any product or vendor mentioned in the book. The contents of this work are intended to further general scientific research, understanding and discussion only. Readers should consult with a specialist where appropriate.

Every effort has been made to trace the owners of copyright material used in this book, if any. The author and the publisher will be grateful for any omission brought to their notice for acknowledgement in the future editions of the book.

Published by : **Daya Publishing House®**
A Division of
Astral International Pvt. Ltd.
– ISO 9001:2015 Certified Company –
4736/23, Ansari Road, Darya Ganj
New Delhi-110 002
Ph. 011-43549197, 23278134
E-mail: info@astralint.com
Website: www.astralint.com

Digitally Printed at **Replika Press Pvt. Ltd.**

Foreword

Food processing industry in India is considered sunrise industry, as our country is rich in diversified agri-resources and offers vast market to the food processors and traders. Smaller family size, women entering into the work force, urbanization, large proportion of younger population, increasing knowledge and concern about hygiene and health are favorable factors for the rapid growth of Indian food processing sector in coming years. The food processing industry as a whole is expected to grow by 15-20 per cent per year during the next decade. To achieve this growth, the farmers, the processors, the academicians and the national agencies have to work together to develop, innovate and continuously improve food products and processes. The synergy between the farmers and the processors will lead to major agro-industrial revolution in India. This will definitely make the processed food products affordable by the common man and the Indian consumer will have access to variety of hygienic and nutritious foods.

The book on Food Science and Technology has been published at such an appropriate time when Indian food processing industry is on its rapid growth path. This book is designed to offer useful information to the food manufacturers, professionals, consumers and students. The editor B.S. Khatkar deserves appreciation and compliments in bringing out this book by careful selection of topics and authors and nicely editing the book. The book is classified into six parts covering 52 topics. Each part comprises of a number of interesting chapters covering different aspects of food science and technology. It provides interesting and updated information on processing of cereals, legumes, fruits and vegetables, meat and meat products, milk and milk products, food biotechnology, food packaging, nutritional quality of foods, food additives and emerging techniques of food processing and preservation.

In summary, the book on Food Science and Technology is well written and informative. Each chapter is well contained. The main strength of the book is that it provides a comprehensive review of various interesting topics of food science and technology. This book is a valuable reference tool for researchers, students and professionals engaged in the area of food science and technology.

Dr. R.P. Bajpai

Vice-Chancellor
Guru Jambheshwar University
Hisar - 125 001, Haryana

Preface

The food processing industry in India is underdeveloped. We process only 3-4 per cent of raw agri-resources into value added products. On the contrary, the value addition on raw foods in developed countries is in the range of 80 to 85 per cent. The main reason why the food processing sector remained underdeveloped so far was the existence of various food laws and multiple authorities. Other reasons are the lack of infrastructure, poor power supply, inadequate transport facilities and inefficient and inadequate post-harvest food handling, processing, distribution and storage facilities. Recently, Indian government has taken some steps to support growth of food processing industry. The Ministry of Food Processing Industries, Government of India has proposed a new Food Safety and Standards Bill to encourage manufacturers and retailers of processed foods. The Bill proposes a powerful food safety and Standards Authority of India, which will be the single window agency for setting rules and regulating the sector. Efforts are also underway to establish food parks in the country to enable entrepreneur to set up cluster of food industries. Establishment of centre of excellence for R&D and testing laboratories are also required to develop and manufacture globally competitive processed foods.

It is equally important to establish and support the training institutes for preparing the technical manpower to manage production and quality of the processed foods. The specialized short-term courses and the books are the medium through which the professionals and students can avail opportunities to learn about appropriate technology and ward off obsolescence.

The book on Food Science and Technology is written to acquaint the readers to various functional areas of food science and technology. Greater emphasis is focused on processing of grains, milk and milk products, meat and meat products, food biotechnology, food packaging, nutritional quality of foods, food additives and emerging techniques of food processing and preservation. The book is divided into six parts and covers 52 chapters. Part I deals with grain processing consisting of eight chapters. It covers convenience foods, extrusion cooking technology, status of flour milling, structure and functionality of wheat gluten proteins, processing of legumes, coarse grains and identification of cereal foods using protein and DNA profiling techniques. Part II gives a detailed account of milk and. milk products. This part covers advances in the production and preservation of traditional milk products, milk concentration by reverse osmosis, dehydration of milk, quality assessment of milk products, quality management in dairy industry and advances in milk processing. Part III is devoted to fruits and vegetable processing. It covers fifteen chapters. This section discusses several important aspects of fruits and vegetables such as the post harvest handling and management, packaging of fresh and processed products, freezing and dehydration, nutritional and medicinal value, shelf life enhancement and potato chips quality. Part IV covers in detail the use of antioxidant vitamins and modified atmosphere packaging in meat and meat products. A detailed account of food additives is presented in Part V of the book. This section deals with role of enzymes, antioxidants and biopreservatives in processing of plant foods. It also covers functional properties of foods and explains ways to remove antinutritional and toxic factors from foods. Several other important chapters are covered under Part VI of the book. In this section sixteen chapters are included covering interesting topics such as status of food processing industry in India, processing, packaging, nutritional and medicinal value of honey, nutritional and medicinal value of mushroom, utilization of food industries wastes, evaluation of pesticide residues in foods, integrated pest management in stored grains, nutraceuticals and its implications on human health, role of dietary fibre in human health, and management of food processing units.

The book can be used as a valuable reference text for the undergraduate and post-graduate level courses in the discipline of food science and technology. The book is also useful for the teachers and food industry professional for understanding various important aspects of food processing. Finally, I express my appreciation to the contributors of this book for providing chapters well in time and for their valuable contributions.

B.S. Khatkar
Department of Food Technology,
Guru Jambheshwar University
Hisar - 125 001, Haryana

Contents

Part IV: MEAT AND MEAT PRODUCTS

Part V: ADDITIVES

Part VI: GENERAL

Part I
GRAIN PROCESSING

1

Convenience Foods from Cereals and Legumes

S.S. Arya
B-431, Kendriya Vihar, Sector 51, Noida (U.P.)

The term "Convenience Foods" is used for variety of heterogeneous group of products, which vary in composition, shape, size, method of preparation and processing and even with regard to their function in the diet. These literally range from simple fried to roasted nuts, to ready mixes to canned and frozen foods to sophisticated warm and serve type dinners. Convenience foods can be defined as those products in which all or a significant portion of their preparation has been transferred from the consumer's kitchen to the processing plant. The major emphasis in this definition is on significant and should provide adequate savings in time and labour in the consumer's kitchen.

All over the world during the last three decades, the convenience food market has witnessed breath-taking changes in quality and quantity of products available and the packaging and technology employed for their processing. In India too, multinational companies in attractive packages now market many of the products, which were hitherto marketed by food service establishments or small-scale artisans. Though the market has not grown to commensurate with the population, the changes are quite impressive keeping in view the low level of industrialization. In the present chapter emphasis has been laid on the factors governing the quality of the products and the technological constraints encountered in large scale production and marketing of convenience foods.

Shelf-stable Fried Products

Fried products form the largest group of convenience foods marketed in India. A large number of fried items are made from flours of wheat, Bengal gram and black gram throughout the country. Frying not only aids the cooking, but it helps in the destruction of antinutritive factors, aids moisture evaporation and thereby imparts shelf stability. Above all, it imparts crispness and fried aroma, which make the fried products highly acceptable. The moisture contents in these products range from 2-6 per cent and products equilibrate to 0.30 a_w. The fat contents vary from 25-37 per cent. As a result of low moisture, the products do not support proliferation of microorganisms and products remain stable for long periods provided storing in moisture proof packaging prevents moisture ingress. The products are highly susceptible to auto-oxidation of fats and concomitant development of off-flavours and rancidity.

Both sweet and savory fried products are very much liked and consumed throughout the country. Despite their high popularity, their commercial marketing has not taken place to a large extent. Only recently a few firms like M/s Haldiram have started marketing some of the fried products in attractive packages at international level successfully. Also high fat content in fried products seems to be impediment as high income groups have become conscious of the calorie intake and obesity. There is need to develop fried products with lesser fat contents without affecting their crispness. Use of lesser fat contents will also reduce their cost of production and thereby improve their market spread. Development of low cost fried foods with higher proteins will also enable their introduction in mid-day meal school programmes as these can be easily distributed among children. A few traditional fried products are discussed below:

Namkeenparas and Shakarparas

These are salty and sweet products prepared from stiff wheat doughs, vanaspati ghee, salt or sugar. The dough is flattered and cut in to desired shapes and fried in oil at 140-160°C. The moisture contents in these products vary from 2-5 per cent and fat content from 27 to 37 per cent. In an another method plain fried paras without salt or sugar are added to concentrated sugar syrup and mixed to enable sugar syrup to dry up. These contain higher moisture contents and are more prone to mould attack. Thakur and Arya (1990) have reported that Shakarparas and Namkeenparas remain stable for one year if packed in paper–Al. foil–polyethylene packs.

Fried Products from Bengal Gram

These are salted and spiced crisps in various shapes prepared from Bengal gram flour dough by extruding through orifices and deep fried in vegetable oils. Products termed Sev, Ganthiya, Papdi, Boondi are highly relished as snacks and are also used in the preparation of curries from curd. Due to rising cost of Bengal gram flour, some attempts have been made to replace part of Bengal gram flour with defatted soyflour and maize flour with encouraging results (Murthy *et al.*, 1980). Horsegram flour is also used to prepare Bhujia. Despite good internal demand and also good export potential, progress in automation of the process is rather slow. Mechanized extrusion of the dough seems to harden the dough and decreases the fat absorption and crispness of the Bhujias.

Fried Products from Rice and Legumes

Chakli, Murruker, Tengolai, Muchorai and Kodbale are some of the popular fried products prepared from rice and legume flours. While Chakli and Tengolai are based on rice and black gram in the ratio of 4: 1 and 3: 1, respectively. Muchorai is prepared from rice and green gram dhal flours in nearly equal proportions and Kodbale is prepared from rice flour and roasted Bengal gram flours having salt and spices and optionally coconut powder. Stiff dough is extruded through a hand-operated press and fried. Crispness and soft texture are the most desirable attributes and these are mostly governed by the dough composition and frying temperature.

Fried Dhals

Fried dhals and whole legumes are prepared by soaking in water along with salt and sodium bicarbonate, frying in vegetable oils and mixing with spices. Bengal gram, greengram, blackgram and horse gram dhals are most frequently used. Fried dhals are highly crisp. Storage studies conducted in various packaging revealed oxidative off-flavours as the major cause of storage deterioration though increase in peroxide value did not correlate with the off-flavours. Sprinkling with antioxygenic salt recently developed (Arya *et al.*) will definitely help in the enhancement of shelf stability without much addition in the cost.

Fried Moist Products

Products like Samosa, Cutlets, Vada, Pakoras, Kachori, Bhaji etc are consumed as snacks very widely in India. These are prepared and marketed by food service establishments or confectioners or street vendors. Total sales of these products may surpass any other snack food but because of their high moisture content (15-40 per cent), they are susceptible to mould growth, ropiness and fermentation. The products continue to be produced by the labour intensive traditional methods. Some plants have been designed for the automatic production of samosas in western countries but these have yet to reach in India. There is also lack of information on the kinetics of heat and mass transfer in these products during frying operation.

Popped and Puffed Cereals

A variety of popped or puffed cereals like popcorn, puffed rice (Kheel, aralu, nalpuri), puffed sorghum (Kheel), puffed ragi and puffed barley (Dhani, Satu) are marked and consumed as snacks by all segments of population. While popcorn is popular in urban elite, other puffed cereals are more popular among middle and lower middle segments of the population in regions where the particular cereal is grown. Also, while all types of popcorn are produced in electrically operated mechanized poppers and marketed in polyethylene packs, the other popped cereals are manufactured by small scale artisans by heating conditioned cereals in hot sand and marketed in loose by shopkeepers. All popped cereals are highly crisp, fragile and except ragi and barley, have fluffy white attractive appearance. But they are highly hygroscopic and tend to absorb moisture and become leathery, chewy and soggy- a drawback in the traditional method of marketing. Blending them with different flavours and marketing them in moisture proof plastic pouches will provide enormous opportunities for increased acceptance and usage of puffed cereals.

Among other factors like colour, flavour, texture and other subjective characteristics, the expansion ratio *i.e.* the relative amount of Kernel expansion on popping is given the largest to weightage in selecting varieties for preparing popped cereals. The relative expansions of various cereals on puffing along with optimum moisture and temperature of puffing have been studied. For efficient puffing, the temperature just before explosion must be sufficient enough to create a high water vapour pressure without burning the pericarp. The temperature increase must be fast enough to build up the required pressure before the water evaporates. Moisture content of the Kernel has a pronounced effect. The Kernels which are too dry pop-up better.

In the preparation of puffed rice, the paddy is equilibrated with moisture and then roasted in sand. The maximum expansion ratio has been observed to be only around 15-fold while optimum moisture varied between 13.5 to 14.5 per cent and optimum temperature was found to be 190-210°C for sand puffing and 215-235°C for air puffing. Soaking paddy in salt solution improved expansion but parboiling and pressure parboiling considerably reduced puffing. Though Antonio and Julanio (1973) have reported that high amylase varieties puffed poorly, other workers could not confirm the above findings. Puffed ragi is not as widely used as puffed rice and popcorn. Some work has been reported on the puffing quality of various varieties of ragin vis-à-vis their physico-chemical characteristics. Wide varietal variations in puffing quality of ragi have been observed but no consistent relationship could be found with grain amylase, protein content and bran thickness.

Despite wide range of puffed cereals in our country, data on their nutritional values and flavouring components are scanty. Also, no efforts have been made to mechanize their large-scale production and market them in moisture-proof packaging. Major constraint seems to be the non-availability of low cost roaster. Developmental effort in this direction as well as exploration of possibility of marketing puffed cereals with innovative flavours will go a long way in increasing their acceptance by a wider section of population.

Expanded Cereals

Expanded rice (murura, puri, muri) is another traditional convenience food widely consumed in India either as such or with

Jaggery, roasted Bengal gram and shredded vegetables and spices. The product is mostly produced in cottage and small scale sector by skilled artisans. In the traditional process, the paddy is soaked in water, preferably overnight until saturation, drained and then either steamed or dry roasted in sand for parboiling. The parboiled paddy is milled, salted and again roasted in sand for expansion. Recently, pressure-parboiling method has been tried in which paddy is soaked for short periods, drained and steamed under pressure to bring about gelatinization. The parboiled rice is then roasted for sand after salt and bisulphate treatment. At present dry heat parboiled rice is in vogue under commercial conditions. Switching over to pressure parboiling treatment will involve establishing additional boiler and retorting facilities involving considerable amount of additional expenditure. Also, pressure parboiling treatment (2.5 cm^2 for 15 min.) results in amber coloured parboiled rice and consequently yellowish colour of expanded rice. Among other factors, the moisture content of parboiled rice, roasting temperature, rice to sand ratio and rice variety also influence the expansion ratio.

Expanded Wurled Wheat

Morgan and Graham (1967) have reported that wurld Wheat prepared by lye peeling and acetic acid bleaching can also be expanded considerably by equilibrating to 20 per cent moisture and then drying in a sprouted bed dryer. In this dryer, air at 400-500°F and at 100 ft/sec carries the heat to the fluidized bed of grain. Expanded wurled wheat takes much shorter time for cooking compared to wurled wheat. At present biggest constraint in upgrading the quality and production of expanded cereals seems to be the non-availability of a suitable low cost mechanized roaster and lack of suitable packing for retail marketing. The present method of production is laborious and is not easily amendable to large-scale production and the product is liable to contamination under retail marketing. Therefore, the product does not create enough confidence of quality product among consumers.

Beaten Rice

Flaked or beaten rice (Poha, Avallaki, Chivada) is also a very popular traditional product which is consumed either as snack after toasting or frying and spicing or after soaking in water and seasoning spicing and vegetables as an item of breakfast. Beaten rice is consumed relatively in larger quantities in western or southern parts. Thinner flakes having whitish colour are preferred as compared to shorter and thicker flakes with darker colour. Commercially beaten rice is prepared by soaking paddy in warm water (40-50°C) for 18-24 hours. After draining off the soak water, the paddy is roasted in sand (220-240°C) and flaked in edge-runner flaking machine. The yield of the flaked rice is rather low (61-64 per cent) and some attempts have been made to increase the yield by altering paddy curing and flaking conditions. There is scope for reduction in breakage during manufacture, and transport besides improvement in colour, flavour and texture. Some progress has been made in the manufacture and sale of mixed snacks from fried beaten rice and nuts and legumes.

Extruded Foods

In India many extruded foods are prepared either at home or by skilled artisans from rice or rice and legume flours and served as snacks. These are mostly prepared by cooking the rice flour with water and extruding the gelatinized mass through a hand operated press in the form of noodles, chords or ribbons and drying them in sun. Dried products remain stable for months and just before serving, these are fried in oils resulting in many-fold expansion and a soft and crisp texture. Alternatively, extruded foods are also prepared by making a dough of rice and legume flour (blackgram, greengram or Bengal gram flour) and extruding in hand operated press into different shapes and directly frying in vegetable oils. Fried products as already stated remain stable for 1-4 months depending upon the nature of frying oil used. Both the types of extruded products are generally prepared in homes but confectioners and street vendors also market small quantities. With the advent of mechanical cooker extruders, a few companies have started marketing extruded pellets, which need frying before use and ready to eat expanded and spiced products in attractive packs. Both the types of products are receiving good consumer response in the market.

Extruded Pellets

These are double extruded products prepared from wheat, rice or maize flours or their blends. After proper conditioning and equilibration the ingredients are cooked in a primary extruder. The cooked material is then cooled, densified and shaped on a secondary extruder. In the secondary stage, the temperature of the extrudate is kept below boiling point of water to eliminate product expansion. Immediately after extrusion, the moisture content in pellets is around 20-25 per cent which is brought down to 7-8 per cent by drying in a tunnel dryer. The dried pellets are packed in attractive moisture proof packs of plastic films or aluminized polyester-polyethylene packs. Before consumption these have to be fried in oil which imparts them crispness and softness and about 4-10 times expansion. The products are not fragile and therefore there is no breakage in transportation and marketing. They are also less susceptible for autoxidation and off-flavour developments during storage.

Ready to Eat Extruded Products

These products are manufactured in high pressure cooker extruders. The products processed by this method, such as crax or corn curls or balls are merely toasted and dried and then coated with flavour or sugar either by dusting or by spraying. Advantages of this type product are that the level of the oil/fat in the product is kept low and the products are ready to eat and can be marketed in different designs having different flavours.

Corn and Tortilla Chips

Corn and tortilla chips are popular snacks in American continent. Both these products are prepared from corn which is cooked with dilute solution of lime by bringing to near boiling. The cooked corn is then cooled and steeped under water for 8-16 hours. The cooking and steeping process loosens the outer hull and germ, which are removed by washing. The washed kernels are ground between milling stones to produce a mass having 50-60 per cent moisture. To make corn chips, the mass is extruded through a slit to give ribbon of dough which is cut-off by a knife. These fall directly in hot oil and deep-fried to a moisture content of around 2-4 per cent. Tortilla chips are produced by rolling the mass into a thin sheet, cutting into desired shapes and finally the chips are transferred into open flame. The heat sets the structure, drives offs the moisture and generates a milled caramelized flavour. The chips are then briefly fried to create a crisp texture, salted and packed. Being low in moisture these are susceptible to rancidity development during long term storage. Though Indian public is not accustomed to these products, the technology of manufacture is simple and can be adopted even in cottage sector.

Fermented Products

Idli and Dosai are the two most popular fermented foods used either as snack or in breakfast especially in southern parts. These are prepared from fermented batters prepared from rice and black gram. At present these are prepared in homes or food service establishments. Dhoklas are prepared from rice and Bengal gram batters in Maharashtra and Gujrat. There have been some attempts to preserve Idli in ready to eat state by thermal processing in cans and retort pouches (Kadkol *et al.*, 1983). However, this concept has not caught up and there is hardly any demand for thermally processed Idli and Dosai.

Secondly, attempts have also been made to develop instant mixes for Idlis and Dosais. The concept of instant mixes is quite attractive and many firms are already marketing instant mixes for Idli, Dosai and Dhoklas. However, the Idlis and Dosais prepared from these mixes are not soft and fluffy and also lack fermented flavour. Some attempts have been made to characterize the microorganisms responsible for fermentation of Idli batter. Surprisingly, even after characterization of microorganisms involved in fermentation of Idli batter, no attempts have been made to develop instant mixes having lyophilised or micro-encapsulated cultures. This aspect definitely needs more attention.

Thirdly, in fermented batters surface active proteins and polysaccharides are well conditioned to retain large volume of gases to give soft fluffy texture. In the ready mixes, this conditioning does not take place and additional hydrocolloids may help in providing support to gel structure in retaining gases generated by the action of chemical leavening. This aspect also needs further evaluation.

Traditional Sweets

There are many traditional sweets prepared from grains, which are consumed as snacks besides biscuits and cookies. Prominent among these are Chikkies, Gajjak etc which are prepared from puffed or expanded rice or Bengal gram, roasted ground nuts and Jaggery. When freshly prepared these are crisp and dry and have a desirable caramelized flavour. These are mostly prepared by small-scale vendors and marketed in the form of slabs, laddus or bars without any packaging. Moisture contents in these products vary from 4-8 per cent and the products are highly hygrospic and tend to become soggy and sticky besides possibilities of microbial contamination (Arya and Kalpatri). The product is a concentrated source of protein and calories and is well suited for mid-day school programmes. At present these are mostly prepared manually in small batches by preparing Jaggary syrup, mixing with roasted grains and nuts and rolling and cutting in to small slabs. Control of syrup concentration is very critical in maintaining the product crispness. Design of suitable machinery for mechanized rolling and cutting is desirable in upgrading their technology.

Sweets like Laddu, Boondi, Jalebies, Imarti and Jhangiri are prepared by deep fat frying Bengal gram, blackgram or fermented wheat flour batter after passing through the orifices to impart different shapes. The fried products are dipped in sugar syrup for varying periods of time to impart sweetness. These products have moisture range of 15-20 per cent and equilibrate between 0.70-0.85 water activity. Because of intermediate moisture range, these are susceptible to mold and yeast spoilage and remain stable under ambient temperature only for a few days. Incorporation of sorbate or benzoate in syrup would definitely help in extending their shelf-life. Non availability of continuous deep fat fryers and machinery for shaping seems to be the major constraints for organized growth in their production and in bringing about uniformity in quality. Some more sweets like Mysore pack, Besan Burfi, Firni, Chiroti, Sohan papdi and Halwas are also important items among traditional sweets from grain products. Information on their composition and causes of spoilage is rather scanty. Recently, Sohan Papdi has been introduced for export in attractive packing.

Papads

These are thin wafers like products prepared from a variety of base ingredients such as black gram, Bengal gram, greengram or their combinations with or without spices. There are mostly prepared at homes or in cottage scale industry. A stiff dough made from legume flour is rolled into thin round discs by rolling pins and dried to a level so that it still remains pliable. The moisture content in papads varies between 14-18 per cent and their water activity is quite near to critical limits of mold growth. Before consumption, papads are either deep fried or toasted on open flame. Recently there has been considerable growth in papad production because of their demand from overseas market. Mechanization of the unit operations of dough kneading rolling, and drying have also been attempted. Though technologically desirable, complete mechanization of papad production may displace thousands of unskilled woman workers from grainful employment. Therefore, it would be more desirable to design semi-manual gazets by which quality of the product and productivity of the worker and their earnings get enhanced without displacement of labour by machine.

In addition to the legume based products, papads are also made from tapioca, sago, jackfruit seeds, gelatinized rice and ragi flavours. Even wheat flour is also used for preparing papads.

Instant Mixes Based on Chemical Leavening

Instant Gulab Jamun, Cake, Pancake, Idli and Dosai, Pakoda, Vada, Bonda and Dhokla mixes belong to this category. These products are prepared by simple blending various ingredients in required proportions along with chemical leavening in a planetary or any other type of dry mixer. Control of granularity, moisture and adequate mixing are essential for product quality and shelf-life. Infestation is one of the major problem faced by most of the manufacturers. Heating at 60°C in a cabinet dryer has been found to be useful in controlling infestation as well as caking and lumping. Low dose (50-100 Krads) in pack irradiation has also proved useful in controlling infestation in mixes. Among various mixes, Gulab Jamun mix is most popular and has the largest sale.

Instant Mixes Based on Precooked Dehydrated Products

Precooked dehydrated convenience foods like instant Pulav, Khichdi, Bisibelebdat, Curried dhals, Rice, Curried Chholey, Halwa, Rasam and Sambhar, Rava Idli and Upma mixes are prepared by precooking of ingredients blending with spices, salt and fat. Depending on the dehydration method and the nature of ingredients, these mixes can be cooked in 2-15 min. Reconstitution of these foods, by mere mixing with cold water has remained the goal of technologists in the area. Freezing or flaking before drying has been found to be highly effective in reducing the reconstitution time of precooked and dried rice and dhals (Patki and Arya, 1994). Recently, "no time" cooking curried dhals which can be reconstituted by mere addition of hot water have developed by Patki and Arya. These have been tried by troops and have been highly liked by them.

Instant Kheer mixes based on roasted semolina, dehydrated rice and wheat vermiciti have also been developed but their commercialization has yet to take place. But procurement of 3-4 lakh pack rations based on these instant foods has catalyzed their large scale production by industry. Potential for instant soup mixes having different flavours and body consistencies is enormous in addition to specific mixes of Indian curries like Chholay, Aloo-chholay and Peas-paneer and vegetable–legume vadi.

Some attempts have been made to develop instant cereal products by freeze drying of prepared whole dishes lemon rice, peas-paneer and wheat porridge mainly for space expeditions where temperatures above 70°C are not obtainable for reconstitution. Because of highly porous structure of freeze-dried products, these can be reconstituted in packages by the introduction of required amount of warm water with the help of a pressure syringe and leaving them for 10-15 min. after which the products will be ready for consumption. These products are however, very costly due to large investment on freeze drying plants and long drying periods.

Ready to Eat Products Stabilised by Antimycotic Agents

Chapati, Parantha and Poori are most popular traditional Indian foods but these have hardly received attention they deserve. These are mostly prepared in homes or in food service establishments. Their organized production and marketing has not taken place due to non-availability of chapatti making plants.

Freshly baked chapaties are highly susceptible to fungal growth and ropiness and also become brittle due to starch retrogradation. Despite many attempts, a satisfactory method of chapatti stabilization emerged when a combination of sorbic acid pH control and in pack heating at 90-100°C was used (Arya *et al.*, 1977). Using this method chapaties could be preserved for more than 6 months under ambient conditions. Long keeping chapaties have also received good acceptance from troops who are being supplied long keeping chapaties regularly in the rations of marine commandos and MRF rations for Army and Navy. Considerable efforts have also been made in the characterization of flavouring compounds in chapaties and their stabilization in long keeping chapaties. Permission to use sorbic acid in long keeping chapaties has also been accorded under PFA.

Some attempts have also gone in to develop machineries for mechanized production of chapaties and paranthas. Prototypes models of leg operated stamping machine for chapatti rolling has given encouraging results during initial trials. Design of a baking oven and synchronization of baking and rolling operations still remains a cherished goal. It is rather surprising that this problem has not attracted machinery-making industry.

Bread, bun and cakes are other convenience food prepared from wheat flour and marketed in ready to eat state. This industry is growing at around 8 per cent annually. Though both sorbet and propionates are permitted under PFA, only propionates are used by baking industry to achieve a shelf-life of 3 days. On the 2g sorbic acid per square meter and heating at 90°C in polypropylene packs has been found to keep bread upto one month. In cakes, sorbic acid is permitted in higher concentration (0.15 per cent) which ensures considerably longer shelf-life.

Retort Pouch Foods

In India work on retort pouch processing started in 1966 to preserve stuffed paranthas which could not be preserved by incorporating sorbic acid or other antimycotic agents. Earlier attempts were made to sterilize stuffed parothas by packing in cellophane which was further encashed in cost polypropylene pouch. However, because of delicate nature of the polypropylene film, pouches required support during retorting and during further handling. Therefore, the polypropylene pouch was held in an aluminium mould for keeping it in a upright state. The products like Rajma in Sauce, Suji Halwa, Potato-peas curry and Koftacurry could be processed by this method. Because of thin geometric profile of the retort pouch, the quality of the retort pouch processed food was far superior than the corresponding canned foods especially solid and semi-solid products because of less thermal damage during retorting. The retort pouch processed Suji Halwa and potato-peas curry has undergone extensive consumer acceptance trials and has been highly liked. These have been supplied in Antartic Expedition and mountaineering Expeditions. Only problem faced in polypropylene based retort pouches is the heavy loss of volatiles and changes in colour and flavour as a result of high oxygen transmission rate of the polypropylene. Recently with the availability of polyester–Aluminium foil-polypropylene (cast) laminate, this problem has been overcome successfully. The technology of retort pouch processing of a large number of legumes based curries, suji-halwa, pulav and palak-paneer curry with a shelf-life of about one year. The industry has also started marketing retort pouch foods both in Indian and export markets. Special retorts for this purpose have also been indigenously fabricated. The only problem faced by the industry is the non-availability of a suitable form-fill-seal machine for handling particulate and liquid products.

Canned Convenience Foods

Besides processing in retort pouches, some attempts have also been made to preserve some of the traditional foods like Upma, Kheer, Halwa, Parotha, Chapaties and whole meal Parothas, Subji and dhal curry Idlis, Pulav (vegetarian and non-vegetarian) peas-paneer curry in cans mainly for supply to troops. Among these canned Upma and Kheer were regularly supplied for Antartica and other expeditions. But the quality of instant Upma and Kheer has improved considerably, the supply of canned items has been discontinued. Though considerable efforts have gone in to optimize the canning of parothas, Idli and whole meal, these products did not find acceptance from troops due to excessive caramalisation and texture hardening during canning operation. Nevertheless these studies do provide a means to preserve foods for troops for emergency situations though these foods may not find utility in commercial sector mainly because of high cost of cans and relatively poor quality of the processed products.

Though canning of rice has posed many problems due to formation of congealed mass of pasty grains after retorting, this problem is not faced in thermal processing of Pulav. High fat content in Pulav prevents congealing of rice grains. Addition of tomato puree and 10 per cent fat and equilibration of rice to 35 per cent moisture before cooking seems to improve the texture and flavour of the stored product. Similarly, coating of Idlis with fried Bengal gram powder has been found to improve the texture of canned Idlis.

Breakfast Cereals

Important products commercially marketed as breakfast cereals include Dalia (Farina), rolled oats and corn and wheat flakes. While Dalia and rolled oats require cooking before use, the corn and wheat flakes are ready to use. In India, however, the use of corn and wheat flakes is limited to affluent sections because of their high cost.

Dalia

Dalia or farina is nothing but middling fraction obtained during roller flour milling. These are chunks of wheat endosperm free of bran amid germ. In the manufacture of Dalia, it is beneficial to use hard wheats since Dalia milled from soft wheats becomes excessively pasty. An instant porridge based on precooked Dalia and milk solids has also been developed for astronauts.

Rolled Oats

In the manufacture of rolled oats, the cleaned grains are heated to 100°C for one hr. This reduces the moisture of about 6 per cent and partially dextrinises the starch. In addition, the roasting step renders the hulls fragile which can be easily removed by passing through a pair of circular stones. The dehulled grains (grouts) may be flaked directly or these may be first cut into pieces by rotary granulators. Thinner the flake, shorter the time it takes for cooking. The quick cooking oats are produced from ½ to ¼ the size of the whole grouts and these require about 5 min. boiling. Recently, the oat bran has received considerable importance because of its beneficial effect in reducing lipids in blood especially harmful cholesterol.

Cereal Flakes

Flaking is relatively a simple process consisting of cooking whole or fragments of cereal grains, flattening the tempered grains by passing through rollers and toasting the resultant flakes at high temperature to impart caramelized flavour. Both wheat and corn flakes are commercially produced in India but total production is limited because of relatively high cost and limited consumer demand. The most frequently encountered problems are excessively breakage during transportation and oxidative off-flavours generally associated with commercially produced products. Recently breakfast cereal flakes from sorghum have been developed at DFRL Mysore. The product is crisp and reconstitutes well in hot milk. The process know-how is yet to be released to industry, but there is potential for its large scale production because of large production of sorghum in our country and lack of value added product from it.

Other Breakfast Cereal Products

There are many other cereals based breakfast items like shredded wheat biscuits, cereal granules and sugar coated puffed cereals commercially produced in western countries. But these have yet to enter in Indian market.

Conclusions

It is evident from the foregoing account that a large number of convenience foods are traditionally marketed in India either by small scale skilled artisans or by cottage scale industry. Many of these products are prepared at home or served by food service establishments. The technology of their production has remained more of an art and little emphasis has been laid on infusing new technological developments in their large scale production. Though some information has been generated in R&D institutions in scientific evaluation of processing parameters and physico-chemical aspects of the products, R&D efforts in design of equipments and machineries for up scaling of technology have been meager. Also, whatever information has been generated in the research institution has not percolated to industry and lack of interaction between R&D institutions and industries. It is therefore suggested that technological needs of this sector be identified and more stress be laid on design/fabrication of indigenous machinery to upgrade the technological base of convenience and snack food industry. Also there is urgent need to inculcate the hygiene and sanitation consciousness in the industry for creating confidence in the consumers. There is also need for filing patents on the new developments in upgradation of technologies so that small scale industry is insulated from exploitation by multinationals companies and industries from developed countries.

REFERENCES

Antonio, A.A., Julanio, B.G. (1973). Amylose Content and puffed volume of rice. *J. Fd. Sci. Technol.*, 18: 15.

Arya, S.S., Thakur, B.R. (1990) Packaging requirement and storage stability of fried wheat snacks. *J. Fd. Sci. Technol.*, 27(2): 76-81.

Arya, S.S., Kalpatri, M.D. (1973). Survival Rations–A Comparative Evaluation, Technical Report No. SL–P_3–69 DFR, pp. 69.

Arya, S.S. Vidyasagar, K. Pasihar, D.B. (1977). Preservation of Chapaties. *Lebenan. Wiss. Technol.*, 10: 203.

Kadkol, S.B., Ramnathan, D.A., Pitchumutha, P., Bhatia, B.S. (1973) Canning of Idlies. *Indian Fd. Packer*, 27(3): 17.

Morgan, A.L., Jr., Graham, R.P. (1967) Expanded convenience products. *Cereal Sc. Today*, 12: 472.

Natraja Murthy T, Tathi, L.S., Rao, J.K.M., Lakshminarayana, G. (1980) Shelf Life of ready to eat fried noodles. *J. Fd. Sci. Technol.*, 17: 281.

Sharma, G.K., Semwal, A.D., Narasimha Murthy, Arya, S.S. (1977) Suitability of antioxygenic salts for stabilization of fried snacks. 60: 19-24.

2

Applications of Extrusion Cooking Technology

Narpinder Singh* & A.C. Smith**

**Department of Food Science and Technology, Guru Nanak Dev University, Amritsar – 143 005*

***Institute of Food Research, UK*

Introduction

Extrusion is simply the operation of shaping a plastic or dough-like material by forcing it through a orifice or die. Extrusion is a versatile high temperature, short time (HTST) process that has become established for the continuous manufacture of new and traditional products. Extrusion cooking normally involve mixing, melting and application of intensive energy to the product at high pressure and within short period of time. Extrusion cooking applications in the food industry cover a wide range of food products based on starches, cereals, proteins and sugars. Extrusion cooking has the capacity to produce a large variety of products from numerous raw ingredients. It can produce products with a variety of shapes, textures, colours, and appearances, whih is not easily feasible using other production methods. It is often characterized by low production costs resulting from the small area required, reduced energy consumption and low numbers of personnel that are necessary. Saving of raw materials, labor and capital investment of 19 per cent, 14 per cent and 44 per cent, respectively has been reported while using expension process (Darrington, 1987). Extrusion produces little or no waste stream. It has challenged traditional ways of fabricating food products and also satisfied the industry goal of new products. High temperature used during extrusion destroy antinutritional factors, undesirable enzymes. In many cases, the amount of added water can be reduced, which eliminates the costly evaporation of surplus water at a later stage in the process. It can also use cheaper feed stock than conventional processes. The extruder can be used to perform many functions such as dehydrations, agglomeration, gelatinization, mixing, shaping, expansion, homogenization, degassing, enzymes inactivation, pasteurization, etc. The extruder may also be described as a complex biochemical reactor and it is in this mode that the greatest scope for new development exists. Supercritical carbon dioxide extrusion is one of the more recent technological developments to broaden its portfolio of applications.

Equipment

Extrusion can be simply defined as the process of forcing a material through a specifically designed opening. The basis of extrusion cooking is a screw pump using twin or single screws. The single screw extruder is still used in industry and its origins are in pasta manufacture, which is a low temperature, high moisture process (Mercier and Cantarelli, 1986). A single screw extruder may be classified based on the extent of shear generated and source of heat generation. Classification based on the extent of shear described by Farrell (1971) and Harper (1981) include: Cold forming extruders, high pressure forming extruders, low-shear cooking extruders, collet extruders, and high-shear cooking extruders. Classification based on heat generaion include: adiabatic extruders, isothermal extruders, and polytropic extruder. Single screw autogenous or adiabetic extruders generate heat through friction. It is a high-shear machine in which high temperature can be achieved in a short time (less than 20 seconds). These are low cost machine can achieve high compression ratios, temperatures and various degrees of puffing. The single screw extruder has a wide residence time distribution and is generally a poor mixer. The single screw extruder may be operated with a pre-conditioner, which may be used to add water either directly or as steam. There is, nonetheless, considerable interest in developing the single screw extruder as a modular design with mixing elements, such as the cavity transfer mixer. A greater variety of twin screw devices is available with intermeshing or non-intermeshing co-or counter-rotating extruders. The counter-rotating intermeshing twin screw extruder is cited for use with low viscosity materials and generally operates at lower screw speeds than its co-rotating relative such as gum, jelly, and licorice confections (Wiedmann and Strobel, 1987). Counter-rotating extruders are not widely used in food industry although they are excellent conveyors. However, the most commonly employed type is the intermeshing, co-rotating extruder comprising self-wiping, conveying and reversing elements, orifice discs and mixing paddles, which may be assembled to convey forward or backward. In these extruders, the flight of one screw penetrates the channels of the other screw. Liquids and other solid or semi-solid ingredients are often fed into the extruder at different stage along the barrel length. The twin screw extruder has the capability for good mixing and narrow residence time distributions. The versatile design of the screw means that the mixing stage can be synchronized with the temperature zones and the pressure may be designed to follow a complex history involving a section at atmospheric pressure in order to vent the extruder or to inject materials or gases. The most recent development in extrusion technology is to combine it with pressures. The combined "supercritical fluid extrusion" process introduces the supercritical fluid, usually carbon dioxide, carrying soluble micronutrients, flavours and colours into the melt stage of the extruder. It has been demonstrated for breakfast cereals, pasta and confectionery production.

A number of innovations are possible as the material leaves the screw section and passes through the die. Multiple dies may be designed with geometries to reduce pressure more or less rapidly to atmospheric. Co-extrusion in which filler is pumped into the center of a hollow extrudate has been popular for a number of years and has enhanced the product types available through this technology.

The occurrence of flow defects and the viscoelastic nature of food materials dictates that the material exiting the extruder do not confirm to the shape of the die. The loss of superheated water as steam has been the basis of many novel extruded products and in these cases, the extruded diameter is markedly greater than the die section. At its simplest, extrusion cooking may be used as a single stage process to produce pasta or expanded cut pieces. However, most extrusion technology is used with other downstream processes for rolling, crimping, cutting, toasting and coating (enrobing).

Further technological innovations are possible with the twin screw extruder in handling wet materials. The extruder may be used for multiple feeds such as dehydrated potato and potato mash (Ferdinand *et al.*, 1989). Similarly, nuts and fruit slurries may be added at appropriate points to avoid high shear or high temperature.

Cereal and Bakery Products

This is a traditional area where extrusion cooking has made the greatest impact. One of the original developments of the food extruder was to rival the pasta press. High protein wheat semolina is mixed to dough with water and is then formed through dies, followed by appropriate drying. Related processes use maize or dehydrated potato, the latter often in combination with potato starches to form collets or dense shaped pieces, which are dried and may be transported in this state of minimum volume. Expansion to create "third generation" snacks may be carried out using hot air, microwaves or frying, which rapidly heat the collet so that it softens and is expanded as water turns to stream. These products may also be manufactured by "sheet and cut" processes, which involve the formation of rolled sheets from which product shapes are cut. Directly expanded or "second generation" snacks and some expanded cereals are made by direct steam- expansion of the extrudates. The water in the food material dough or melt in the extrude is superheated, but flashes off as steam, when the pressure falls at the die. The food material is sufficiently plastic under elevated temperature conditions that it deforms. Cooling and loss of moisture cause the material to solidify in the expanded state.

One approach, which has been advanced to better understand extrusion processes is to take a biopolymer science approach, which considers the polymeric nature of starch and protein molecules

(Slade and Levine, 1993). Polymers melt in the extruder under hydrated conditions and solidify, when they pass through their glass transition. The glass transition varies with water content and is also lowered by other small molecules such as sugars. It can be concluded that the initially solid material melts in the extruder, but remains sufficiently mobile in the flaking step to be flattened without breaking. Removal of further moisture on toasting, allows ductile deformation of the surface of the flake, causing blistering before the flake solidifies. The final flake owes its crisp or crunchy texture to this loss of mobility. From thermal analysis or calorimetry, it is possible to measure glass transition data for particular compositions (Georget and Smith, 1995).

Expanded ready- to-eat cereals are manufactured from mixtures of cereal flour and starch combined with small amounts of malt, fats, sugars and salt (Harper, 1981). Extrusion can be used to produce directly formed or expanded cereals and also as part of a total process, for example, for making cereals flakes. In the latter case, the extruder produces filaments of slightly expanded cereals, which are cut, rolled and then toasted. The alternative process is pressure-cooking combined with rolling and toasting. While pressure-cooking uses whole grains, extrusion uses grits and flours as feedstock with consequent cost savings. It is clear that the extruder subjects the material to shear in a HIST operation, which contrasts with the conventional shear- loss, comparatively long time process. The conventional and extrusion- based processes can be compared and contrasted at a number of structural levels, which underpin consumer properties in terms of flavour generation and texture, including subsequent bowl life in hot or cold milk. The microstructure of starch affects its solubility and expansion properties. In general, high specific mechanical energy extrusion results in destruction of starch microstructure leading to high expansion and high solubility (Smith, 1992). It is difficult to obtain highly expanded cereals, which have desirable low solubility, which reduces perceived stickiness of these products.

One approach to overcome the problem of expansion while low solubility, is to operate an extruder under low temperature and low specific mechanical energy conditions and to use the injection of gas into the extruder barrel to solubilise it so that it is released, when the pressure falls on exit from the extruder (Ferdinand *et al.*, 1992).

More recently, supercritical fluid extrusion has been developed to produce a structured product under low temperature conditions (Rizvi *et al.*, 1995).

Crispbreads may be produced by extrusion cooking and these are formed into sheets, then creased so that they may be broken cleanly after drying and toasting. This compares with a traditional baking process.

The first part of the brewing process involves mashing of the malted cereal, which gelatinizes the starch and degraded into fermentable sugars by amylolytic enzymes. Malted cereals are relatively expensive and this leads to its supplementation by adjuncts. Extrusion-cooked cereals have potential as brewing adjuncts (Briggs *et al.*, 1987; Wiedmann and Strobel, 1987). Extrusion destroys the starch granular structure so that it is more readily hydrolyzed by enzymes in the mashing process. It has potential advantages over the use of cereal flakes in the high yield of fermentable extract. Expanded rice cakes of fine structure can be produced by extrusion cooking compared to the harder texture and longer processing time typical of conventional hot air drying. Extrusion cooked rice flour has also been reported to improve the quality of *Idli,* a traditional south Indian food (Singh *et al.*, 1995).

Corn- soy blends with optimized nutritional properties have been studied extensively, often using low cost extrusion technology (Harper 1981; Lorenz and Jansen 1980; Jansen and Harper, 1980). Extrusion can be used to treat starches and flours to alter their functional properties, particularly water holding capacity and cold and hot aqueous viscosities. Soup ingredients and infant foods are based on modified cereal and vegetable flours, which have a high cold water viscosity. The extruded material is basically dried, cooled and milled. An alternative conventional process is drum drying, which involves metering the aqueous flour dispersion onto hot rotating drums. The water is vaporized and the flour films removed from the drums by scraping. Again, difference between the two processes may occur at the microstructure level. Drum drying is a traditional starch gelatinization process whereas the starch microstructure may be disrupted by shear in the extruder. The resulting flours may have very different solubility and rheological characteristics, when reconstituted.

Texturised Proteins

One of the early innovations of extrusion technology was to make fibrous meat analogues from soya, which were originally marketed as meat substitutes, but are more commonly used as meat extenders. Many of these extrudates need to be rehydrated before use. More innovatively, long cooling dies are used with extruders to produce structured meat analogues at higher moisture contents (Noguchi, 1989). Injection moulding of this material is also possible in an entirely analogous way to the process used to make thermoplastic products. Whey proteins may be heat- denatured and textured using shear. The extrusion of animal proteins has also been discussed (Areas,1992). The alternative processes use spinning processes. Wet spinning uses high pH solutions of proteins, which are spun into coagulating acid salt baths. Stabilization of the spun fibers may be achieved by heating or the use of cross-linking agents. Dry spinning involves drawing fibers from a protein melt followed by solvent evaporation.

Confectionery

Confectionery items are another important type of products from the extrusion cooking process (Best, 1994). Sucrose and glucose may be processed at low water contents to make boiled sweets and strands, making optimal use of the die shapes and co- extrusion. The residence time must be sufficient to allow melting of crystals and the temperature- time history must be short to avoid formation of off- flavours. The traditional approach involves the energy-intensive vaporization of large quantities of water. Liquorice and fruit gums make use of the venting capability of a twin screw extrude and the downstream addition of flavours and colours. Toffee, caramel and fudges of various types may also be made, replacing batch boiling pans. Aerated confectionery is also possible making use of gas injection.

Chocolate may also be manufactured using a twin screw extruder, making use of volatile removal and addition of cocoa nibs at the full length and of other ingredients nearer the die. The process removes off- flavours and provides sterilization followed by development of desirable flavours. Traditional chocolate production involves roasting of beans, grinding the cocoa nibs to form cocoa mass, which is mixed with cocoa butter and sugar over long periods to produce chocolate.

A related area is the use of the extrusion cooker to produce flavours, for example, by controlling Maillard reactions, roasting and caramelisation (Wiedmann and Strobel, 1987). A comparable process is drum- drying, but the extrusion process uses lower moistures. The Maillard reaction has nutritional implications in loss of available lysine and also changes the colour of an extrudate through non- enzymatic browning.

Extrusion cooking of chewing gum offers advantages of sustained release of flavour over traditional processes. The flavour is added early in the extrusion process to ensure adequate dispersive mixing with the base gum.

Animal Feeds and Pet Food

Extrusion cooking is an excellent method for combining waste feedstock of the production of complete feeds. Dry extrusion is used to process high moisture by-products of the animal, food and marine industry into high quality feed ingredient. They are used to process complete feeds such pig starters, fish feed, dog food, cat food, mink and fox feed, poultry feed and calf starter feed. Fish and meat wastes are often combined with soybean meal. Fish foods with designed flotation characteristics are formed as flakes or pellets. Again, the microstructure may be manipulated to have the appropriate wetting and solubility characteristics to float or sink.

Pet foods production is a major product type of the extrusion cooking process (Rokey, 1994). They encompass hard dry varieties based on cereals, protein meals and fat to semi- moist types with added humectants. Co- extruded and multi- layered pieces and biscuits are refinements of the basic product.

The Extruder as a Reactor

Cationic starches are used in the paper industry because of their affinity with cellulose. Chlorohydrine reacts with starch in the presence of an alkaline catalyst. The highest degree of reaction occurs with the lowest water contents, which may only be achieved in extrusion cookers. Carr and Cunningham (1989) reported the formation of glycol glucosides, which are chemical intermediates in the production of urethane foams, from starch and ethylene glycol by twin screw extrusion cooking. Starch phosphates may be produced with a similar degree of substitution and less reagents, compared to the conventional method (Salay and Ciacco 1990; Chang and Lii 1992).

As much as 80 per cent of the protein in milk consists of the casein complex. Separating casein from milk involves coagulation followed by purification and concentration by spinning and drying. Mildly acidic milk albumen becomes water soluble through neutralization with caustic soda. Extrusion cooking may be used rather than agitator vessels in combination with drum dryers and flaking equipment to form powdered caseinate (Wagner, 1989). Caseinate is used in baking and as a texturising agent.

Sucrose esters are powerful non- calorific fat replacers and extrusion cooking has been cited to implement the necessary chemical synthesis.

The use of thermostable α-amylase in the extrude and mixed to saccharify starches has been demonstrated (Linko, 1989), using alpha-amylose and glucoamylase. This has been suggested as a first step in ethanol production. The possibilities of modifying other renewable resources to materials or ingredients with desirable and functional properties is a future area for growth of extrusion or related processes. Vegetable waste is being examined by extrusion and other processes to provide useful functional ingredients. The extruder offers enormous scope to provide selective physical, chemical and enzymatic breakdown of natural materials. It is perhaps in this regard that the extruder has yet to make its greatest contribution to effect reactions, which rely less on chemicals.

Another application of extrusion cooking in the brewing industry is to treat hops to facilitate the transformation of bittering precursors to bittering components (Westwood, 1994). Hops are introduced into the wort boiling stage of the brewing process in order to impart bitterness to beer. Alpha acids present in hops, which are not particularly bitter or soluble in beer, undergo isomerisation into iso-alpha acids on boiling. The rate of isomerisation is increased under alkaline conditions and hence the extruder is used to react hops and alkaline salts with water addition.

Nutrition, Enzyme Inactivation and Sterilization

All heat treatments cause modification or loss of nutrients. In general, the high temperature of the extruder is destructive to some nutrients, but this is countered by the short time of the process. Dietary fiber is defined by Asp and Bjorck (1989) as the polysaccharides and lignin that are not digested in the human small intestine. Processing affects the physical, chemical and physiological properties of dietary fiber. Extrusion alters the ratio of insoluble to soluble dietary fiber in favour of the soluble type.

Many food plants contain substances, which inhibit nutrient absorption and processes need to remove these anti- nutritional factors. Protease inhibitors are inactivated in the extrusion cooker, which increases the digestibility of the proteins, although this is offset by loss of essential amino acids and cross-linking, which occurs on heating. Lysine reduction is least at high extrusion moisture, but is comparable with that occurring in baking processes. The loss in protein nutritional value is of most importance in infant foods. Extruder can be used to detoxify peanuts or cotton seed through the reduction of aflatoxin from both the seeds and the reduction of the free gossypol from cotton seed.

The young are very sensitive to anti-nutritional factors in soybeans and other seeds. Legumes have substantial toxins and extrusion cooking of bean flour destroys anti- nutrient components and also improves palatability. Beans contain lectins, trypsin and alpha- amylase inhibitors, which are inactivated and in addition, starch and protein digestibility is increased. It is important that mineral availability should not be reduced by processing. Absorption of iron and zinc does not differ between conventionally and extrusion cooked starch- bran mixtures. Most vitamins are process sensitive and the extrusion cooking process destroys them in proportion to the process severity, as might be expected. Enrichment of extruded products with vitamins compensates for many of these losses.

Many enzymes in food materials cause deteriorative reactions during storage, if they are not inactivated in processing. Lipase and alpha- amylase can be particularly thermostable, the latter being exploited in starch liquefaction. Rice bran oil is an important lipid resource in our country. Majority of rice milling plants are small in scale and scattered throughout the country. The collection of rice bran from rice milling plants and its transportation to an oil extraction company takes a long time, and the bran deteriorates during this period, mainly by lipase action. Extrusion of rice bran reduces the endogenous lipase. Oil extraction from rice bran was favoured by preliminary extrusion. In general, oilseeds can be efficiently pre-treated by extrusion cooking prior to solvent extraction. The extruder cook the oilseeds, rupture the cell and partially dehydrates the seeds. Lipoxygenase is also inactivated in full fat soybean extrusion. Extrusion cooking is also used to sterilize spices and pharmaceutical materials.

Extrusion cooking is also used for sterilization of herbs and spices. It has been reported to be offer an alternative method to sterilization by irradiation. Spices like poprika can be extruded under nitrogen over to avoid undesirable changes in colour. Extrusion cooking reduced bacterial count while retaining flavour characteristics of herbs and spices because of short processing time.

Recycling of By-products

Majority of by-products contain high moisture and required to be dried. The blends are extruded and processed to desired moisture and shape in extruder.

Non Food Uses

There is a growing interest in biodegradable plastics. Foamed loose fill packaging material may be produced by more conventional extrusion of starches material as a competitor with non-degradable expanded polystyrene. Starch is also modified chemically and physically in its own right for food and non-food uses and extrusion cooking is exploited for these purposes.

REFERENCES

Areas, J.A.G. (1992). Extrusion of food proteins. *Crit. Rev.Food Sci. Nutr.* 32: 365.

Asp, N.G. and Bjorck, I. (1989). Nutritional properties of extruded foods. In: Extrusion Cooking (C. Mercier, P. Linko and J. M. Harper, eds.) A.A.C.C., St. Paul, USA p.399.

Best, E. T. Confectionery extrusion. In: The Technology of Extrusion Cooking (N. D. Frame, ed.) Blackie, London. p. 190.

Briggs, D. E., Wadeson, A., Statham, R. and Taylor, J. F. (1986). The use of extruded barley, wheat and maize as adjuncts in mashing.*J. Inst. Brew.* 92: 469.

Chang, Y. H. and Hii, C.Y. (1992). Preparation of starch phosphates by extrusion. *J. Food Sci.* 57: 203.

Carr, M.E. and Cunningham, R. L.(1989). Glycol glucosides from starch by continuous twin- screw extruder processing. *Cereal Chem.* 66: 238.

Darrington, H. (1987). A long running cereal. *Food Manual,* 3: 47–48.

Fanta, G. F., Burr, R. C. and Doane, W. M. (1980). Polymerisation of alkyl acrylates and alkyl methacrylates with starch. *J. Appl. Poly. Sci.* 25: 2285.

Farrell, D. (1971). "Extrusion equipment-types, functions and applications". Symposium on extrusion: Process and product development. American Association of Cereal Chemists, St. Paul, MN.

Ferdinand, J.M., Kirby, A.R.,Senouci, A. and Smith, A. C. (1989). The extrusion cooking of potato products. In: Process Engineerimg in the Food Industry (R. W. Field and J. A. Howell, eds.) Elsevier, London. p.59.

Ferdinand, J.M., Clark, S.A. and Smith, A. C. (1992). Structure formation in extrusion cooked starch- sucrose mixtures by carbon dioxide injection.*J. Food Eng.* 16: 283.

Georget, D.M,.R. and Smith, A. C. (1995). The mechanical properties of wheatflake components. *Carbohydr. Polym.,*28: 305.

Gry, P., Holm, F. and Kristensen, K. H. (1984). Degermination of spices in an extruder. In: Thermal Processing and Quality of Foods (P. Zeuthen *et al.*, eds.) Elsevier, London. p. 185.

Harper, J. M. (1981).Extrusion of Foods, Vol. 1 and 2. CRC Press, Boca Raton, Florida.

Jansen, R. and Harper, M. (1980). Application of low cost extrusion cooking to weaning foods in feeding programmes. *Food and Nutrition* 6:2.

Kaur, K. and Singh, N. and Singh, H. (2000). The effect of extruded flour and fermentation time on some quality parameters of idli. *Journal of Food Quality* 23: 15.

Linko, P. (1989). Extrusion cooking in bioconversions. In: Extrusion Cooking (C. Mercier, P. Linko and J.M. Harper, eds.) A.A.C.C. St. Paul, USA.p.235

Lorenz,K. and G. R. Jansen (1980). Nutrient stability of full-fat soy flour and corn- soy blends produced by low- cost extrusion.*Cereal Food World* 25: 161.

Mercier, C. and Cantarelli, C. (1986). Pasta and extrusion cooked foods: Some Technological and nutritional Aspects, Elsevier, London.

Noguchi, A. In: Extrusion cooking (C. Mercier, P. Linko and J.M. Harper, eds.) A.A.C.C., St. Paul, USA, p.343.

Parker, R. and Smith, A.C. The mechanical properties of starchy food materials at large strains and their ductile-brittle transition. In: The Glassy State in Foods (J. M. V. Blanshard and P. J. Lilliford, eds.) Nottingham University Press. p. 159.

Rizvi, S. S. H., Mulvaney, S. J. and Sokhey, A.S. (1995). The combined application of supercritical fluid and extrusion technology. *Trends Food Sci.Tech.* 6: 232.

Rokey, G.J. (1994). Petfood and fisfood extrusion. In: The Technology of extrusion Cooking (N. D. Frame, ed.) Blackie, London.p.144.

Salay, E. and Ciacci, C. F. (1990). Production and preparation of starch phosphates produced by the extrusion process. *Die Staerke* 42:15.

Singh, N., Bawa, A. S. and Sekhon, K. S. (1995). Quality improvement of *Idli* using extruded rice flour. *J. Food Quality*, 18: 193.

Slade, L. and Levine, H., The glassy state phenomenon in food molecules. In: The glassy state in Foods (J. M. V. Blanshard and P. J. Lilliford, eds.) Nottingham University Press. p. 35.

Smith, A.C. (1992). Studies of the physical structure of starch based materials in the extrusion cooking proces.In: Food Extrusion Science and Technology (J. L. Kokini, C. T. Ho and M. V. Karwe, eds.) Marcel Dekker, New York. p. 573.

Smith, A. C. and Singh, N. (1996). New applications of extrusion technology. *Indian Food Industry*, 15: 14-23.

Swanson, C.L., Shogren, R.L., Fanta, G.F. and Imam, S.H. (1993). Starch- plastic materials–preparation, physical properties and biodegradability. *J. Environ. Polym. Degradation* 1: 155.

Wagner, J. (1989). Making caseinate by extruder. *Food Eng.Intntl.* March: 42.

Westwood, K. T. (1994). Extrusion of Brewers' hops. In: The Technology of Extrusion Cooking (N. D. Frame, ed.) Blackie, London. p.237.

Wiedmann, W. and Strobel, E. (1987). Processing and economic advantages of extrusion cooking in comparison with conventional processes in the food industry. In: Extrusion Technology for the Food Industry. (C. O.' Conner, ed.) Elsevier, London. p. 132.

3

Flour Milling Industry Status in India

Vinod Kapoor

Kapoor Brothers Roller Flour Mill,
64, Industrial Area, Panchkula – 134 109, Punjab

Wheat grinding is probably the oldest practice after hunting and dates back to the existence of human civilisation. It started with primitive methods of manual cleaning and pounding it between rocks, to Slaves and animal powered grinding. Windmills were the first known mechanical power grinders in early 17th century. After windmills, the power was derived by steam, and on the coast by pools of water captured at high tide and allowed to run out and turn the mill wheel during low tide. At the start of the American Revolution, the wheat was cleaned of chaff and ground by means of revolving mill stones. After grinding the flour was bolted to remove as much of bran and germ as possible.

The first revolution in flour milling is attributed to inventive genins of Oliver Evans who moved from every thing with water. All opertions were mechanised. The grain moved from Receiving hopper–the bucket elevator–cleaners–mill stones–screw conveyor–

elevator–upper floor for cooling–bolting–flour chests–flour packing. This machinery sounds very familiar to the present day millers. Todays versions have improved form machines only. Water power however gave way to steam power. Flour purifier air blast was introduced around 1870 and new process patented by George Christian. This new process enabled millers to mill wheat, remove bran, manufacture gluten rich flour. From "low" grinding method to make as much flour as possible had given way to the principle of "gradual reduction". In this wheat was granulated on the first grinding–purified–middling ground still finer to fines.

The second revolution in the milling industry came around 1875 introducing the use of rollers for grinding which reportedly came from Hungary. The porcelain rollers were used to convert purified middling into flour. Immediately thereafter Mr. John Stevens used hard weaving chilled steel to make a roller and also devised a system of corrugations.The millstone became obsolescent at this stage. The subsequent stages of inventions saw in the increase in the number of stages in the gradual process system to reach a dozen reduction steps. It however finally settled on the five or six reductions.Around this time,dust collectors and tempering of wheat before grinding too was introduced.Beyond 1890 and present days there has been refinement in processes and machines only besides the application of improved power devices, automation and computerisation. There has not been any significant invention of the new machinery or process.

Initially bran and offals were burned, dumped or thrown away. During 1880's its use as feed was introduced, other grains like Rye and Oat were used for flour too. Process control systems were introduced. Pneumatic conveying added to the improvements in yields, better hygiene in the plants, lesser infestation and improved handling practices. Automation of machine and computerisation of processes have resulted in reduction in energy consumption and manpower at all levels.

Roller Flour Milling Technology

Presently, the basic flour milling technology of converting wheat into wheat products remains the same in the entire world, *i.e.*, pre-cleaning, cleaning, tempering or conditioning, milling and packaging. The Indian Flour Milling Industry is presently confronted with the problem of inconsistent raw material supply, over capacity, lack of economies of scale, low margins, poor extractions and inconsistent quality. A major capacity in the industry is in the small scale sector. The technology adoptions by these units is based on the indigenous small scale machinery manufacturers limitations. Process controls are manual and the processing system machines performance is low and energy consumption is high. Modernisation involves high capital cost. The storage system of raw materials and products is bag based and there are no silos for storing raw materials and end products. Packaging systems at the raw material as well as end products stage are manual. Since all the machinery manufacturers are in the small scale sector they have lesser technical back up and mostly have copied the machines from the imported ones.All these factors contribute to low productivity and high energy costs. Energy consumption in Indian mills with indigenous machinery is higher by 30-50 per cent over the International levels. Maintenance costs and break downs are very high which contribute to the increased production costs.

The technology in the world is taking place to improve productivity, reduction in tempering time, greater flexibility, grinding process shortening, achieving better environment,pollution control,automation and control of milling operations by computers and improved on line equipments in processing, weighing and packaging to produce a consistent quality products for meeting the requirements of the end use industry.

New process has been developed for removal of most of the wheat bran layer from the wheat kernel before the first break. This is likely to reduce power consumption and give a better colour flour. The space required for such like plant will be small. Technology has also been developed for setting up small capacity short flour mill plants to produce speciality products. Technology has been developed for computer control, batch blending, adding additives mixing for functional improvements in flour for end use applications, enriching and fortifying flour with vitamins, iron and other minerals. Containerised flour mills have also been developed which are compact and can be erected in very short time.Bulk silos systems including grain handling and loading have been developed.

Flour Milling Industry in India

Background

In India, there are almost 1000 Roller Flour Mills with an installed capacity of more than 20 Million tonnes. The industry is well spread throughout the country and can reach the entire Indian population.

The average installed capacity per unit is around 70 MTs per day. The capacity utilisation in the industry is around 50 per cent only.

Out of total wheat consumption in the country, 85 per cent is consumed in the form of Atta primarily through chakki route. Roller Flour Milling Industry processes around 12-15 per cent of the total wheat consumed in the country, the balance being consumed through stone chakkis.

A major part of the industry is in the small scale sector belonging to trader turned entrepreneur. The expansion taking place is not the result of market survey or analysis but out of sheer entrepreneurial urge to follow.

Roller flour milling industry is a mass based, high volume, low added value, low margins industry. It operates in a volatile raw material pricing regime and a price sensitive environment with the threat of Government intervention disrupting the operation of free market environment. The industry's performance varies from year to year and region to region and from state to state in the region.

From the era of regulation and control, we have passed through deregulation of the industry, liberalisation in terms of raw material supplies to free market economy and globalisation.

The industry's only raw material is wheat. Wheat is consumed to provide basic food after processing. Wheat being an essential food item is always an area of concern for the Government. We have now achieved self sufficiency and are on the threshold of entering export markets.

Wheat, the golden grain behaves speculatively like Gold in the world market. In India the problem is accentuated due to absence of reliable market information and accurate statistical data on production, market surplus and its spatial distribution, procurement during the year; wheat stocks built up with the Government;

Government policies in vogue on fees, levy, sales tax, movement. stock restrictions, selective credit control directives, marketing, distribution and import-export policy; minimum support price offered to the farmer and psychosis of plenty/shortage in a year etc. Wheat available with us today is not available as per specific end use application as per International standards. Wheat procured/ marketed on the basis of FAQ standards in India has excessive foreign material admixture, varietal admixture, low protein quantum, poor quality protein, poor postharvest technology adoption.

Presently the industry is passing through a vicious circle. To convert into a virtuous circle, efforts need to be put in jointly by growers, processors, the Government, Research Institutions and the industry in the developmental perspective.

Wheat Production

Wheat production per hectare in the country is low and is stagnating at the present levels because of small holding, low mechanisation, low nutrient consumption, low seed replacement, improper water use etc. There is urgent need for increasing productivity by introducing proper replacement,varietal diversification, to develop specific end use varieties, developing disease resistant hybrid seeds, adopting better agronomic practices. Public and private sector seed companies be encouraged to take up contract farming through their extension services. There is urgent need to encourage economies of scale by land reforms/cooperative farming as well as developing a self sustained technology for small farmers

Varietal segregation and improvement of specification for specific end usage conforming to International standards need to be adopted.

Investment in postharvest storage facilities creation is essential to enable farmers realise a better price.Field extension and education for growing specific end use varieties be developed regionally.

Wheat Marketing

In the world, wheat production and marketing is done by the wheat growers whereas wheat marketing in India has a large number of intermediateries in the chain from the farmer to trader and finally to the processor. The system was necessary in times of shortages necessitating stricter Government Controls. In the present times, brain storming needs to be done to reduce the number of intermediaries. Procurement specifications of Fair Average Quality parameters donot match with the international classifications. Global challenges and surplus status should motivate us to work towards adopting international specifications in domestic trade too.

Agricultural Produce Marketing Act and Rural Development Act were enacted to regulate markets and improve Rural Development for the benefit of farmers. The application of the Act is not uniform in terms of its levies in the country. Different perceptions exist in different states. Market fee and Rural Development Fund on agricultural produce brought to the mandis is levied at the time of auction and is paid by the buyer at different rates in different states.This increases the number of operations unnecessarily.

Agricultural Produce not brought to the mandis (market yards) by the farmers does not give any realisation to the state Government. This produce is sold by the farmer or in the name of the farmer without payment of market fee and RD Fund outside the State of produce to buyers of other states.

Wheat, a Declared Commodity under CST Act is subjected to sales tax once in each state wherever tax is applicable. Wheat consumption is possible only after processing. Since wheat products by virtue of Supreme Court judgement have been classified different than that of wheat, some states have imposed tax on wheat products also. Since wheat products are consumed only after processing wheat, the above situation leads to multiple taxation on wheat and wheat products. Besides this wheat moved out by the farmer or in the name of the farmer does not attract any sales tax. Different states have different tax rates.This creates difficulty for the honest entrepreneurs.The suggestions to provide a transparent and an equal opportunity to all the following sugestions need to be considered:

1. There should be uniformity in the levy of market fees, under the Agricultural Produce Markets Act, Rural Development Fund under the Rural Development Act, Sales tax under the States Sales Tax Act and classification of "wheat and wheat products" as a single entry as goods of special importance under the Central Sales Tax Act 1956.

2. If farmers are undertaking trading from their state of produce, they should pay all the taxes as applicable to the trader in their states and no special exemption should be given from levy of fees, fund and taxes.
3. Procurement specifications should be brought on par with the International specifications.
4. Specific end use varietal marketing needs to be conceptualised.

Processing

Major part of the industry is in the small scale sector as per the then Government policy. The industry lacks economies of scale and the process technology is less efficient, high power consuming and less productive too. The basic technology the world over is the same, *i.e.,* Pre-cleaning, cleaning, conditioning, milling and packing storage system is in godowns and not in Silos. Process technology system machines differ in terms of performance in production capacity, power consumption and consistent end product quality.

Process controls are manual whereas in the developing world they are electronic based,. Packaging systems are manual both at raw material stage as well as end product stage. Loading and unloading is in bags and is manual. No reliable manufacturer of indigenous equipment is available. Machine manufacturers are mostly in small scale sector.

Potential areas of investment and modernisation are:

1. Transportation

   Raw Material Stage: Development of bulk handling systems
2. Storage and Milling Stage: Transportation systems

   Process Equipment

   Process Controls

   High Efficiency Electrical Equipment.
3. Packaging and: Packaging mode change from bag to bulk Storage SystemSilos storage systems at mill stage as well as end user industry level like Bread and Biscuit.

4. Infrastructural Development: At cargo handling Railway Projects: networkdevelopment. Silos.

Government Policies

The Government policies play a significant role in the roller flour milling industry since it produces products of daily food items which are essential in nature. The present need for consideration of various matters requiring changes in the Government policies are–

Futures Trading

In the past three or four years in the free market economy, volatility in prices has been witnessed in the market. Such volatility becomes speculative and psychosis related. The world over futures trading exists in wheat trade. There is no futures market. In absence of futures market, the industry has become more trade related than production and performance related. No long term planning can be resorted to by the industry in terms of raw material procurement and end products marketing in absence of futures price projections in a given period. Industry cannot enter into a long term contract with a foreign buyer. Future market needs to be developed.

Exports and Imports

The present adhocism in the policy should give way to long term planning. We need to plan for a long term import-export policy. In view of the World Trade Organisation agreement, free import and export should be permitted.

Commodity Management

Wheat management in times of shortages and plenty vary since free import export was not allowed. Whenever Government intervenes in the open market in case of a rising price trend, the Government gets sensitive. Commodity markets behave on a demand supply situation. Even though volatility in pricing is there in absence of a futures markets and sentiments, yet the policies announced are adhoc and are not transparent. Government prices disturb the free market economy parameters too.

The Government needs to develop an all embracing attitude of confidence in the industry rather than coercive action planning and not disturb the free market economy parameters specially in decisions pertaining to open market sales scheme.

Prevention of Food Adulteration Act

Needs to review the provisions hurting the industry with special reference to distinction between Adulteration and Substandard category.

Essential Commodities Act

Currently the E.C.Act is in force. In the present global village environment of pleading there is no justification of retaining it on the statute book and only create harassment/nuisance value for trade and industry. The E.C.Act needs to be scrapped in view of the globalisation of trade. If however the Government wants to continue with it, at least practical difficulties should be ameliorated like accepted variations in physical stocks, incomplete record for the day etc.

Selective Credit Control Directive

RBI in times of shortage had been imposing selective credit control directive limiting quantum of advances, fixing interest rates, stipulating higher margins etc. It is also imposed on wheat and wheat products. Even though all products are perishable in nature still it is imposed on the flour milling industry only whereas user industries are exempted. Even though in the present scenario there may not be any likely eventuality of imposing such kind of restrictions yet whenever imposed, these should be restricted to wheat which is hoardable. It should not be applied to wheat products which can neither have a long shelf life nor hoardable.

Annexure–I

Year-wise Number and Capacity–Roller Flour Mills in India

Year	*Number*	*Yearly Capacity (million tonnes)*
1968	202	4.53
1969	206	5.13
1970	211	5.41
1971	218	5.50
1972	221	5.68
1973	232	5.95
1974	232	5.95
1975	232	5.95
1976	232	5.95
1977	232	5.95
1978	232	5.95
1979	232	5.95
1980	234	5.95
1981	306	7.41
1982	310	7.49
1983	365	7.85
1984	426	8.59
1985	454	8.91
1986	464	8.99
1987	496	9.80
1988	557	10.21
1989	580	10.39
1990	605	11.00
1991	623	12.89
1992	702	16.00
1993	710	16.00
1994	750	17.72
1995	774	19.10
1996	800	21.00
1997	812	21.80

Annexure–II

Sl.No.	Name of the State	No. of Roller Flour Mills
1.	ANDHRA PRDESH	52
2.	ASSAM	39
3.	BIHAR	66
4.	CHANDIGARH	7
5.	DELHI	18
6.	GUJARAT	45
7.	HARYANA	45
8.	HIMACHAL PRADESH	9
9.	JAMMU and KASHMIR	44
10.	KARNATKA	39
11.	KERLA	55
12.	MADHYA PRADESH	35
13.	MAHARASHTRA	70
14.	MANIPUR	1
15.	MEGHALAYA	3
16.	MIZORAM	1
17.	NAGALAND	3
18.	ORISSA	45
19.	PONDICHERY	2
20.	PUNJAB	55
21.	RAJASTHAN	17
22.	SIKKAM	3
23.	TAMILNADU	49
24.	TRIPURA	2
25.	U.P	8
26.	WEST BENGAL	43
27.	GOA	2
	Total	**837**

4

Advances in Structure and Functionality of Wheat Gluten Proteins

Bhupendar Singh Khatkar
Department of Food Technology,
Guru Jambheshwar University, Hisar – 125 001

Introduction

Wheat is cultivated in all parts of the world except in the hottest tropical regions. World wheat areas is currently about 220 million hectares and its production is around 550–600 million tonnes per annum. Most wheat is consumed in the form of various types of baked goods in most countries of the world (Khatkar and Schofield, 1997). The technological importance of wheat flour is attached mainly to its gluten proteins, *i.e.* the water-insoluble complex proteins. The paramount importance of gluten proteins among wheat constituents for bread making quality is attributed to its unique physical properties such as viscoelasticity, water absorption capacity, film forming ability and thermosetting properties. Recently, wheat protein research has gained considerable momentum to understand how wheat gluten acquires these special physical properties and also efforts have been made to understand how these properties help to retain gas in bread making process. In this article, relationship of physical properties and structural aspects of wheat gluten proteins with bread making quality will be discussed.

Classification of Wheat Gluten Proteins

Proteins of wheat flour can be separated into non-gluten forming and gluten forming groups, when a wheat flour is wetted and mixed with water. The term gluten refers to a viscoelastic protein mass, which is recovered after aqueous washing out of starch and water soluble components from wheat flour, when dried it yields a cream coloured free flowing powder of high protein content (75–80 per cent) and bland taste. When rehydrated, it regains its viscoelastic properties, *i.e.* rubber like characteristics. The non-gluten forming proteins are albumins and globulins, which account approximately 10–20 per cent of the total flour proteins and their amounts are relatively constant in wheat flours.

According to Osborne (1907) gluten forming proteins, which represent 80–90 per cent of the total proteins of wheat flour, have been classified into two major groups, *viz.*, gliadin and glutenin, based on their extractability and unextractability, respectively, in aqueous alcohol. The terms, gliadin and glutenin, are the specific names for the wheat proteins, corresponding to the generic terms prolamin and glutelin of all cereal proteins. Neither of these groups are considered to consist of pure proteins and a considerable amount of overlap between these groups may occur, depending on the exact extraction conditions, such as time, temperature and type, as well as concentration of alcohol used. In view of this disadvantage of Osborne classification, an alternative classification of gluten proteins has been proposed by Shewry and Colleagues (1986) that reflects biological, chemical and genetic relationships among component polypeptides of the gluten complex. They divided gluten proteins into three main groups, *viz.*, sulphur-poor prolamins, sulphur-rich prolamins and high molecular weight (HMW) prolamins (Figure 4.1). All these gluten protein groups are considered typical prolamins, because they are rich in proline and glutamine

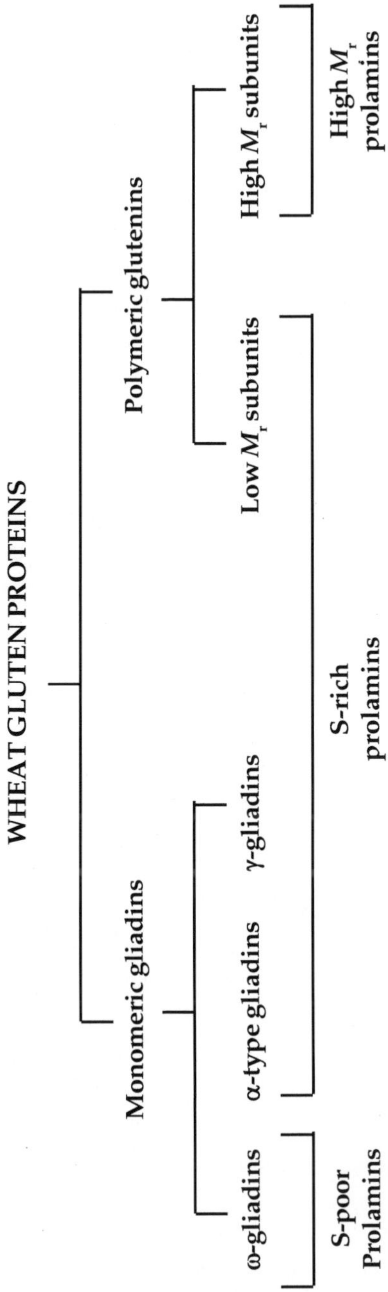

Figure 4.1: Classification and Nomenclature of Wheat Gluten Proteins (Shewry *et al.*, 1986)

amino acid residues and are extractable, at least partially in aqueous alcohol, particularly in the presence of reducing agent.

The gliadin and glutenin fraction can be divided into subclasses. The gliadins are usually separated by lactate polyacrylamide gel electrophoresis (PAGE) into four main sub-categories, α-, β-, γ- and ω-gliadins in decreasing order of electrophoretic mobility under acidic conditions and increasing order of relative molecular mass. Amino acid sequencing has revealed that the α- and β-gliadins are structurally closely related polypeptides and, therefore, both of these gliadin polypeptides have been classified into one group, *i.e.* alpha-type gliadins. The relative molecular masses of α-type and γ-gliadins are between 30000–45000, whereas ω-gliadins generally have relative molecular mass of 44000–80000. The γ-gliadins, in general, have somewhat higher relative molecular masses than alpha-type gliadins.

According to their sodium dodecyl polyacrylamide gel electrophoresis (SDS-PAGE) under reduced conditions, the glutenin fraction can be divided into two sub-categories, *i.e.* low and high molecular weight subunits. The high molecular weight (HMW) subunits, also known as A subunits of glutenin are the largest polypeptides and contain the highest level of glycine residues among gluten proteins. Their molecular weights measured by SDS-PAGE range from 80000-160000. The HMW subunits are further divided into x- type and y-type. The x-type subunits are somewhat large polypeptides, while the y-type subunits are smaller .

The HMW subunits of glutenin are encoded at the *Glu-1* loci on the long arms of the chromosomes IA, IB and ID. These loci are designated as *Glu-A1, Glu-B1* and *Glu-D1*, respectively. *Glu-1* quality scores are assigned to the individual subunit or subunit pair based on their values determined by the SDS-sedimentation test, in indirect measure of bread making quality (Table 4.1). The alleles of some HMW subunits are complex in that they contain genes that encode for two subunits, and those subunits are often found together, for example, subunits 2+12, 5+10 from *Glu-D1* and subunits 7+9, 6+8, 17+18 from *Glu-B1. Glu-1* scores are assigned to submit pairs rather than individual subunits for such alleles. The *Glu-1* score of a cultlvar is calculated by summing the scores of the subunits it contains (Payne *et al.*, 1987). The *Glu-1* scores of bread wheats usually range from 3 to 10 for poor and good bread making quality, respectively (Table 4.1).

Table 4.1: *Glu-1* Quality Scores Assigned to Individual High M_r Subunits or Subunit Pairs of Glutenin

Locus			Glu-1 Score
Glu-A1	*Glu-B1*	*Glu-D1*	
-	-	5+10	4
1	17+18	-	3
2*	7+8	-	3
-	13+16	-	3
-	7+9	2+12	2
-	-	3+12	2
Null	7	4+12	1
-	6+8	-	1
-	20	-	1

Scores are ranked from 1 to 4, indicating poor and good bread making quality, respectively (Payne, 1987).

Structural Properties

In recent years, our knowledge concerning the structure of gluten proteins has improved enormously. The amino acid sequences of purified individual gluten proteins are now available. Comparison of these sequences reveals that all types of gluten protein polypeptieds, in general, have at least two or three distinct structural domains, *i.e.*, a central repetitive domain, flanked by non-repetitive C-terminal and N-terminal domains (Shewry *et at.* 1994). The S-rich prolamins are characterised by a shorter repetitive domain and a longer C-terminal domain compared with HMW prolamins and S-poor prolamins. S-rich prolamins have their cysteine residues in the C-terminal domain, the only exception being the LMW subunits of glutenin, which also contain one cysteine residue in the N-terminal domain, S-poor prolamins have no cysteine residues, as indicated from omega-secalin of rye, a homologue of wheat omega-gliadins. On the contrary, HMW prolamins have the majority of their cysteine residues (3 and 5 in x-type and y-type subunits, respectively, in the N-terminal domains and only one in the C-terminal domain (Khatkar and Schofield, 1997). Several y-type subunits and x-type subunit 5 contain an additional cysteine residue in the central repetitive domain.

Total numbers and relative positions of cysteine residues are important for polymer size and the different polymerisation behaviour of gliadins and glutenin subunits. Inter-molecular disulphide: linkages are considered responsible for giant size of glutenin polymer. According to Ewart (1979) the subunits of glutenin are cross-linked in a head to tail fashion to form a linear macropolymer of glutenin known as 'concatenations'. Based on Ewart's findings and recent understanding of the structure of glutenin subunits, in particular HMW glutenin subunits, many hypothetical models for glutenin polymer have been proposed. All models propose a central linear polymer of either HMW glutenin subunits only (*e.g.* model of Graveland *et al.*, 1985) or a mixture of HMW and LMW subunits of glutenin (Kasarda 1989).

Unfortunately, none of these models describes the underlying mechanism involving interaction of glutenin with gliadin and for non-gluten components, such as lipids and carbohydrates, which might be essential to explain the complete model of the gluten network. Hence, the models proposed so far are highly simplified and obviously suggest that much remains to be investigated to arrive at a comprehensive model for gluten network (Khatkar and Schofield, 1997)

The Basis of Gluten Viscoelasticity

Wheat flour dough is largely composed of starch, proteins, lipids, water and air cells. If the gluten proteins are removed from the flour, then the property of forming a viscoelastic dough is lost. It is generally agreed, therefore, that the gluten proteins form the framework of the dough structure. When flour is wetted, the formation of gluten is mainly caused by a complex interaction between the endosperm storage proteins, *viz.* gliadin and glutenin.

Usually the elastic properties of gluten are ascribed to the glutenin fraction, whereas the viscous properties come from the gliadin fraction. It is understood that the gluten of wheat owes its unique viscoelastic behaviour to an appropriate balance in the amounts of gliadin and glutenin proteins, but variation in the composition and elastic properties of the glutenin proteins appear to be largely responsible for the differences in the gluten viscoelasticity among wheat cultivars. For this reason, it is widely believed that the viscoelasticity of gluten is controlled by the glutenin proteins.

The 'linear glutenin hypothesis' proposed by Ewart has received wide acceptance in explaining the viscoelasticity of gluten and flour doughs. This hypothesis has passed through different stages of sophistication. According to Ewart, the glutenin polypeptide chains form long linear 'concatenations' with two S-S bonds connecting each chain in a head-to-tail fashion to the next chain. The 'concatenations' of glutenin molecules are assumed to participate in the formation of a three-dimensional entangled network structure.

The entanglement may be purely physical and/or at the point of entanglement, known as nodes, non-covalent interactions might occur to form cross-links (branching). The branching of glutenin polymers by S-S bonds is also not ruled out. The effectiveness of entanglement and cross-linking will depend on the length of the glutenin 'concatenations', which in turn depends on the number and distribution of covalent bonds and non-covalent interactions.

Under stress conditions, the entangled and non-covalently cross-linked structure in glutenin offers resistance to deformation, which is manifested in increased elasticity (elastic modulus, G′) during controlled stress rheometry. It also enables glutenin polymers to recoil after stress conditions are withdrawn. Viscous flow depends predominantly on molecular slippage at nodes and the labile nature of weak secondary forces acting between the glutenin polymers. Sulphydryl-disulphide bonds interchange and mechanical scission of S-S bonds may also contribute to viscous flow. Mechanical scission of S-S bonds probably occurs when the rate of deformation exceeds the rate at which molecular slippage or SH/SS interchange can occur resulting in glutenin polymers being subjected to stress values more than their elastic limit. The viscous nature of gliadins may be attributable to the absence of effective entanglements owing to their low M_rs. Gliadin polypeptides help molecular slip acting analogously to ball bearings.

Spiral structure hypothesis has also been suggested to explain the elasticity of glutenin. This hypothesis is based on the structural analogy of high M_r subunits of glutenin to the mammalian connective tissue protein, elastin, in which the elasticity is attributed to the presence of β-Spiral structure. Circular dichroism spectroscopy and structural prediction studies have indicated that the central repetitive domain of high M_r glutenin subunits are rich in β-turns. The β-turns are hypothesised to occur so frequently and regularly that they have been suggested to form a β-spiral structure, as in elastin. The β-spiral structure is assumed to be intrinsically elastic.

It has been suggested that under stress conditions, the β-spirals undergo deformation and on release of stress, the β-spiral structures resume their original energetically favourable conformation. Experimental evidence for the β-spiral conformation of the high M_r glutenin subunits has been obtained using scanning tunnelling microscopy of a purified high M_r glutenin subunit (1Bx 20) from durum wheat. There is no direct evidence as yet for the elastic nature of β-spiral, however.

Recent experiments have shown that the high M_r subunits of glutenin are not elastin-like in their interaction with water, however. It has also been argued that the spiral structure of high M_r of glutenin is based on γ-turns rather than β-turns. Thus, there is still a need for a more complete structural model to explain the viscoelastic behaviour of' gluten and flour doughs.

Importance of Gluten Viscoelasticity in Gas Retention and Bread Making

The technological significance of wheat flour is related to the viscoelastic nature of its flour-water dough, which retains gas produced during fermentation and in the early stages of baking. The ability of a viscoelastic dough to retain gas results in a light porous crumb textured bread. Wheat gluten proteins are believed to be primarily responsible for the viscoelastic behaviour of dough and hence for the uniqueness of wheat flour. Proteins of other cereals such as corn, rice, rye, etc. do not have the ability to produce such viscoelastic dough. Flours from other cereals, therefore, fail to retain gas and give very dense and small loaves.

The bread making process is accomplished through three basic operations: mixing, fermentation and baking. Mixing transforms the flour and water into cohesive viscoelastic dough. Another key function of the mixing operation is the incorporation of air. The air is essential as it introduces the gas cells into which the CO_2 produced by the yeast fermentation diffuses. The importance of these gas cells emerges from the fact that yeast cannot produce new gas cells. Therefore, if no air cells were present, the crumb grain of the final bread would be very coarse with only a few large cells.

Dough is developed as a result of mixing to high work input levels and at high speed in mechanical dough development systems, such as the Chorleywood Process, or by mixing to low work inputs at low speeds plus long fermentation in bulk in 'traditional' bread making systems. Such developed dough is capable of retaining the CO_2 produced by the yeast during fermentation and in the early stages of baking. Baking rheologically transforms an expanded viscoelastic dough foam into an elastic bread sponge. Flour-water dough retains gas owing to its discontinuous discrete cellular foam structure, whereas baked bread is a continuous open-celled sponge that is permeable to gases.

Understanding the structure of the gluten proteins, and how they interact with themselves and other flour constituents to convert a wheat flour into a viscoelastic system, still largely remains a challenging problem. Also, much remains to be learned about how a viscoelastic dough retains gas and how the expanded dough is transformed into bread. However, some progress has been made in this area of research, and in the discussion that follows the role of gluten proteins in gas retention is explained. The discussion below, although based on available evidence, may be somewhat speculative.

From a colloidal point of view, in a mixed flour-water dough the hydrated gluten proteins form the continuous phase with starch and air cells as the discontinuous phase embedded in it and the yeast cells are dispersed throughout the aqueous dough phase. The 'free' water in the dough represents the aqueous dough phase, which dissolves the water-solubles and, acts as the medium for chemical reactions and for dissolving CO_2 up to its saturation point.

Yeast ferments sugars and continuously produces CO_2 in the aqueous dough phase. When the aqueous dough phase is saturated with CO_2, most of the CO_2 diffuses into the air cells that are formed in the dough during mixing. This is attributed to the fact that the rate of diffusion of CO_2 in the dough and hence its ultimate evaporation into the surrounding atmosphere is slow. This is due to the presence of continuous gluten protein film, pentosans and lipids.

The diffusion of CO_2 into gas cells increases the pressure within gas cells that provides the driving forces for dough expansion. The viscous flow properties of dough owing to its monomeric proteins (mainly gliadins) allow the gas cells to expand. The expansion of gas cells releases the pressure within gas cells, however, a little pressure does exist in the dough system during fermentation. The pressure in gas cells has been reported to be slightly greater than the atmospheric pressure, *i.e.* about 1.01 atmosphere. This small over-pressure results from the surface tension at the gas dough interface and the resistance of dough to deformation, *i.e.* expansion.

During baking, the temperature increases, water evaporates from the liquid dough phase into the gas cells, CO_2 and ethanol produced by the yeast and dissolved in the liquid dough phase also diffuse into the gas cells, which together result in an overall increase in the pressure within the gas cells. The viscous components, *i.e.* monomeric proteins (mainly gliadins) allow gas cells to expand to equalise the internal gas cell pressure, whereas the elastic components, *i.e.* polymeric proteins (mainly glutenins) provide strength to prevent the gas cells from over expanding, and thus preventing the rupture of the gluten film network enveloping the gas cells.

In a flour-water dough containing insufficient amounts of good quality polymeric proteins, the gluten network would be too extensible, allowing uncontrolled expansion of gas cells, which would ultimately rupture the continuous gluten film network and, thus continuity between gas cells would be established. The continuity between gas cells or the discontinuity in the gluten film network is seen as pinholes at the dough surface. This results in the rapid loss of CO_2 even at very low temperature during baking and very early end of the oven spring. Consequently, bread of low loaf volume and dense open crumb is produced. On the other hand, good quality flour-water dough with enough strength retains CO_2 until more advanced stages of baking, *i.e.* ~70°C. This produces a good bread loaf volume and even crumb texture on completion of baking.

Differences in the loss of CO_2 between poor and good quality flour doughs are not related to changes in the starch or to changes in the gluten proteins but appear to reside in the inherent differences in the gluten proteins. The ratio of monomeric to polymeric proteins and composition as well as properties of glutenin subunits in flours appear crucial to the gas retention and dough expansion. Dynamic rheological studies showed that gluten-water doughs from poor quality flours had lower elasticity and greater viscosity than those from good quality flours (Khatkar *et al.*, 1995).

The rapid loss of CO_2 at ~70°C is assumed to be the result of the increased elastic character of the dough due to starch gelatinisation and polymerisation of the glutenin proteins via disulphide cross-linking. This leads to a sharp increase in the tensile stress in the dough system, thus causing rupture of the gluten film network at the gluten-starch interface and subsequently loss of gas retention. More recently, rupture of the liquid film surrounding the gas cells has been considered to contribute to the rapid loss of gas during baking. The liquid film may have an additive effect with the gluten film network in gas retention.

Associations of Wheat Gluten Proteins with Mixing Properties and Bread Making Quality

It has long been recognised that the mixing properties and bread making quality of wheat flours are governed by the quantity and quality of their proteins. A recent study conducted by Khatkar and co-workers (1996) indicates that the mixing characteristics of flours from wheat cultivars in the medium-strong range are significantly influenced by their gluten protein contents. On the other hand, the mixing properties of 'extra strong' or weak flours are relatively less affected by their gluten protein contents, but gluten protein quality primarily controls their behaviour during mixing. Weak gluten develops quickly, where as strong and 'extra strong' glutens need a longer time to mix to peak dough development (PDR). This clearly suggests that the mixing requirement of flour is related to its gluten protein composition. It has been demonstrated that gliadin fraction generally decreases mixing time, whereas certain glutenin fraction increases the mixing requirements of a wheat flour. Addition of different gliadin subgroups to a base flour decreases mixing time in the order omega → gamma → beta → alpha-gliadins in accordance with the decrease in their molecular size and/or the increase in their charge (Khatkar and Schofield, 1997).

The type and composition of HMW glutenin subunits have also been considered important for determining the mixing characteristics and bread making performance. It is generally considered that subunits 5+10, 17+8, 1 and 2* are related to longer mixing times, dough strength and superior bread quality, whereas subunits 6+8, 2+ 12, 3+ 12 and 20 are related to shorter mixing times and dough weakness (Khatkar *et al.*, 1996; Khatkar and Schofield, 1997). Both quantitative and qualitative effects may contribute to quality difference associated with specific HMW glutenin subunit alleles. According to Gupta and MacRitichie (1994), the alleles at the *Glu-B1* and *Glu-D1* loci produce HMW subunits (17+18 and 20 or 5+10 and 2+12) in similar amounts. Their results indicate that the qualitative differences in HMW subunits may be responsible for the differences in the mixing characteristics and bread making quality among wheat cultivars. This is consistent with the previous observations that the superior subunit 5 has one additional cysteine residue and has a greater molecular mass than the inferior subunit 2. The superiority of the subunits 17+18 over subunit 20 in the case of genome B has been attributed to the number and position of the cysteine residues. Furthermore, on the basis of amino acid sequence differences in the regions towards the C-terminal ends of the repetitive domains of HMW glutenin subunits 10 and 12, it has been postulated that subunit 10 may form a more regular beta-spiral structure than subunit 12, and that this in turn may lead to improved elastic properties (Khatkar and Schofield, 1997). Beta-spiral structure has been proposed to be intrinsically elastic.

It has been argued that it is the quantity of glutenin subunits and not the composition that is important. Studies have been reported in which the HMW subunit pairs 5+10 and 2+12 failed to show the expected differences in dough strength and bread making performance in many genotypes. Bread wheat, such as Egret, contains superior subunit pair 5+10 yet produces weak dough, whereas inferior subunits 2+12 are found present in many Australian good bread making quality wheats, for example, Cook, Gabo, Oxely and Timgalen and also in the European wheats, such as Courtot, Hardi and Capitole (Khatkar *et al.*, 1996).

REFERENCES

Ewart, J.A.D. (1979). Glutenin structure. *J. Sci. Food Agric.*, 30: 482-492.

Graveland, A., Bosveld, P., Lichtendonk, W.J., Marseille, J.P., Moonen, J.H.E. and Scheepstra, A. (1985). A model for the molecular structure of the glutenins from wheat flour. *J. Cereal Sci.*, 3: 1-16.

Gupta, R.B. and MacRitchie, F. (1994). Allelic variation at glutenin subunit and gliadin loci, *Glu-1, Glu-1,* and *Gli-1* of common wheats. II. Biochemical basis of the allelic effects on dough properties. *J. Cereal Sci.*, 19: 19-29.

Kasarda, D.D. (1989). Glutenin structure in relation to wheat quality. In: *Wheat is Unique,* (Ed.) Y. Pomeranz. AACC, St. Paul, MN, p. 277-307.

Khatkar, B.S., Bell, A.E. and Schofield, J.D. (1996). A comparative study of the inter-relationships between Mixograph parameters and bread making qualities of wheat flours and glutens. *J. Sci. Food Agric.,* 72: 71-85.

Khatkar, B.S., Bell, A.E., & Schofield, J .D. (1995). The dynamic rheological properties of glutens and gluten sub-fractions from wheats of good and poor bread making quality. *J. Cereal Sci.,* 22: 29-44.

Khatkar, B.S. and Schofield, J.D. (1997). Molecular and physio-chemical basis of breadmaking properties of wheat gluten proteins: A critical appraisal. *J. Food Sci. and Technol.,* 34: 85-102.

Osborne, T.B. (1907). *The Proteins of the Wheat Kernel.* Carnegie Inst., Washington, DC.

Payne, P.I., Holt, L.M., Harinder, K., MacArtney, D.P & Lawrence, G.J. (1987). The use of near-isogenic lines with different high M_r glutenin subunits in bread making quality and glutenin structure. In: *Proc. of the 3rd Intl. Workshop on Gluten Proteins,* (Ed. R. Lasztity and F. Bekes. World Scientific, Singapore, p. 216-226.

Sherwy, P.R., Halford, N.G. and Tatham, A.S. (1994). Analysis of wheat proteins that determine bread making quality. *Food Sci. Technol. Today,* 8: 31-36.

Shewry, P.R., Tatham, A.S., Forde, J., Kreis, M. and Miflin, B.J. (1986). The classification and nomenclature of wheat gluten proteins: A reassessment. *J. Cereal Sci.,* 4: 97-106.

5

Identification of Cereals and Cereal Derived Foods Using Protein and DNA Profiling Techniques

Santosh Dhillon, Dharam Singh and Anita Ahlawat

Department of Biochemistry
CCS Haryana Agricultural University, Hisar – 125 004, India

Cereals such as wheat, rice and maize account for a major part of all human diets and the range of cereals and the foods produced from them continues to diversify. Identification of cereals and cereal derived foods is important from the point of view of plant breeders, farmers, traders, food processors and consumers. Some closely related cereal genotypes may have very different processing properties resulting in significant differences in the value and potential uses of grain varieties that cannot be distinguished by their appearance. For example, the noodle making properties of wheat and the brewing characteristics of barley differ greatly for varieties that show no morphological differences. This has necessitated the development of molecular techniques which are fast, relatively inexpensive, reproducible and free from variations due to environment. These are finding applications not only in fingerprinting varieties/ cultivars but also in testing of parentage, genetic mapping of loci governing economic traits and measurement of genetic diversity.

Electrophoresis of Proteins

Proteins and enzymes are direct gene products which reflect the genomic composition of varieties accurately. Seed proteins in general and storage proteins in particular are widely exploited, although a range of vegetative proteins and enzymes have also been used in the identification of varieties. Cereals have been routinely identified by protein electrophoresis and occasionally by HPLC of proteins. The first report of the successful separation of wheat seed proteins (Bietz 1983) has been followed by the unequivocal demonstration that HPLC can be used for varietal identification in a range of crops, primarily cereals such as wheat, barley, rice, oats, maize and sorghum but also in soybean and other legumes (Smith 1986). Alcohol soluble proteins (prolamins) such as gliadins of wheat and zein of maize have been most commonly analyzed though there are published methods involving the separation of albumins, globulins and glutelins. Reverse-phase HPLC has been commonly employed for cereal proteins, although ion-exchange columns can also be used. Despite high discrimination power of HPLC, its use has been limited.

Electrophoresis of seed or seedling proteins and isoenzymes is one of the simplest and routine laboratory techniques used for cultivar identification (Smith and Smith 1992). Seed storage proteins which are encoded at several loci, are present in comparatively larger amounts, are readily extracted and exhibit considerable polymorphism in almost all species. Also they are not influenced by the environment. Hence, seed storage proteins have been widely used in characterizing plant genotypes and have proved extremely useful for variety identification in cereals.

Several types of electrophoretic methods such as paper, starch gel, polyacrylamide gel and agarose gel electrophoresis are available (Dhillon 2000), but polyacrylamide gel electrophoresis (PAGE) is by far, the most commonly employed technique for protein and enzyme analysis. However,starch gel electrophoresis (SGE) has also been used in the analysis of isoenzymes. A number of methods and modifications of PAGE, which differ in type and composition of buffer pH, pore size of the gel etc. are available. The choice of the method depends largely on the crop species in question. Separation of gliadins by Acidic–PAGE has proved useful in wheat variety identification. White and Cooke (1992) found that in a collection of 353 European barley varieties, 70 different groups could be recognized on the basis of hordein composition. Similar methods have been used with modifications to reveal polymorphism of seed storage proteins in oats, rice, maize, triticale and sorghum (Cooke 1995). Polyacrylamide gel electrophoresis in the presence of sodium dodecyl sulfate (SDS-PAGE) has been used in identification of genotypes of ryegrass and was adopted by International Seed Testing Association (ISTA) as a standard reference method for identifying commercial seed lots of ryegrass varieties and species (ISTA 1992).

Isoelectric focussing (IEF) and ultra thin layer isoelectric focusing (UTLIEF) are other very high resolution electrophoretic techniques and have been used in identification of varieties and hybrids. Recently, capillary electrophoresis (CE) also known as high performance capillary electrophoresis (HPCE) has shown much promise for protein analysis and cultivar identification. Gliadin proteins from seventeen wheat cultivars were separated by CE, gliadin electrophoretograms of all cultivars showed clear qualitative and quantitative differences (Yan *et al.*, 1998).

Isozyme polymorphism is the other most widely used marker system in plants. Isozymes are multiple molecular forms of enzymes with the same substrate specificity but differ in electrophoretic mobility. Isozymes are revealed when tissue extracts are subjected to electrophoresis in various types of gels and subsequently immersed in a solution containing enzyme specific stain. Analysis of isozymes has been used for cultivar identification in crops such as maize, wheat and barley (Smith and Wych 1986: Cooke 1995).

One of the problems in electrophoretic techniques is of multiplicity of methods. For example, 27 or more published methods are available for wheat variety identification. This renders results of different laboratories non-comparable. However, various organizations such as International Association of Cereal Science and Technology (ICC) and the International Seed Testing Association (ISTA) are rationalizing the problem by adopting standard methods. Some of these methods have proven to be reproducible across the laboratories and have been included in the ISTA rules (ISTA 1999) for verification of varieties of *Triticum, Hordeum, Pisum, Lolium, Zea mays and Avena sativa* (Table 5.1). In addition to the variety committee of the ISTA, the Biochemical and Molecular Techniques group of the UPOV (International Union for the Protection of New Varieties of Plants) is engaged in developing protocols for testing distinctness of varieties. Knowledge of genetic control for each of the protein/isozyme markers is required by the UPOV for inclusion in the guidelines for DUS testing. Electrophoresis protocols have been included as supplementary tests in the UPOV guidelines for identification of wheat and maize (UPOV 1994)

Table 5.1: Electrophoresis Methods Recommended by ISTA (1999) for Various Crop Species

Crop Species	*Method*
Triticum and *Hordeum*	Acid-PAGE of alcohol soluble proteins (gliadins from *Triticum* and Hordein from *Hordeum*) using 100 individual seeds.
Pisum and *Lolium*	SDS-PAGE of Tris-HCl soluble seed proteins (100 individual seeds are used in *Pisum* while in *Lolium* bulked seeds are analyzed).
Zea mays	Alcohol soluble proteins (zeins) separated by UTLIEF (200 individual seeds analyzed)
Avena sativa	Urea/ethylene glycol soluble proteins (avenins) separated by acid-PAGE

DNA Profiling Techniques

DNA polymorphism provides an extremely powerful tool for variety identification as variability at DNA level is much more compared to proteins or enzymes due to presence of large portion of non-coding regions. DNA profiling of cereals has not been adopted rapidly in industry because of the extensive experience and investment in more conventional approaches to sample identification. However, conventional techniques are relatively time consuming and do not always permit distinction between closely related cereal varieties. DNA based methods have recently been investigated as more generally applicable techniques for cereal identification (Ko and Henry 1994). These methods are broadly of two categories a) probe based techniques such as restriction fragment length polymorphism (RFLP) and b) PCR amplification technologies.

Restriction Fragment Length Polymorphism (RFLP)

This technique makes use of a unique class of restriction endonucleases (Type II) which have the ability to cut DNA in a precise and reproducible manner. These enzymes recognize specific base sequences usually 4-6 bp in length and cleave the DNA at these or adjacent sites. Due to inherited mutational changes in DNA, nucleotide base sequence differs between the individuals, which results in a loss or gain of restriction enzyme sites in DNA. Hence, digestion of total DNA of plant with a restriction enzyme will generate a unique restriction profile for an individual. Restriction enzyme digested DNA from two closely related individuals will appear as different sized fragments on agarose gel electrophoresis when visualized under UV light after staining with ethidium bromide. This difference in length of fragments obtained when DNA of two or more individuals is digested with a restriction enzyme is called restriction fragment length polymorphism.

In the most widely used procedure for detection of RFLPs, total DNA from the plant of interest is digested with a restriction enzyme and separated by electrophoresis in agarose gels. The DNA is then transferred from the gel to a nylon membrane by Southern blotting and RFLPs are detected by hybridization with labeled probes followed by autoradiography. The pattern of restriction fragments detected by one or more cloned probes can often be used as a taxonomic character (DNA fingerprint) to identify a cultivar or even an individual plant. RFLP analysis using randomly selected single copy or low copy genomic and cDNA clones as probes has revealed differences between varieties in a number of crop species such as wheat, barley, rice, maize and oats. Wang and Tanksley (1989) used this approach to fingerprint cultivars of rice. They analyzed 10 single copy RFLP markers on 70 rice cultivars selected to represent the total genetic diversity present in cultivated rice. Fifty eight of the 70 cultivars could be uniquely determined by combination of the probes used. Dallas (1988) used a human minisatellite probe to distinguish rice varieties, and Liu and coworkers (1992) isolated a moderately repeated sequence from wheat to produce hypervariable DNA fragments in common wheat. Vaccino and Corbelline (1993) reported the isolation of highly polymorphic RFLP probes specific for high molecular weight glutenins and γ–gliadins to identify more than 50 wheat cultivars from Italy and other parts of Europe. Drawbacks of this technique include its high cost, relatively low resolution and intensive labor requirements.

PCR Amplification Techniques

PCR is a procedure for the *in vitro* enzymatic amplification of a specific segment of DNA (Mullis and Faloona 1987). The capability of PCR to amplify specific nucleotide sequence millions of times facilitates the use of technologies for the analysis of DNA that would be difficult with conventional molecular biology techniques. PCR methods are rapid (6-8h), need very little quantities (ng) of target DNA, avoid the need for labeling, blotting and hybridization steps and are more amenable to automation.

In PCR, two oligonucleotide primers are used to hybridize to opposing DNA strands flanking the desired sequence in the target DNA so that DNA polymerase (Taq DNA polymerase, thermostable) can synthesize a copy of the intervening target DNA. The procedure results in an exponential amplification of the target fragment, approximately 2^n times where n is the number of cycles of amplification performed. The exponential amplification mediated by cycles of heating (94°C), cooling (35°C) and extension (72°C) can produce 10^6-10^9 copies of the target DNA sequence. Amplification products are separated and recorded by a variety of detection methods. Variations in primer sites on the target DNA and length variations between primer sites generate molecular polymorphism. Several variations of basic PCR are available among these randomly amplified polymorphic DNA (RAPD), amplified fragment length polymorphism (AFLP) and simple sequence repeat (SSR or microsatellites) are more commonly employed for DNA profiling and cultivar identification.

Randomly Amplified Polymorphic DNA (RAPD)

The discovery that arbitrarily chosen primers, made with no knowledge of any specific target DNA, can act as templates for the amplification of several fragments of genomic DNA (Williams *et al.*, 1990) has led to new generation of diagnostic techniques known as RAPDs (randomly amplified polymorphic DNA) and AP-PCR (arbitrary primed PCR). RAPD markers are based on the use of single short synthetic obligonucleotide (8-10 nucleotides long) primers of arbitrary sequence. The nature of the amplified fragments depends upon the primer sequence and on the DNA being analyzed. Different primers give rise to different amplified bands and polymorphism at the priming sites results in the disappearance of an amplified band. RAPDs in recent years have successfully been applied to distinguish between varieties, genotypes, lines and species of several crops and to find markers for important traits in cereal breeding.

While considerable work has been done using RAPD marker as a varietal identification method in numerous species, primarily because of its simple and rapid nature, however, this method is finding less favour now as more reliable methods in terms of reproducibility and amount of polymorphism detected between cultivars are available.

Amplification Fragment Length Polymorphism (AFLP)

ALFP technique is a recent innovation (Vos *et al.*, 1995) that has proven both powerful and reliable in fingerprinting cultivars. The technique has several advantages over earlier marker technologies– very little DNA is required, many polymorphisms can be generated per reaction and no prior knowledge of specific DNA sequences is needed. The following three steps are involved in the technique:

1. Digestion of genomic DNA with two restriction enzymes.
2. Ligation of adapter sequences to the restriction ends and
3. Selective amplification of sets of restriction fragments by two successive PCR reactions.

The PCR products are separated in denaturating PAGE and detection is done by silver staining or autoradiography. The technique has great potential for distinguishing cereal varieties because of large number of markers generated. AFLP has been used to differentiate genetically populations or individuals within numerous species including wheat and rice (Maheshwaran *et al.*, 1997; Ridout and Donini 1999).

PCR with Specific Primers

Amplification of specific sequences by PCR has many applications in cereal analysis. The advantage of working with a known gene or sequence is that the identity of the sample can be confirmed by comparing the PCR product with the size predicted for the target sequence, by sequencing the PCR product, or by hybridization with the PCR product. This technique has been applied for the identification of cereal species using a number of specific primers such as 5S rRNA primers, α-amylase gene primers and primers for storage protein genes (Table 5.2) and discrimination between very closely related varieties has been achieved (D'Ovidio 1993; Weining *et al.*, 1994; Ko and Henry 1996 and Ko *et al.*, 1996). For example, primers designed to be homologous to both conserved and variable regions in the α–amylase genes distinguished between barley varieties chebec (feed) and Schooner (malting). D'Ovidio (1993) used PCR of low molecular weight glutenin genes to select durum wheat genotypes with good quality. Hordein genes have been analyzed in a similar way in barley (Chee *et al.*, 1993).

Table 5.2: DNA Profiling in Cereals Using Polymerase Chain Reaction

Gene(s)	*Primers*	*PCR Products*	*Application*	*Reference*
Unknown	Arbitrary	Many (small) (10 mers)	Variety identification	Ko *et al.* (1994)
				Analysis of genetic
diversity				
Unknown	Consensus (intron/exon)	Many	Variety identification	Weining *et al.* (1991)
5S RNA	Specific	One to many	Species identification	Ko and Henry (1996)
α-amylase	Specific	One to many	Variety identification	Weining *et al.* 1994), Ko *et al.* (1996)
Storage proteins (glutelins, gliadins hordeins)	Specific	One to many	Variety identification	D'Ovidio(1993) Chee *et al.* (1993)
Short sequence Repeats	Specific	One to few	Variety identification	Zhao and Kochert (1993); Roder *et al.* (1995)

Simple Sequence Repeats (SSRs or Microsatellites)

Simple sequence repeats (SSRs) or microsatellites are DNA sequences that consist of two to five nucleotide core units such as $(AT)_n$, $(CTT)_n$ and $(ATGT)_n$ which are randomly repeated. Cereal genomes consist of at least 70 per cent repeated sequences. The dinucleotide repeat (AT)n is abundant in the barley genome, while the trinucleotide repeat (GGC)n has been found to be widely distributed in the rice genome and wheat contains high levels of (CAA)n and (CAG)n. The regions flanking the microsatellites are generally conserved among genotypes of the same species. PCR primers to the flanking regions are used to amplify the SSR containing DNA fragment, length polymorphism is created when PCR products from different individuals vary in length as a result of variation in the number of repeat units in the SSR.

Human geneticists first demonstrated the highly polymorphic nature of SSRs (Litt and Luty, 1989). In contrast to RAPD and AFLPs, SSR or microsatellite technology requires DNA sequence data that are species specific and development of primers for SSRs requires significant resources. However, once SSRs are developed they can be used to provide profiles that are highly discriminative and have been studied in several cereal systems such as maize (Senior and Heun, 1993); wheat (Roder *et al.*, 1995), rice (Zhao and Kochert 1993), barley (Beckman and Heun 1995) and sorghum (Smith *et al.* 2000).

In most plant species the level of polymorphism with microsatellites is considerable higher than found with RFLP markers. Given this high level of informativeness, abundance and random distribution in plant genomes combined with reliable amplification via PCR, SSR markers are likely to become an important and widely used DNA marker system in plants.

Conclusion

Interest in methods/technologies for varietal identification will remain exciting from a range of scientific and user point of view. Electrophoresis continues to remain the method of choice. However, the procedures available are required to be standardized for a universal acceptance. The analysis of electrophoretic polymorphism is likely to be based on more than one method or component and results will have to be matched by computer assisted digital gel image processing softwares. The newer molecular tools/techniques based on nucleic acid polymorphism may prove to be the ultimate tools for establishing the identity or purity of seed lots in commercial production, in segregating grain with differing processing qualities and associated market values and in the verification of labeling of foods derived from cereals. Technology for the utilization of DNA markers is evolving rapidly at the present time, and further advances are sure to occur soon. Some of these will involve making the process of developing and utilizing DNA markers technically simpler, less expensive and more capable of automation.

REFERENCES

Becker, J. and Heun, M. (1995). Barley microsatellites: allele variation and mapping. *Plant Mol. Biol.*, 27: 835-845.

Bietz, J.A. (1983). Separation of cereal proteins by reversed phase high performance liquid chromatography. *J. Chromatography*, 255: 219-238.

Chee, P.W., Pederson, L., Matejowski, A., Kanazin, V. and Blake, T. (1993). Development of polymerase chain reaction for barley genome analysis. *J. Am. Soc. Brew. Chem.*, 51: 93-98.

D'Ovidio, R. (1993). Single seed PCR of LMW glutenin genes to distinguish between durum wheat cultivars with good and poor technological properties. *Plant Mol. Biol.*, 22: 1173-1178.

Dallas, J.F. (1988). Detection of DNA fingerprints of cultivated rice by hybridization with a human minisatellite DNA probe. *Proc. Nat. Acad. Sci., USA*, 85: 6831-6835.

Dhillon, S. (2000). Electrophoretic Techniques. In: *Introductory Practical Biochemistry*, (Eds.) S.K. Sawhney and R. Singh. Narosa Publication, p. 237-264.

ISTA (1992). *Handbook of variety testing–electrophoresis testing*. (Ed.) Cooke, R.J., Zurish, p. 1-42.

ISTA (1999). International rules for seed testing. *Seed Sci. Technol.*, 27: 251-270.

Ko, H.L., Cowan, D.C., Henry, R.J., Graham, G.C., Fox, G.P., Chadbone, D.A. and Haak, I.C. (1994). Identification of cereals using the polymerase chain reaction. *J. Cereal Sci.*, 19: 101-105.

Ko, H.L. and Henry, R.J. (1996). Specific 5S ribosomal RNA primers for plant species identification in admixtures. *Plant Mol. Biol. Rep.,* 14: 33-37.

Litt, M. and Lutty, J.A. (1989). A hypervariable microsatellite revealed by *in vitro* amplification of a dinucleotide repeat within the cardiac muscle action gene. *Am. J. Hum. Genet.,* 44: 391-401.

Liu, Y.G., Ikeda, T.M. and Tsunewaki, K. (1992). Moderately repeated, dispersed and highly variable (MRDHV), genomic sequences of common wheat usable for cultivar identification.*Theor. Appl. Genet.,* 84: 535-543.

Maheshwaran, M., Subudhi, P.K., Nandi, S., Xu, J.C., Parco, A., Yang, D.C. and Huang, N. (1997). Polymorphism, distribution and segregation of AFLP markers in a double haploid rice population. *Theor. Appl. Genet.,* 94: 39-45.

Mullis, K.B. and Faloona, F.A. (1987). Specific synthesis of DNA in vitro via a polymerase chain reaction. *Methods Enzymol.,* 155: 835-850.

Ridout, C.J. and Donini, P. (1999). Use of AFLP in cereals research. *Trends Plant Sci.,* 4: 76-79.

Roder, M.S., Plachke, J., Konig, S.U., Borner, A., Sorells, M.E., Tanksley, S.D. and Ganal, M.W. (1995). Abundance, variability and chromosomal location of microsatellites in wheat. *Mol. Gen. Genet.* 246: 327-337.

Senior, M.L. and Heun, M. (1993). Mapping maize microsatellite and polymerase chain reaction confirmation of the targeted repeats using a CT primer. *Genome,* 36: 884-889.

Smith, J.S.C. (1986). Biochemical fingerprints of cultivars using reversed phase high performance liquid chromatography and isozyme electrophoresis: A review. *Seed Sci. Technol.,* 14: 753-768.

Smith, J.S.C., Kresovich, S., Hopkins, M.S., Mitchell, S.E., Dean, R.E., Woodman, W.L., Lee, M. and Porter, K. (2000). Genetic diversity among elite sorghum inbred lines assessed with simple sequence repeats. *Crop Sci.,* 40: 226-232.

Smith, J.S.C. and Wych, R.D. (1986). The identification of female selfs in hybrid maize: a comparison using electrophoresis and morphology. *Seed Sci. Technol.,* 14: 1-8.

Smith, J.S.C. and Smith, O.S. (1992). Fingerprinting crop varieties. *Adv. Agron.*, 47: 85-140.

UPOV (1994). Guidelines for the conduct of tests fot distinctness, uniformity and stability of maize (*Zea mays* L.).

Vaccino, P., Accervi, M. and Corbellini, M. (1993). Cultivar identification in *T. aestivum* using highly polymorphic RFLP probes. *Theor. Appl. Genet.*, 86: 833-836.

Vos, P., Hogers, R., Bleeker, M., Reijans, M., Lee, T.V.N., Homes, M., Frijters, A., Pot, J., Peleman, J., Kuiper, M. and Zabeau, M. (1995). AFLP: A new technique in DNA fingerprinting. *Nuc. Acids Res.*, 23: 4407-4414.

Wang, Z.Y. and Tanksley, S.D. (1989). Restriction fragment length polymorphism in *Oryza sativa* L. *Genome*, 32: 1113-1118.

Weining, S., Ko, H.L. and Henry, R.J. (1994). Polymorphism in α-amyl gene of wild and cultivated barley revealed by the polymerase chain reaction. *Theor. Appl. Genet.*, 89: 509-513.

Weining, S. and Langridge, P. (1991). Identification and mapping polymorphism in cereals based on polymerase chain reaction. *Theor. Appl. Genet.*, 82: 209-216.

White, J. and Cooke, R.J. (1992). A standard classification system for the identification of barley varieties by electrophoresis. *Seed Sci. Technol.*, 20: 663-676.

Williams, J.G.K., Kubelik, A.R., Livark, K.J., Rafalski, J.A. and Tingy, S.V. (1990). DNA Polymorphisms amplified by arbitrary primers are useful as genetic markers. *Nucleic Acids Res.*, 18: 5631-5635.

Wu, K.S. and Tanksley, S.D. (1993). Abundance, polymorphism and genetic mapping of microsatellites in rice. *Mol. Gen. Genet.*, 241: 225-235.

Yan, Y., Liu, G.T., Prodanovic, S. and Zoric, D. (1998). Discrimination between some Chinese and Yogoslav winter wheat cultivars by capillary electrophoresis of gliadins. *Seed Sci. Technol.*, 26: 839-843.

Zhao, X. and Kochert, G. (1993). Phylogenetic distribution and genetic mapping of a (GGC)n microsatellite from rice (*Oryza sativa* L.). *Plant Mol. Biol.*, 21: 607-614.

6

Processing and Food Uses of Grain Legumes

Umaid Singh

Department of Foods and Nutrition,
CCS Haryana Agricultural University, Hisar – 125 004, Haryana, India

ABSTRACT

Nutritional quality of grain legumes with respect to their chemical composition and anti-nutritional factors have been the subject of numerous investigations in the past two decades or so. Relatively less attention has been paid to the processing aspects that affect the overall utilization of food legumes. Two very important processing considerations are the product development and consumer acceptance of the product. These are greatly influenced by the primary processing practices such as dehulling, decortication, pearling etc. and the inherited grain characteristics such as the size, shape and colour and, chemical composition of the grain. The storage conditions and any pre-treatment received during processing also influence the product quality. Under adverse storage conditions, grain legumes can develop hard-to-cook defects, depending upon genotypes and cultural practices. The secondary processing technologies such as roasting, steaming, germination, fermentation and extrusion cooking, will play an important role in determining the product quality of pulses. Greater emphasis should be given for marketing the new product of grain legumes.

Introduction

Grain legumes, also called pulses, are increasingly becoming important in human nutrition, because they are less expensive sources of proteins than the animal proteins especially in the developing countries. In addition, they are valuable sources of carbohydrates, minerals and vitamins. Their protein quality for food is low, however, when mixed with cereals the total diet quality can exceed 70 per cent of casein or lactalbumin. In India, judiciously mixing cereals and legumes to provide a good quality diet was in practice long before people understood the nutritional importance of this practice. Food legumes or grain legumes are frequently referred to as pulses in the Indian subcontinent.

The majority of pulses are consumed in the regions where they are grown after processing by suitable methods. Like other countries, consumers in India are becoming increasingly aware of the processing and grain quality factors that affect the nutritive value of the food they consume. This article reviews and highlights the work on various aspects of processing quality of pulses with special emphasis on food uses as influenced by traditional and modern methods of processing.

Primary Processing: Dehulling of Dry Seeds

The fibrous seed coat, or testa ('husk or skin') of pulses, with a high content of cellulose and hemicellulose, is often indigestible, and sometimes has a bitter taste probably due to the presence of tannins. In view of this, not only does primary process improve palatability and digestibility of legume seeds, it also reduces their cooking time. Proper dehulling of pulses is essential for marketing and human consumption. Two major steps:

1. Loosening and removal of seed coat, and
2. splitting the cotyledons, are involved in a typical dehulling operation.

Methods of Dehulling

Dehulling methods can be broadly classified by scale of operation into small-scale processing, generally adopted by households in villages and large-scale processing adopted by the commercial mills or agro-processing industries in urban areas. For small-scale processing, wooden pestle and mortar, and stone chakki (quern) are used for dehulling chickpea, field pea and lentil in several Asian countries. For large-scale processing, carborundum rollers and discs are employed by the commercial agro-processing units. These large-scale processing operations are carried out by the dhal mills located in cities in urban areas of the country.

Attrition-type dehullers and roller mills are particularly suitable for dehulling and splitting legume grains with loose seed coats; whereas abrasive-type dehullers are suitable for dehulling grains with more tightly adhering seed coats. The use of an emery-coated roller is a common practice in commercial dhal mills in India (Singh 1995). The emery-coating, also called carborundum, is made of silicon carbide (carbon + crystallized alumina) and used for abrasive or refractory action. Some millers use a carborundum roller for both dehusking and splitting, whole others use a roller and disc sheller alternatively for this purpose to reduce the dehulling losses.

The disc sheller is generally used for wet processing and works on the principles of attrition, which removes the husk and splits the cotyledons simultaneously resulting in excessive breakage of legume cotyledons. The disc shellers are generally used for dehusking rice.

Dhal yields of chickpea are considerably higher than other pulses (Table 6.1) indicating that dehulling losses in terms of by-products such as broken chips, powder, and husk of pulses are highly variable. The seed characteristics and agronomic factors that would affect these losses in pulses when grown in different environments need to be examined carefully because they influence the market price of the commodity. In some countries, legume decorticators are typically used for dehulling legumes in large-scale processing. The outer layers of legume cotyledons are invariably scarified by the abrasive action of the dehulling process resulting in considerable nutrient losses (Singh *et al.*, 1992a). Considerable amounts of nutritionally important nutrients such as protein, calcium, and iron are lost in the form of powder fraction of chickpea dehulling process (Table 6.2). These losses of dehulling operation could be reduced by developing better dehulling methods and by selecting for more suitable genotypes of pulses. This will have significant beneficial effects on market price of processed product and the nutritional of status of people who consume these legumes in their daily diets.

Table 6.1: Processing Yields (Decorticated Dry Split Cotyledons) of Different Legumes Obtained by Commercial Processing Mill[a]

	Seed Coat (%)[a]	*Dhal Yield (%)*	*Theoretical Yield*
Chickpea	14.2	80.0	85.8
Pigeonpea	15.6	72.5	85.4
Field pea	10.0	78.0	90.0
Lentil	8.4	73.0	91.6

[a] Average of values for different sources.

Table 6.2: Effect of Dehulling on Chemical Constituents (g/100 g) of Dhal and Powder Fractions of Chickpea cv. Annigeri[a]

Dehulling Time (min)	*Dhal*					*Powder*				
	Protein (%)	*Sugar (%)*	*Starch (%)*	*Calcium*	*Iron*	*Protein (%)*	*Sugar (%)*	*Starch (%)*	*Calcium*	*Iron*
0	18.6	6.8	56.2	43.7	5.7	–	–	–	–	–
2	18.0	6.5	57.8	39.5	5.0	23.6	12.1	48.0	85.0	12.0
4	17.5	6.3	57.8	38.0	4.8	21.8	10.5	50.3	65.5	10.5
8	17.5	6.0	58.0	35.5	4.3	19.8	9.5	52.0	45.0	8.5
12	16.4	6.1	60.8	35.0	3.8	18.9	8.6	55.4	45.0	7.0
SE+	0.18	0.21	0.31	0.78	0.21	0.21	0.13	0.51	1.90	0.32

[a] For each treatment, results are averages of two samples obtained as in Table 6.1 and analysed separately. Results are expressed on a moisture-free basis. Source: Singh *et al.* (1992a).

Factors that Affect Quantitative and Qualitative Dehulling Losses

Two factors are extremely important for dehulling; (I) Firstly, it is conditioning of the seed material and (II) Secondly, the seed characteristic itself that would influence the dehulling losses.

Conditioning

Loosening of seed coats achieved by suitable pre-conditioning pre-treatments generally results in higher dhal yield from chickpea and field pea (Table 6.3). Among chemical pre-treatments, soaking in sodium bicarbonate followed by oven-drying may produce the highest dhal yield both in chickpea and pigeonpea (Table 6.3). This pre-treatment not only loosens the seed coat but also reduces the cooking time of dhal. Soaking in water (2-14 hr) followed by sun-drying is common practice followed by the consumers for dehulling small quality of pulses in India. This is primarily followed to loosen the seed coat and subsequently increase the dhal recovery. It may be emphasized that dehulling can be rendered more easy by prolonged soaking in water but the dhal remains uncooked and tough even on prolonged boiling. Pre-heating the seeds at higher temperature help in loosening the seed coat and increasing the dhal yield, but excessive heating may result in breakage of cotyledons and reduce the dhal yield. Therefore, the preconditioning treatments should be carefully standardized and followed to increase the dhal yield of cool season legumes. Some interactions between pre-conditioning treatments and legume genotypes may be expected. In this context, the newly developed high yielding genotypes of cool season legumes should be examined for their pre-conditioning effects on dhal yield.

Table 6.3: Effect of Preconditioning Treatments on Dhal/Split Yields of Chickpea and Field Pea[a]

Treatment	*Dhal/Split Yields*	
	Chickpea	*Pigeonpea*
Control	71.1	65.4
Water	73.6	66.3
Sodium chloride (1.0 per cent)	75.2	–
Sodium bicarbonate (1.0 per cent)	76.1	71.1
Edible oil (0.5 w/w)	–	70.5
Preheating	–	71.0

[a] *Source*: Singh (1995).

Seed Characteristics

Several characteristics affect the dehulling efficiency in terms of increasing the dhal yield reducing nutrient losses in different legumes. Dhal yields primarily depend on the content of seed coat of legume genotypes (Table 6.1). Genotypes with lower seed coat content lead to higher dhal yields. In addition, seed size, shape and hardness also influence the dhal yield; very small-to-small seeds are generally difficult to dehull. Greater than 75 per cent of the variability in dehulling efficiency or yield of legumes could be accounted for by grain hardness and resistance to splitting into individual cotyledons (Reichert *et al.*, 1984). That would suggest that losses in terms of broken and powder fraction would be greater, if grains of legumes are too hard, requiring more abrasive force during the dehulling operation. On the other hand, if grains are too soft, they would tend to break during the operation resulting in more losses in terms of broken fraction and powder. This characteristic needs to be carefully evaluated for monitoring and improving the dehulling quality of grain legumes.

William *et al.* (1993) reported that efficiency of dehulling and splitting of lentil is favoured by large seed size, thin testa, short storage period and correct wetting (conditioning) practices. Further they reported that very angular seeds lose excessive amounts during the dehulling, because the dehulling process attacks sharper edges preferably and more seed mass is removed from flatter seeds. Generally, rounder seeds split more readily than the flatter seeds. Variability in dehulling characteristics of pulses is influenced by genotype dhal yield ranging between 71.1-85.7 for field pea and for chickpea (67.7-84.8 per cent) and lentil (62.1-80.2 per cent). However, the role of agro-climatic conditions in influencing the dhal yield of food legumes is not clearly defined. Faba beans are usually consumed as whole seeds after soaking and cooking, or as bean flour, and are not usually dehulled. But such legumes are suitably procured by other methods such as germination, roasting, fermentation etc. before being consumed as human foods. These have been discussed in the section that follows:

Secondary Processing and Food Uses

Secondary processing practices for grain legumes generally include soaking, boiling, frying, roasting, steaming, baking, germination, fermentation, extrusion etc. Such processes are directly involved in making food products or value-added snack items depending on the type of food and region of consumption. All these processes improve appearance, texture, cooking quality and appearance. Some important food products of pulses prepared by such practices and consumed in different regions.

The grain legumes are often soaked in water as a pre-treatment to reduce their cooking time. Soaking has the additional beneficial effects of either completely or partly reducing content of undesirable substances such as phytate, trypsin inhibitors, tannins and flatulence-forming carbohydrates. Boiling is an indispensable process and most commonly used to cook chickpea, faba bean, field pea and lentil products. Among several market demands, the cooking time is an important consideration. Large differences exist in cooking time of these legume grains. Cooking time is a fairly heritable characteristic in that if genotypes differing widely in cooking time are grown at different locations, the differences between genotypes will persist. Generally, cooking time is positively and significantly correlated with seed size in some pulses. The cooking time is greatly reduced by dehulling of grains. Cooking has been reported to improve the carbohydrate digestibility, possibly by changing the content and chain length of the amylose components.

Soaking and frying processes could prove to be very helpful in expanding the snack market for these legumes. The usual process is to soak dry field peas and chickpeas overnight, and fry them in hot oil. Sometimes peas could be coated with other materials such as rice flour before frying to provide different flavouring characteristics. The product is then seasoned, packaged and marketed.

Germination has specific effects in terms of reducing antinutritional factors such as protease inhibitors, and oligosaccharides in legumes. In addition, it helps develop desirable flavours for certain food products. Germinated and boiled faba beans make a very good sauce. Very specific snack items can be prepared by the roasting process alone and there are clear-cut differences in the roasting quality of chickpea genotypes. Although India is the largest producer of chickpea in the world, there are only a limited number of genotypes that are suited for making roasted products such as 'phutana', which is very commonly consumed as snack item in the Indian subcontinent. The market demand for 'phutana' type chickpea is very high and development of genotypes for this purpose will play a rich dividend.

Except for mungbean, fermented products of food legumes are not popular with traditional consumers. By using some pulses, tempe, natto, dhokla, kiyit injera being the important fermented products, could be prepared and popularized, particularly in the country. Extrusion is relatively new process involving the application of both pressure and heat. This process has wider application in making value-added products from legumes, particularly from dry peas and chickpeas. Peas are ground into fine flour that is passed through an extruder under pressure to create different sizes and shapes. The extruded shapes may be fried, seasoned and packaged.

Greater emphasis be given for marketing the new products of legumes. For example, quick-cooking dhal, also called instant dhal of grain legumes has good market potential. The most serious drawback in the utilization of food legumes is their long-cooking time. Even though dehulling and splitting into dhal reduces the cooking time considerably, the cooking process is time and energy consuming. Special soaking solutions containing inorganic salts have been employed for quick-cooking of legumes. The cooking time of field peas could be remarkably reduced by soaking in sodium carbonate and bicarbonate solutions (Singh *et al.*, 2000). However, such treatment adversely affects the product quality. Pectinase enzyme treatment significantly decreased the cooking time as compared to salt solutions and it also improved the general acceptability of the product (Singh and Rao 1995). These legumes can be used in various foods including noodles, breads, snacks, soups, stews, and purees. Legume flour can be blended up to 20-30 per cent in sausages and beverages with acceptable results. Chickpea flour can be used in tortilla and unleavened bread. Protein concentrates by pin-milling and protein isolates by precipitation of these legumes can be prepared containing approximately 50-60 per cent and 80-90 per cent protein, respectively. They should have a place in second generation functional foods, with and without blending with cereals.

Storage Practices

Storage conditions affect the processing and grain quality of food legumes which are generally stored at farmer, trader and government levels in various types of storage structure. The quantitative and qualitative losses occur during storage. The influence of storage on processing and grain quality is the result of interactions among storage conditions, physiological changes during aging and the activity of storage microorganisms and insects. Pulse beetle popularly known as bruchid is an important storage pest of chickpea, lentil and field pea. Sometimes, the insect begins its infestation in the field and is subsequently carried with the grain into the stores after harvesting. The dehulling yields of pulses are considerably reduced as a result of insect infestation. This happens due to physical damage to the seed by the bruchid. The greater susceptibility of large-seeded cultivars of soybean to field deterioration was used to explain their poor storability in comparison to small seeded cultivars.

Noticeable differences in susceptibility of pulse cultivars to bruchid attack have been observed. However, seed characters such as size, colour and texture have not been related to bruchid preferences, but seed-coat thickness may influence the bruchid influence. Breeders need to develop legume varieties with increased resistance to postharvest infestation.

Storage under adverse conditions of high temperature and high humidity may render these legumes susceptible to a hardening phenomenon also known as hard-to-cook (HTC) defect. Particularly, beans with this defect are characterized by extended cooking times, are less acceptable to the consumer and are of lower nutritive value. The mechanism causing the HTC phenomenon in food legumes could possibly be due to (*i*) limited hydration of intercellular protein; (*ii*) pectin insolubilization in the middle lamella by Ca and/or Mg; and (*iii*) cross-linking of phenolics (lignification) and/or protein in the middle lamella of cell wall. None of these phenomenon has been systematically studied in cool season food legumes. The accumulation of uric acid in pulses stored for 4-5 months generally make them unacceptable to the consumers. Protein quality is also adversely affected when pulses are stored for long period.

The harmful effects of fungi and bacteria on stored pulses have been established. Chickpeas can be contaminated with mycoflora and mycotoxins as a result of storage. Aflatoxin B1 and B2 are also were present in some pulses when stored under adverse climatic conditions. The low germination of pea seeds following commercial storage has been attributed to *Aspergillus* infection. Storage effects on some quality parameters such as colour and hydration capacity of faba bean, lentils and field peas have been noticed. As the storage period increased, the time taken for cooking also increased irrespective of the type of pulse stored and the treatment given. The loss of protein solubility and digestibility, off-flavour development, and oxidative rancidity as a result of storage of cool season food legumes are the other parameters that need to be studied.

REFERENCES

Singh, U., Williams, P.C. and Petterson, D.S. (1997). Processing and grain quality of pulses to meet the market demands. *Int. Food Legumes Conf.*, 26-28 Sept. 1997, Adelaide, Australia.

Williams, P.C., Erksine, W. and Singh, U. (1993). Lentil processing. *Lens Newsletter, ICARDA, Syria,* 20: 3-13.

Singh, U., Rao, P.V. and Seetha, R. (1992a). Effect of dehulling on nutrient losses in chickpea.

Singh, U., Santosa, B.A.S. and Rao, P.V. (1992b). Effect of dehulling on nutrient losses in chickpea.

Singh, U., Santosa, B.A.S. and Rao, P.V. (1992b). Effect of dehulling methods and physical characteristics of grains on dhal yield of pigeonpea. *J. Food Sci. Technol.*, 29: 350-353.

Reichert, R.D., Oaman, B.D. and Young, C.G. (1984). Factors affecting the abrazine type of dehulling of grain legumes investigated with a new intermediate-sized batch dehuller. *J. Food Sci. Technol.*, 49: 267-272.

Singh, U., Sehgal, S. and Tomer, Y.S. (2000). Influence of dehulling, soaking solution and enzyme treatment on the cooking quality of improved varieties of pulses. *J. Food Sci. Technol.*, 37: 627-630.

7

Processing and Utilization of Coarse Grains

Asha Kawatra

Department of Foods and Nutrition, CCSHAU, Hissar

Coarse grain occupy important place in the grain economy of the world. Their agricultural importance arises from the fact that they are hardy crop as compared to wheat and rice, could be grown in regions with comparatively low rainfall and can stand to greater extent vagaries of weather and other environmental changes. Coarse grains generally available and used in country like maize, barley, rye, sorghum and various millets like pearl millet, fingermillet, foxtail millet, poromillet etc.

The coarse grains including millets are comparable and even superior to major cereals with respect to protein, energy, vitamins and minerals. These grains are highly nutritious and in many respects are superior to rice and wheat. For example, dehusked proso millet has a protein content of 12.5 per cent whereas wheat has about 11.8 per cent. Thiamine present in foxtail millet is approximately three times more than that of rice while carotene content is sixteen times higher than that of rice. Finger millet is known for its high calcium content of 344 mg/100 g edible grain compared to wheat with only about 41 mg/100 g grain (Hegde and Linge Gowda, 1989). Beside they are rich in fibre and contain phytochemicals, micronutrients, neutracenticals and hence now a days lightly termed as nutricereals (Desikachar, 1977). In addition higher level of non-starchy polysaccharides as well as uniqueness of nature of starch of millets, makes its digestion slower as compared to rice starch. These qualities give them status of health cereals (Lakshmi and Vimala, 1996).

The coarse grain and millets are generally consumed by the people from socio-economically weaker section of society. Poor grain quality characteristics such as rough texture, high fibre content, lack of gluten and typical flavour limit their uses in various food preparations. The outermost layer of husk is tough and irritating to tongue. They also require more cooking time and have relatively poor digestibility and low bio-availability of minerals due to the presence of inherent antinutritional factors.

The food use of these crops is very much restricted to traditional consumers mainly due to their non-exposure to these food grains as well as nonavailability in ready to use or ready to eat forms and other novelty products (Malleshi and Desikachar, 1985).

The simple processing method which could improve their acceptability without seriously affecting their nutritive value and diminishing the total food availability are however necessary. Malting, blanching, soaking, sprouting, cooking, roasting may be good choice in this context as there are reports indicating that these processing enhances the nutritional quality and utility of these grains.

Exploratory studies on popping and extrusion cooking of sorghum and millets are encouraging. Such products have shown better shelf life. Among tropical cereals, millets have a pride place for malting. Millet malt is store house of digestive enzymes and when slurry made out of malt flour, mixed with water is heated to boiling the starch gets hydrolysed to dextrin and lower molecular weight carbohydrates which reduces the water holding capacity of foods. Malt is also being used by the brewing industries. Millets are small sized grain cooked to soft texture. This unique characteristic could be used to develop high fibre flakes.

According to scientific literature some research work has been carried on formulation of weaning. A supplementary food prepared from the coarse grains in combination with other cereals and pulses.

Maize

Maize (*Zea mays*) knows as 'corn' is utilized in more diversified ways than any other cereal. Maize grains may be white, yellow or reddish in colour. With its high percentage of carbohydrates, lipid and protein, it is nutritious for human consumption. A high percentage of maize grown in developing countries is used for food, in India it is 80-90 per cent. The principal maize varieties are flint corn, dent corn, sweet corn, pop corn, flour corn and waxy corn. The classification is based on the nature and distribution of starch in the endosperm.

Processing and Utilization

Milling

Maize is milled by a dry or wet process. In both processes, the germ is separated from the grain in order to extract and recover germ oil. Wet milling involves a steeping stage and complete disintegration of the endosperm in order to recover starch and protein. In dry milling, the object is to recover the maximum amount of grits with the minimum amount of flour. In India, maize is ground for wholemeal atta in power driven or hand operated chakkis. In large scale milling, maida and suji are produced. Grits are used in the preparation of porridge, to make corn flake, as a brewing adjunct and to manufacture glucose by hydrolysis. One is extracted from germ, while bran and germ meal are utilized as animal feed. Maize is wet milled to obtain starch, oil, cattle feed and the products of starch hydrolysis, *viz.* Liquid and solid glucose and syrup.

Table 7.1: Corn Based Foods Made with Whole Corn or by Traditional Stone Grinding

Type	*Specific Products*
Raw corn	Corn salad
Fermented corn	Pickled miniature corn
Fried products	Corn and mush, corn fritters, hush puppie

Sorghum

Sorghum (*Sorghum vulgare*) is a unique cereal due to its drought tolerance and adaption to tropical conditions. So it is major staple food in Africa and India. In India, it is known as 'jowar'. The whole grain or broken may be cooked like rice or the whole grain ground to flour and used to make chapaties.

Table 7.2: Uses for Dry-milled Corn Products

Products	*For use*	*Industrial use*
Flaking grits	Breakfast cereals	
Coarse grits	Breakfast cereals	
	Snacks	
Medium grits	Breakfast cereals	
	Snacks	
Fine grits	Brewing	
	Maize porridge	
Meal	Brewing	Wall paper paste
	Snacks	
	Pancake, waffle, muffins	
Cones	Snacks	Dusting agents
	Combread	Handsoap
	Breakfast cereals	
Flour	Baby foods Frozen foods	Fermentation media Explosives
	Biscuits	Wallboard
	Wafers	
	Pancake mixes	
	Meat extender	
	Breading, batters	
Speciality products	Confectionery toppings Jam and Jelly extenders	Foundry binders label adhesives
	Fig paste extenders Coconut extenders	Edge paste
	Thickening agents	
	Dough conditioners	
	Brewers flakes	
	Meat extenders	

Processing and Utilization

Sorghum grains are processed by dry milling, wet milling and by fermentation. The products of milling are chiefly starch and feed products. Dry milling is used to obtain products low in fibre, fat and ash, and wet milling to make starch and its derivatives. The residue from wet milling is used as a feed.

Sorghum starch is indistinguishable from corn starch and can be used interchangeably with corn starch in most industrial applications. Sorghum oil obtained from germ fraction, after refining, is used for salads and general cooking. Kernel residues containing bran and gluten are processed as cattle feed.

Sorghum is as good as barley for malting purposes. The sorghum malt is used in the preparation of infant foods and malt extracts produced from sorghum are employed in the pharmaceutical industry. Sorghum malt has been widely used in Africa for brewing "Kafir beer".

Sorghum and wheat flour blends have been used to produce an array of baked products including yeast leavened breads, cakes, muffins, cookies, biscuits, flour tortillas and others. The level of sorghum substituted for wheat flour depends upon the strength and quality of gluten in the wheat flour, the baking procedure, the definition of acceptable bread, the colour and particle size of the sorghum flour and the use of gums, emulsifiers and other additives. Usually 5-10 per cent sorghum flour is substituted for wheat flour.

Sweet sorghum varieties which produce extra quantities of sugar in the juice are used to produce sorghum syrup or molasses in the southern United States.

Barley

Barley is an important crop worldwide and it belongs to genus Hordeum which includes both wild and cultivated forms of barley. In both types of barley, there are hulled and hulless varieties. It was used as a bread grain in early times. Nowadays, barley is extensively used as food, cattle feed and for malting of brewing.

Processing and Utilization

Barley is milled to make blocked barley, pearl barley, barley grits, barley flakes and barley flour. Barley is used worldwide for animal feed, brewing and human food. In India, it is mainly used as human food. Its nutritional quality is equal to or better than wheat. Barley is eaten as popped barley, barley flakes, sprouts, barley starch and as sweeteners, barley malt flour supplements, malted milk, infant food syrups, barley tea or coffee substitute, and rice extender. Barley flour is generally mixed with wheat or gram flour for preparing chapaties. Wheat flour mixed with barley or barley and rye or oats flour is used in making bread. Barley is steamed, milled and supplemented with vitamins and minerals, it will produce an infant baby food where all the protein and carbohydrates come from barley.

Another important use of barley is in malting, brewing in the manufacture of beer and for distilling in the manufacture of whisky. Barley is used in preference to wheat and rye for malting because of its specific physical and chemical properties. In addition to brewing beer, the remaining 5 per cent of barley malt is used in the US for baking, malt extracts (powders and syrups), diastate (energy rich powder), whisky, gin, vodka, alcohol and malt vinegar.

Other uses of malt include textile designing, pharmaceutical preparations, breakfast cereals, malted milk concentrates, infant foods, bakery products and candies.

Oats

Oats (*Avena sativa*) forms one of the most nutritious cereals for human use and rank sixth in world cereal production after wheat, maize, rice, barley and sorghum. The chief disadvantage is the difficulty in removing large proportions of hulls from the grain. Oatmeal obtained from the removal of husk is a common article of breakfast. Oatmeal flour is not suitable to bread making.

Processing and Utilization

The steps involved in the milling are cleaning, drying, or light roasting of oats to a moisture content of 6 per cent and dehulling, cutting and flaking and oat flour.

Most oats used as food are consumed as whole grain products. The major food uses of oats are in hot cereals, cold cereals, bread products, cookies and infant foods. Oats products are often used in bakery items to produce desired texture or to increase moisture retention. The oat bran is rich in soluble fibre that is active in lowering blood serum cholesterol levels. Oat flakes are manufactured from pinhead meal by cooking this meal in a steamer, rolling the cooked product while hot, moist and plastic, between heavy rollers and drying the flakes so formed.

Rye

Rye (*Secale cereale*) is cultivated extensively in Europe and it is the bread grain of the poor people.

Processing and Utilization

Rye milling industry is in Europe. The rye is cleaned and conditioned and then passed through successive break rolls. The husk is aspirated and finally the purified middlings are passed through reduction rolls. Rye flour thus obtained is dark in colour.

The products of rye milling are flour for soft breads, coarse rye meal for hard breads and rye flakes for hot breakfast cereals. Rye flours are used occasionally in snack-type crackers. A very light rye has been recommended for use in soda crackers and some cookies as partial replacement for wheat flour. Rye flours are also used in some pancake and waffle mixes to produce a special flavour. Rye flours are used as fillers for sauces and soups. Rye is also flaked in the United States to make a hot breakfast cereal. Rye grain may be fermented to produce alcoholic beverages or industrial alcohol.

Millets

Millets are a set of highly variable small seeded plant species indigenous to many areas of the world. Millets are of value especially in semi-arid regions because of their short growing season and higher productivity under heat and drought conditions. Often millets are produced in areas where maize and sorghum crops usually fail. The grain and forage are valuable as food and feed resources in Africa, Russia, India and China. Important millets grown in India include Pearl millet, Finger millet, Foxtail millet, Kodo millet, Proso millet, little millet, Barnyard millet etc.

Pearl Millet

Pearl millet (*Pennisetum glaucum*) is an important food crop of arid and semi-arid regions of the world. It is produced to a tune of 7.64 MT during 1997-98. Pearl millet forms a major source of dietary protein to larger section of the population and supplies 80-90 per cent of the calories for millions of poor. Its nutritive value especially protein, fat and mineral content is comparable to other cereal food grains. It is known by various common names like cattail millet, candle millet, spiked millet, bulrich millet etc. In India, it is popular as Bajra/Bajri. Processing techniques like sprouting, fermentation, decortication, blanching, popping, extrusion cooking, etc. are used for improving its utilization.

Processing and Utilization

Malting is commonly used process. Malting has been reported to cause a significant reduction in the antinutritional components and increase in the *in vitro* digestibilities of the malt (Chaturvedi and Sarojini, 1996; Archana *et al.,* 1998). Malted flour has been utilized for development of various supplementary foods (Archana, 1998). Malt is also used for alcoholic beverages.

Blanching of pearl millet is done by adding grain to the boiling water in the ratio of 1:5 and boiled for 30 seconds and then dried at 50°C for 60 min. Blanching completely arrests the lipolytic deterioration of flavour and thus enhances the shelf life of the pearl millet flour upto one month (Chavan and Kachare, 1994; Rekha, 1998). The blanched flour can be used for product development.

Germination improves the vitamin content of the grain and lowers the level of lipids, phytate and oxalate. Kumar and Chauhan (1993) reported that sprouting increased total protein, ash, ascorbic acid, total soluble sugars, reducing non-reducing sugar and sprouts incorporated in foods such s salad, baby foods, cake and biscuit were well acceptable.

Popping is a simple traditional technology enables to produce ready-to-eat foods which can be utilized for the production of low cost nutritious supplementary foods. Popped starches in comparison to native starches exhibited higher susceptibility to *in vitro* enzymic digestibility. Fermentation without prior malting, hydrolyzes starch, softens flour particles and lowers the pH which helps bleach the flour and slightly increases protein digestibility.

Product Development

Pearl millet is ground into flour and made into chapaties. It is also used in porridge and the grain is sometimes eaten after it is parched, the product being similar to popcorn. Pearl millet is also used for preparation of traditional and non-traditional recipes like cake, biscuit, dhokla, kheer, shakkarpara, suhali, pakora, chat, cookies, bread etc. It is also used to make weaning food mixtures. Raw and processed pearl millet can be sued for development of extruded products. It is also utilized for making certain diabetic foods.

Finger Millet

Finger millet (*Eleusine coracana*) is cultivated for human food in Africa and Southern Asia. The production of finger millet in India in 1994-95 was 2,430.4 million tonnes (Anon, 1996). In India, finger millet ranks third in area and production after sorghum and pearl millet. It is principal food grain of the rural population of Southern India. Generally the grain colour of finger millet varieties is brown but some white varieties are also available.

Processing and Utilization

Generally the finger millet seeds are powdered and wholemeal is utilized for food preparations or the whole grains are germinated to make malt for direct consumption or for brewing (Malleshi, 1986). Some other techniques for processing of finger millet such as popping, flaking, extrusion, parboiling have also been reported.

Finger millet possess superior malting characteristics (Malleshi and Desikachar, 1986). Finger millet has agreeable flavour with adequate starch hydrolysing enzymes. The amylase activity in malted flours ranges from 75-200 units after 3 days of germination (Malleshi and Desikachar, 1979). Finger millet is also malted for brewing purpose.

Finger millets are also popped to prepare ready-to-eat products. Popped grain is crunchy, porous, precooked product. Popped flour is also obtained which is blended with other legumes to form nutritious supplement.

Parboiling of millet grain is expected to help in dehulling. It completely eleminates the stickiness in cooked finger millet. Dumpling introduces a little grittiness or bite into products and improves consumer acceptability (Desikachar, 1975).

Finger millet flour can be blended with roasted mung bean flour and dried skim milk powder which is then extruded to make precooked ready to eat weaning foods. Extrusion enhances the *in vitro* protein digestibility (Malleshi *et al.*, 1996).

Finger millet can also be utilised for development of various bakery products by partial supplementation of white flour with finger millet flour. Finger millet and its flour ash been utilized in various traditional recipes consumed in southern India. Products developed from finger millet include idli, dosa, poushtik roti, dhokla, chakli, pakoda, halwa, ampli, doughnut, papad etc. These products can be used conveniently with no deviation from day to day practices (Mustar Begum *et al.*, 2001).

Foxtail Millet and Barngard Millet

Foxtail millet is grown as an intercrop with maize or red gram, pigeon pea, cotton etc. in U.P. Hills. The common names for foxtail millet are Kangni, Korra, rala and tenai. Barnyard millet is grown as a kharif crop under rainfed conditions. Its common name is sawan, oodalu etc. It is a short and fastest growing millet. Both these grains are particularly low in phytic acid and rich in mineral particularly iron and calcium.

Processing and Utilization

Both these millets are generally cooked as rice. Foxtail must be pounded or otherwise husked before cooking. Foxtail millet also finds its way to the local specific sweets and savouries. The most common ways of processing these grains are milling, roasting, popping, malting and parboiling.

Traditional milling is done by pounding. Both dry and slightly wet grain are used after which outer bran can be removed by winnowing. The husk yield is about 16 per cent. The debranned barnyard millets are white in colour whereas debranned foxtail has light golden yellow colour. The yield of milled grain is around 77 per cent for both millets. Germination and malting are done as per routine procedure.

Both the millets are generally used as snack food either after spicing or sweetening. Besides popped flour has wide scope for use in various traditional food preparations. Use of popped grain has been found acceptable in weaning mixtures. Foxtail millet is also utilized for development of diabetic foods alongwith other foods like fenugreek, legumes, peanut, etc.

REFERENCES

Anon. (1996). All India final estimates of food grain. *Agricultural Situation in India*. 73: 95.

Archana, Sehgal, S. and Kawatra, A. (1998). Influence of processing on polyphenol and phytic acid content of pearlmillet grains. *Plant Foods for Hum. Nutr.*, 53: 93-98.

Chaturvedi, A. and Sarojini, G. (1996). Malting of pearl millet, its effect on starch and protein digestibility. *J. of Food Sci. and Tech.*, 33(4): 342-344.

Chavan, J.K. and Kachare, D.P. (1994). Effect of seed treatment on lipolytic deterioration of pearl millet flour during storage. *J. Food Sci. and Tech.*, 31(1): 80-81.

Desikachar, H.S.R. (1975). Processing of maize, sorghum and millets for food uses. *J. Sci. Ind. Res.*, 34: 232.

Desikachar, H.S.R. (1977). Processing of sorghum and millets for versatile food uses in India. p. 41-46. In: *Proceedings of Symposium on Sorghum and Millets for Human Food,* (Ed.) D.A.V. Dendy. Tropical Products Institute, London, U.K.

Hegde, B.R. and Linge Gowda, B.K. (1989). Cropping systems and production technology for small millets in India. In: *Proceeding of the First International Small Millets Workshop*, Bangalore, October 29-November 2, 1986. Oxford and IBH Publishing Co. Pvt. Ltd. Pp 209-235.

Kumar, A. and Chauhan, B.M. (1993). Effects of phytic acid on protein digestibility (*in vitro*) and HCl-extractability of minerals in pearl millet sprouts. *Cereal Chemistry*, 70(5): 504-506.

Lakshmi, K.B. and Vimala, V. (1996). Hypoglycemic effect of selected sorghum recipes. *Nutrition Res.*, 16(10): 1651-1658.

Lorenz, Klaus, J. and Karel Kulp. (1991). *Handbook of Cereal Science and Technology*. Marcel Dekker, Inc., New York.

Melleshi, N.G. (1986). Processing of small millets for food and industrial uses. In: *Proceeding of First International Small Millet Workshop, Bangalore,* October 29-November 2, 1986. Small millets in Global agriculture. A. Seetharam; K.W. Riley and G. Harinarayana. Eds. New Delhi, Oxford and IBH. p. 321-324.

Malleshi, N.G. and Desikachar, H.S.R. (1985). Milling, popping and malting characteristics of some minor millets. *J. Food Sci. and Tech.*, 22:400-403.

Malleshi, N.G. and Desikachar, H.S.R. (1986). Nutritive value of malted millet flour. *Pl. Food Hum. Nutr.*, 36(3): 191-196.

Malleshi, N.G., Hadimani, N.A., Rangaswami-Chinnwamy, Klopfenstein, C.F. and Chinnaswamy, R. (1996). Physical and nutritional qualities of extruded weaning foods containing sorghum, pearl millet or finger millet blended with mung beans and non fat dried milk. *Plant Foods for Human Nutr.*, 49(3):181-189.

Manay, N.S. and Shadaksharaswamy, M. *Foods, Facts and Principles*. Wiley Eastern Ltd., New Delhi.

Mushtari Begum, J., Vijayalakshmi, J., Sumangala, S.G. and Shamsad Begum, S. (2001). *Ragi and Nutri-rich Recipes*. U.A.S., Bangalore.

8

Golden Rice: A Future Staple for Developing Countries

Randhir Singh

Plant Biochemistry and Molecular Biology Laboratory, Department of Biochemistry, Chaudhary Charan Singh Haryana Agricultural University, Hisar – 125 004

Rice (*Oryza sativa* L.), the major food staple for more than two billion people, contains neither β-carotene (provitamin A) nor C_{40} carotenoid precursors thereof in its endosperm. Insufficient dietary (pro) vitamin A leads to several clinical symptoms. In Southern Asia alone, it is estimated that five million children develop the eye disease-xerophthalmia every year, of which 0.25 million eventually go blind. Furthermore, vitamin A deficiency exacerbates afflictions such as diarrhoea, respiratory diseases, and childhood diseases such as measles. It is estimated that 124 million children worldwide are deficient in vitamin A and that improved nutrition could prevent 1 million to 2 million deaths annually among children. Oral delivery of vitamin A is problematic, mainly due to lack of infrastructure, so alterations are urgently required. Success might be found in supplementation of a major staple food, rice, with provitamin A.

Because no rice cultivars produce this provitamin in the endosperm, recombinant technologies rather than conventional breeding are required. This approach would have the major advantage that excess dietary β-carotene, in contrast to vitamin A supplements, has no harmful properties.

Carotenoids besides serving as precursors of vitamin A in human and animal diets, also play a key role in protection against various forms of cancer. A number of studies conducted in the past have shown that the diets with green and pigmented vegetables reduce the incidence of certain cancers. Moreover, supplementation with β-carotene in diets reduces the risk of several cancers including those of lung and stomach. β-carotene also offers protection against vascular diseases. In view of these health related characteristics, considerable interest has aroused in the manipulation of carotenoid content and composition in plants to improve the agronomic and nutritional value for human and animal consumption. In the past few years, genes and cDNAs encoding nearly all the enzymes required for carotenoid biosynthesis in green plants have been identified and sequenced, and their products characterized. A brief account of the biosynthetic pathway and the important enzymes involved is given below:

Biosynthesis of Carotenoids

Plastids are the sites of carotenoid biosynthesis in plants. In chloroplasts, they accumulate primarily in the photosynthetic membranes. In chromoplasts of ripening fruit and flower petals and in chloroplasts of senescing leaves, these may be found in membranes, or in oil bodies or other structures within the stroma. The biosynthetic pathway for carotenoids consists of three parts namely the synthesis of isopentenyl pyrophosphate (IPP), phytoene and its subsequent conversion to different carotenoids. Two molecules of acetyl CoA react to produce acetoacetyl CoA and with further addition of one more acetyl CoA, the product formed is β-hydroxy β-methyl glutaryl CoA (HMG CoA). HMG CoA synthase catalyzes both these reactions. In subsequent reactions, esterified carboxylic group of HMG CoA is reduced by two molecules of NADPH to an alcohol group, accompanied by the hydrolysis of the energy rich thioester bond. Mavalonate thus formed is converted to 5-pyrophosphate mevalonate in two successive phosphorylation steps catalyzed by two different kinases. This in turn is converted to isopentenyl pyrophosphate, the central metabolic and building block for all isopemoid compounds by reactions of enzyme pyrophosphomevalonate decarboxylase utilizing ATP.

In the next sequence of reactions, IPP is isomerized to its allylic isomer, dimethyl allylpyrophosphate (DMAPP), which condenses with another molecule of IPP to produce geranyl pyrophosphate (GPP) with the release of pyrophosphate. In this way the chain is further elongated by head to tail condensation with IPP to form famesyl pyrophosphate (FPP) and geranyl geranyl pyrophosphate (GGPP). In subsequent reactions, the enzyme phytoene synthase catalyzes the two step conversion of two molecules of GGPP into phytoene pyrophosphate (PPPP) and then to phytoene. Phytoene, in turn undergoes a series of desatuartion reactions that result in the formation of first phytofluene and then ζ-carotene, neurosporene and lycopene. The four sequential desaturations undergone by phytoene are catalyzed by two related enzymes namely phytoene desaturase, ζ-carotene desaturase. Lycopene is then cyclized by a single gene product, the lycopene β-cyclase to bicyclic β-carotene.

Enzymes of Carotenoid Biosynthesis

IPP Isomerase (EC 5.3.3.2)

The enzyme IPP isomerase located in plastidic and cytosolic fractions of plant cell has been isolated from pepper and daffodil. Plant cDNA for this enzyme was first identified in *Clarkia brewerii* (*Ipi* 1). A second gene (*Ipi* 2) in the same plant has also been reported. Two distinct cDNAs (IPP 1 and IPP 2) in *Arabidopsis, Haematococcus pluvialis* and lettuce have recently been identified. IPP 2 in *Arabidopsis* predicts a polypeptide of 284 amino acids with an N-terminal extension of IPP 2, suggesting a cytosolic location. The N-terminal extension in IPP 2 may serve to target this enzyme to chloroplast. Introduction of any of a number of different plant, algal or yeast IPP isomerase cDNAs, or of additional copies of *E. coli* gene for this enzyme, indicating the possibility that activity of this enzyme might limit the biosynthesis of carotenoids and other isoprenoids in plants.

GGPP Synthase (GGPP, EC 2.5.1.29)

The enzyme is one member of a closely related family of prenyl transferase enzymes that are distinguished by the length of the final end product. The enzyme has been isolated as a soluble homodimer from the chromoplasts of pepper, and corresponding cDNA has been identified and sequenced. Five different cDNAs predicting polypeptides with substantial sequence similarity to the pepper GGPP synthase have been identified in *Arabidopsis.* Ggps 1 encodes a 371 amino acid polypeptide with an N-terminal extension of 76 amino acids. However, the specific roles and subcellular locations of the various forms of GGPP synthase in *Arabidopsis* have not yet been ascertained.

Phytoene Synthase (PSY, EC 2.5.1.32)

Biosynthesis of phytoene from GGPP is catalyzed by the enzyme phytoene synthase, which has been isolated from pepper. Furthermore, genes encoding this enzyme have been isolated from pepper, tomato, melon, rice and *Arabidopsis.* The *Arabidopsis* PSY predicts a polypeptide of 423 amino acids.

Phytoene Desaturase

Phytoene undergoes a series of desaturation reactions catalyzed by two related enzymes in plants, *i.e.* phytoene desaturase (PDS) and ζ-carotene desaturase (ZDS). The *Arabidopsis* PDS specifies a polypeptide of 556 amino acids. The cDNAs encoding PDS.have been cloned from pepper. An *Arabidopsis* homologue of pepper ZDS has also been described, which encodes a polypeptide of 558 amino acids.

Cyclases

There are two cyclases involved in the cyclization of lycopene. One catalyzes the formation of β-carotene and subsequently also the synthesis of minor cyclic carotenoids such as ziaxanthin and estaxanthin, while the other is involved in the formation of α-carotene leading to xanthophylls such as lutein. Both enzymes are encoded by related, single copy gene. The gene for Iycopene cyclase (LYC, crtL) involved in the synthesis of β-carotene has recently been cloned by genetic complementation from a mutant strain resistant to the LYC inhibitor. The predicted protein product of this gene contains an apparent plastid transit peptide.

After knowing the details of the biosynthetic pathway leading to the formation of β-carotene, it will now be easy to understand the experiments conducted by Potrykus and his colleagues (1997, 2000) in obtaining transgenic rice rich in provitamin A. Prior to developing a strategy for rice transformation experiments, it was crucial to obtain information on the status of isoprenoid biosynthesis in rice endosperm. *In vitro* incubation of immature endosperm in the presence of radiolabelled isoprenoid precursors yielded information on the biochemical competence of this tissue for prenyl lipid biosynthesis. Incubation experiments with (1–^{14}C) IPP showed that no detectable carotenoids were produced. However, the formation of geranyl geranyl diphosphate (GGPP), analyzed as its corresponding alcohol, geranyl geraniol, was observed. GGPP represents the substrate for the first carotenoid-specific biosynthetic reaction. Furthermore, the formation of farnesol and squalene, both derived from the cytoplasmic C_{15} branch of the isoprenoid biosynthesis, was observed. Incubation with more immediate precursors of β-carotene such as (^{14}C) GGPP and (^{14}C) phytoene gave no indication for the presence of later carotenoid-synthesizing enzyme activities. Based on these results, transformation experiments using a cDNA coding for the first carotenoid-specific enzyme, phytoene synthase, were performed.

For transformation, Burkhardt *et al.* (1997) used a full length cDNA representing the nuclear encoded daffodil *psy* gene. Two constructs were prepared and linked to a CaMV 35S promoter. DNA was delivered to the scutella of 590 precultured embryos of rice variety Taipei (TR) 309 by microprojectile bombardment using a particle inflow gun. From these experiments, 87 transgenic plants were regenerated, of which 47 (54 per cent) were fertile. Among the fertile, Ro plants obtained from transformation experiments with either pCPsyH or pGt1PsyH, 27 contained an intact *psy* transgene when total DNA was digested with the appropriate restriction enzymes and analyzed by Southern blotting. Of these 27, 11 plants were derived from transformation experiments with plasmid pCPsyH and 16 from experiments with plasmid pGt1PsyH.

To determine qualitatively whether or not the integrated and intact psy transgenes were transcribed, RT-PCR was used. Template RNA from all of the transgenic plants examined allowed the synthesis of *psy* PCR products of the expected size, whereas RT-PCR of the untransformed control did not, suggesting that endogenous rice *psy* gene may not be expressed in endosperm of the developmental stage examined. In order to assess phytoene synthase protein levels in transgenic TP 309 rice lines, immunoblots were performed, using polyclonal antisera raised against the daffodil phytoene synthase. Only some of those rice lines that had been transformed with plasmid pGt1PsyH showed and immunoreactive signal with the expected mol. Wt. of 42 kDa, corresponding to a mature, plastid-localized phytoene synthase lacking the transtit peptide. Carotenoid formation in transgenic rice seeds was studied by pigment extraction, followed by HPLC analysis. Phytoene was unequivocally identified in several transgenic rice endosperms, particularly those transformed with the Gt1 promoter construct pGt1PsyH. In one plant, the highest phytoene content of 0.74 $\mu g\ g^{-1}$ dry seed weight was observed. In control plants, phytoene was not detectable or was present in trace amounts probably derived from the seed coat. No carotenoids derived from phytoene could be detected in either the controls or samples obtained from transgenic plants. These results have demonstrated that it is, in principle, possible to introduce carotenoid production into a noncarotenogenic plants. Although phytoene accumulated in the transgenic rice endosperms that over-expressed daffodil phytoene synthase, no later carotenoids could be detected. It is not yet clear whether the other two enzymes necessary to produce β-carotene, ζ-carotene desaturase and lycopene cyclase, are present in rice endosperm. Thus, the synthesis of β-carotene requires the complementation with three additional plant enzymes: phytoene desaturase and carotene desaturase, each catalyzing the introduction of two double bonds, and lycopene β-cyclase, encoded by the *lcy* gene.

Ye *et al.* (2000) used *Agrobacterium*-mediated transformation to introduce the entire β-carotene biosynthetic pathway into rice endosperm in a single transformation effort with three vectors. The vector pB19hpc combines the sequence for a plant phytoene synthase (psy) originating from daffodil with the sequence coding for a bacterial phytoene desaturase (crt 1) originating from *Erwinia uredovora* placed under control of the endosperm specific glutelin (at1) and the constitutive CaMV 35S promoter, respectively. To complete the β-carotene biosynthetic pathway, they co-transformed with vectors pZPsc and pZLcyH. Vector pZPsc carries psy and crt1, as in plasmid pB19hpc, but lacks the selectable marker aphid expression cassette. Vector pZLcyH provides lycopene β-cyclase from *Narcissus pseudonarclssus* controlled by glutelin promoter and the aphid gene controlled by CaMV 35S promoter as a selectable marker. Lycopene β-cyclase carried a functional transit peptide allowing plastid import.

Pre-cultured immature rice embryos (n = 800) were inoculated with *Agrobacterium* and hygromycin-resistant plants (n = 50) were analyzed for the presence of psy and crt 1 genes. All samples analyzed carried the transgenes and revealed mostly single insertions. In co-transformed plants, one to three transgene copies were found. Ten plants harbouring all four introduced genes were transferred to greenhouse for setting seeds. The transformed plants showed a normal vegetative phenotype and were fertile. Mature seeds from to transgenic lines and from control plants were air dried, dehusked, and. in order to isolate the endosperm, polished with emery paper. The transformed endosperms were yellow, indicating carotenoid formation. The goal of providing at least 2 $\mu g\ g^{-1}$ provitamin A in homozygous lines (corresponding to 100 μg retinol equivalents at a daily intake of 300 g of rice per day) seems to be realistic. However, it is yet to be worked out whether the lines producing provitamin A (β-carotene) or lines possessing additionally zeaxanthin and lutein would be more nutritious.

Golden Rice: The Challenge of Technology Transfer

Golden rice so far consists of a series of provitamin-A producing laboratory lines (TP-309). The characters of these lines must be transferred to as many locally adapted varieties and ecotypes in as many rice-growing countries as quickly as possible and this transfer has to be organized in a way that all rules and regulations concerning the handling and use of GMOs are strictly followed. To achieve this objective, a "Golden Rice Humanitarian Board" has been established to help make the right decisions and to provide secretarial support. In India, golden rice will be introduced through the organisational framework of Indo-Swiss collaboration in Biotechnology, which has ten years of experience in technology transfer. World Bank, ICAR and DBT will share the cost in India. Traditional breeding will transfer the trait into locally best adapted lines, and again will make sure that varieties important to the poor will be used. However, Greenpeace and associated GMO opponents are opposing its introduction citing the same old adverse effects of GMO on environment and human health. But golden rice nullifies all arguments against genetic engineering. Some of the clarifications are as (Potrykus, 2001):

1. Golden rice has not been developed by or for industry.
2. It fulfills an urgent need by complementing traditional interventions.
3. It presents a sustainable, cost free solution, not requiring other resources. It avoids the unfortunate negative side effects of the Green Revolution.
4. Industry does not benefit from it.
5. Those whose benefit are poor and disadvantaged.
6. It is given free of charge and restrictions to subsistence farmers.
7. It does not create any new dependencies.
8. It will be grown without any additional inputs.
9. It does not create advantages for rich land owners.
10. It can be resown every year from the saved harvest.
11. It does not reduce agricultural biodiversity
12. There is, so far, no conceptual negative effect on the environment.
13. There is, so far, no conceivable risk to consumer health.
14. It was not possible to develop the trait with traditional methods.

Norman Borlaug (2000) shares the optimism concerning the potential that GMO technology has for improving the living conditions of the poor and under-privileged in developing countries.

References

Borlaug, N.E. (2000). Ending world hunger: The promise of biotechnology and the threat of anti-science zealotry. *Plant Physiol.*, 124: 487-490.

Burkhardt, P.K., Beyer, P., Wunn, J., Kloti, A., Armstrong, G., Schledz, M., vonLintig, J. and Potrykus, I. (1997). Transgenic rice endosperm expressing daffodil phytoene synthase accumulates phytoene, a key intermediate of provitamin: A biosynthesis. *Plant J.*, 11: 1071-1078.

Potrykus, I. (2001). Golden rice and beyond. *Plant Physiol.*, 125: 1157-1161.

Ye, X., Al-Babili, S., Kloti, A., Zhang, J., Lucca, P., Beyer, P. and Potrykus, I. (2000). Engineering provitamin A (β-carotene) biosynthetic pathway into (carotenoid free) rice endosperm. *Science,* 287: 303-305.

Part II

MILK AND MILK PRODUCTS

9

Technological Advances in the Production and Preservation of Traditional Indian Milk Products

Dharam Pal

Dairy Technology Division, National Dairy Research Institute, Karnal – 132 001

Introduction

The traditional Indian milk products are the products of masses being made in India since time immemorial. These products have great social, religious, cultural, medicinal and economic importance. It is estimated that more than half of the total milk produced in India is utilized for the manufacture of traditional milk products (Banerjee 1997). The classification of these products based on the principle of their manufacture is given in Table 9.1.

The traditional milk products provide the means of preserving precious milk solids for a comparatively longer time than the fluid milk whose shelf life under Indian tropical conditions is only 5-6 h at ambient temperature. It is reported that about 10-12 per cent of raw milk becomes unfit for processing due to sourage. The chilling facilities for preserving raw milk are highly inadequate and the use of LP system, which is considered to be rather safe method, has not yet been approved by the statutory agencies. Since the conventional methods of manufacture of traditional milk products are very simple and do not require complicated and expensive equipment, these can be adopted at rural collection centres. This should substantially reduce the transportation, operating and overhead costs. More importantly manufacture of traditional milk products generates lot of employment. Traditional milk sweets have a distinct advantage in that they are value added products. With these advantages, traditional milk products are also encountered with some inherent problems. These are given as below:

Table 9.1: Classification of Traditional Milk Products

Sl.No.	*Principle of Manufacture*	*Products*	*Uses*
1.	Heat desiccation	Khoa	Khoa based sweets (Burfi, Peda, Gulabjamun, Kalakand, Milk Cake, etc.)
		Rabri	Direct consumption
		Basundi	Direct consumption
2.	Acid coagulation	Chhana	Chhana based sweets (Rasogolla, Sandesh, Rasamalai,Chhana murki, Cham-cham, etc.)
		Paneer	Culinary dishes, direct consumption
3.	Fermentation	Dahi	Culinary dishes, direct consumption
		Chakka	Shrikhand
			Shrikhand vadi
		Misti doi	Direct consumption
4.	Fat concentration	Makkhan	Direct consumption, Ghee
		Ghee	Culinary purpose, direct consumption
5.	Frozen	Kulfi/Kulfa	Direct consumption
6.	Cereal based	Kheer/Payasam	Direct consumption

1. Wide variations in chemical, microbial and sensory qualities from batch to batch
2. Small scale batch processes unsuitable for commercial adoption

3. Low heat transfer coefficients causing equipment to be bulky
4. Insanitary operations as these are open to atmosphere
5. Excessive strain and fatigue on the operators.
6. Poor packaging
7. Limited shelf life of the products.

Until recently the surplus milk in flush season has been converted into skim milk powder and ghee by most of the dairies. The skim milk powder is largely used for the recombination purpose during the lean season. Now we are fast moving from an era of scarcity to that of plenty. Thus the demand for skim milk powder and ghee is decreasing. Our dairy industry, therefore, may have to change its strategy, that is, from conventional skim milk powder and ghee to some value added products such as khoa, chhana and sweets made from them. The production of traditional Indian dairy products will enable the dairy plants to become economically viable and also sustain fluid milk, skim milk, etc. being produced by them. These value added products have also great potential for export because of increasing ethnic population in many countries and implementation of the historic GATT agreement which enable free trade through the opening of protected market and reduction of export subsidies.

In order to obviate the limitations and to achieve the goals as stated above, major innovations are needed in manufacture, quality assurance, packaging and process engineering to adopt traditional milk products to global consumers' acceptance and competitiveness. The important technological advances that have taken place in recent years in respect of traditional milk products, particularly that of commercial significance, are discussed here.

Khoa

Khoa, an important Indian milk product, is prepared by continuous boiling of milk until desired concentration (65 to 72 per cent TS) and texture is achieved. It is used as a base material for a variety of popular sweet-meats, such as burfi, peda, gulabjamun, kalakand, etc. According to PFA rules, as amended in 1999, khoa is the product obtained from cow or buffalo milk or a combination thereof by rapid desiccation and having not less than 30 per cent milk fat on dry matter basis. Buffalo milk is preferred to cow milk in making khoa because besides giving a larger turn-out, it gives a product of soft body and smooth texture which is highly desirable for khoa-based sweets. It is estimated that about one million tonne of khoa is produced annually, which is equivalent to 5.5 per cent of India's total milk production.

Scaling Up of the Technology of Khoa Making

Most of the attempts on up gradation of the technology of khoa have been directed towards mechanization of the process, and for developing continuous khoa making plant.The first semi-continuous khoa making machine was developed by Banerjee *et al.* (1968) and its subsequent modification was done by De and Singh (1970). The equipment consisted of scraped surface heat exchanger and two open semi-jacketed pans with reciprocating spring loaded scrapers. The machine was too bulky and the product was lacking in satisfactory organoleptic quality.

Sawhney *et al.* (1987) mechanized the traditional batch process by providing jacketed, shallow open pan with a swinging hanger type scraper for stirring during the desiccation process. In this equipment, the jacket is filled with water upto two third of its height and is placed over the furnace. Steam is generated in the jacket and its temperature and pressure are controlled by adjusting the safety valve, which is provided to adjust the water steam pressure in the jacket in a range from 0 to 4 kg/cm^2. The equipment is developed for manufacture of khoa under rural conditions. More (1987) designed batch type semi-mechanized scraped surface heat exchanger. It consists of jacketed drum with vapour exhaust and scraper assembly. The scraper assembly comprises of spring-loaded blades, rubber boots and a central shaft. In each batch, 4 kg of milk is used. Both these mechanized units are of batch type unsuitable for commercial production of khoa.

A mechanized conical process vat for preparation of khoa was developed by Agrawala *et al.* (1987). This equipment consists of stainless steel conical vat with cone angle 60° and steam-jacket partitioned into 4-segments for efficient use of thermal energy and less heat loss. The mechanism is consisting of 3-equidistant arms supported at two points in the shaft and each arm having three independent spring-loaded blades. A positive displacement screw pump is connected at the outlet of the vat for recirculation and spreading of the product over heat transfer surface. This is also a batch type equipment.

Reverse osmosis process for pre-concentration of milk followed by desiccation in a steam jacketed open pan has been successfully tried by Pal and Cheryan (1987). Such process offers attractive energy saving in the initial concentration of milk. The quality of khoa made by the membrane process was found to be identical to the conventionally prepared product. It is also possible to replace jacketed pan with SSHE to make the process continuous.

National Dairy Development Board has developed an inclined scraped surface heat exchanger (ISSHE) for continuous manufacture of khoa (Punjrath *et al.*, 1990). In this machine, concentrated milk of 42 to 45 per cent total solids is used as feed. The inclination of ISSHE permits formation of a pool of boiling milk critical to formation of khoa. A thin film scraped surface heat exchanger (TSSHE) system has been developed at NDRI by Dodeja *et al.* (1992) for continuous manufacture of khoa. In this unit, two SSHE are arranged in a cascade fashion. Milk is concentrated into first SSHE to about 40-45 per cent TS and finally to khoa in the second SSHE. Unlike ISSHE, the feed for this unit is buffalo milk and thus suitable for organized as well as small dairies/entrepreneurs. The capacity of both these units is about 50 Kgs khoa/h, and many organized dairies have adopted these continuous khoa making plants.

Christie and Shah (1992) have designed and developed three stage unit for continuous khoa manufacture. The machine has three jacketed cylinders placed in a cascade arrangement. This facilitates easy transfer of milk from one cylinder in to other which works as heat exchanger. The heat exchangers are installed with a mechanism of providing inclination. The slope allows the movement of the contents in longitudinal direction. The unit is provided with variable pulley drive so that speed adjustment is possible. The unit is highly bulky requiring too much flooring area.

Extension of Shelf Life of Khoa

Khoa has a limited shelf life of about 5 days at 30°C. In the absence of proper packaging, the egress of moisture from the product adversely influences the texture and enhances the rate of chemical deterioration such as oxidation and browning. Different workers have studied the effects of different packaging materials and the preservatives on the shelf life of khoa. Using four-ply laminated pouches and tin containers, it was possible to increase the shelf life of khoa upto 13 days at 30°C and to 75 days in the cold storage. Sterilization of polypacks with gamma radiation using CO^{60} prior to product filling proved to be beneficial. Addition of 0.3 per cent potassium sorbate at the last stage of khoa making increased its shelf life by another 10 days. Vacuum packaging of khoa could enhance the shelf life upto 120 days in refrigerated storage (Rajorhia *et al.*, 1984).

Ghee

Ghee, a heat clarified butter fat, is most important traditional dairy product as about 28 per cent of total milk production is converted in to ghee, the value of which is equivalent to several hundred crore rupees per annum. The methods of ghee making vary according to the base materials used (milk/cream/butter), intermediate treatment of raw materials, and handling of the semi- or fully finished ghee. The methods are broadly classified as (a) indigenous milk-butter method (*Desi* method), (b) direct cream method, (c) creamery butter method, and (d) prestratification method. Regardless of the volume of ghee production, all these methods are essentially of batch type and suffer from the inherent problems as discussed earlier.

Several technological advances have taken place in terms of scaling up of the methods of ghee manufacture, recovery of fat, characterization and improvement of flavour and textural attributes of ghee prepared in different regions of the country, and extension of shelf life.

Continuous Ghee Making

A prototype continuous ghee making plant of 100 kg/hour capacity was developed on the principle of flash evaporation by Punjrath (1974) which was subsequently improved and being used at Sugam Dairy, Baroda. Abichandani *et al.* (1982) designed a continuous ghee making machine on the principle of evaporation from a scraped surface falling film evaporator at atmospheric pressure. It comprises of (a) continuous butter melter and (b) straight sided horizontal thin film scraped surface heat exchanger for converting molten butter in to ghee. The steam consumption in continuous ghee making plant, when using butter of 80-85 per cent fat at initial temperature of about 10°C and ghee making temperature 115-118°C, was found to be 35 kg per 100 kg of ghee as compared with 68 kg required in jacketed steam kettle (batch process). It was also observed that 10-15 per cent more vitamin A retained in ghee made by continuous method than that of batch method, and the organoleptic quality of ghee prepared by the two methods was similar (Bector *et al.*, 1996).

Preservation of Ghee

Although ghee has a better capacity to resist spoilage by chemical and microbial attack than any other milk product, it is common knowledge that, upon prolonged storage at ambient temperature, it undergoes oxidative changes. Reaction of oxygen with the unsaturated fat, aggravated by metallic contamination and sunlight, is a major cause of spoilage. It gives rise to a typical, strong and disagreeable odour accompanies with the loss of nutritive value.

A number of synthetic antioxidants, such as gallates (ethyl, propyl, octyl), butylated hudroxyanisole (BHA), butylated hydroxytolune (BHT), tertiary butyl hydroquinone (TBHQ), ascorbic acid, -tocopherol, phospholipids, and some natural antioxidants, namely curry leaves, betal leaves, soya bean powder, safflower and *amla (Phyllanthus amblica)* have been tried in small amounts with a view to either completely prevent or partially retard the oxidation of fat during storage (Rajorhia 1993).

Chhana

Chhana, an important base for several varieties of Indian sweets, occupies a prominent place in the dairy economy of India. It is estimated that about 4 per cent of the total milk is converted into chhana and paneer per annum. The product is usually prepared on a small scale by the private sector of India dairy industry by acid coagulation of hot milk. Though chhana and paneer differ in respect of method of their preparation, the PFA rules are same for both the products, *i.e.* the moisture content not more than 70 per cent and the milk fat shall not be less then 50 per cent of dry matter.

Upgradation of Chhana Technology

To overcome the drawbacks of the batch method, *i.e.* high labour cost, more production time and unhygienic quality of product, a prototype machine with a capacity of 40 kg chhana per hour has been designed and developed (Aneja *et al.*, 1982). The equipment consists of various components *viz.* a balance tank, injection chamber, holding coil, cooling chamber and strainer. Standardized cow milk is pumped from a balance tank to an injection chamber where culinary live steam is directly injected into the milk. Steam gets completely condensed in milk and raises its temperature to 90-95°C. Thereafter, milk is brought in contact with sour whey, the quantity of which is regulated manually in proportion to the rate of milk flow. The mixture of milk and whey is circulated through a holding coil to facilitate complete coagulation of milk. The coagulated product, along with the whey is then pumped to a double jacketed cooling tank, where it is cooled down to room temperature. Finally, the product is directed to a mechanical strainer, a double-jacketed inclined sieve, where it is drained thoroughly. Chhana with 55-65 per cent moisture is discharged through the outlet and collected in the basket. Drained whey is transferred to a separate tank for subsequent use.

Cow milk is preferred for chhana making because it yields soft body and smooth textured product that is suitable for chhana based sweet particularly rasogolla. Buffalo milk, because of many inherent differences in physico-chemical make up than that of cow milk, poses many technological problems in preparation of good quality chhana based sweets. Rajorhia and Sen (1988) have reviewed the attempts made to improve the quality of chhana produced from buffalo milk.

Application of ultrafiltration, a membrane process that enables separation of proteins and fat from lactose and soluble minerals (process similar to traditional chhana making) and their partial concentration, has been reported for the preparation of chhana of acceptable organoleptic quality (Sharma and Reuter 1991).

Extension of Shelf Life of Chhana

Chhana has a very limited shelf life of about one day at ambient conditions and 5-6 days at refrigeration temperature. Since fresh chhana is preferred for making sweets of good quality, much research efforts have not been made in extending its shelf life.

Paneer

Paneer, the indigenous variety of soft cheese, is also obtained by acid and heat coagulation of milk, like that of chhana. In the manufacture of paneer a certain amount of pressure is applied for the drainage of whey from the coagulated mass whereas in chhana free drainage of whey is allowed. The demand for the value added culinary dishes and other products based on paneer are growing rapidly throughout the country. Its export potential is also being greatly realized in the present GATT era. Simultaneously, lot of research developments has taken place particularly on the aspects related to technology upgradation and shelf life extension.

Upgradation of Paneer Technology

The processing conditions *viz.*, type of milk, heat treatment to milk, coagulation conditions (temperature, pH, coagulants, etc.), and subsequent operations have been standardized for preparing paneer with higher recovery of solid, improved flavour and body and texture characteristics (Sachdeva and Singh 1988; Rao *et al.*, 1992). Best quality paneer is made from buffalo milk. Cow milk produces an inferior product in terms of body and texture. Many attempts have been made to prepare paneer of good quality from cow milk. These include addition of Calcium Chloride @ 0.08-0.1 per cent (Sachdeva *et al.*1991) or 0.05 per cent Calcium Sulphate (Mistry *et al.*, 1992) to cow milk, coagulation at higher temperature, and nanofiltration of cow milk to about 1.5 fold level (Pal *et al.* 2000).

A process for commercial production of paneer employing UF was developed by Sachdeva *et al.* (1993). It involved standardization and heating of milk, its UF to a concentration level of about 40 per cent TS, followed by cold acidification to get the desired pH. The flowable product dispensed/filled into containers and then subjected to texturization by microwave heating. The resulting product had typical characteristics of normal paneer and yield increased by about 25 per cent. In another approach a long life "paneer-like" product was developed employing vacuum concentration/UF of milk to the TS level as in paneer, packaging in metallized polyester (report) pouches, sterilization at 115°C and in pack texturization with GDL (Rao 1991).

Preservation of Paneer

The shelf life of paneer is reported to be only 6 days under refrigeration though its freshness is lost within 3 days. At room temperature paneer does not keep good for more than one day. The relatively short shelf life of paneer is a major handicap in the commercial adoption of paneer manufacture. The spoilage in paneer occurs due to the surface growth of microorganisms which is due to post manufacturing contamination. A greenish yellow slime forms on the surface and the discoloration is accompanied by an abnormal odour. It is only the surface that gets spoiled early while the interior remains good for a longer time. To curb the surface growth of microorganisms and thereby increase the shelf life of paneer, the use of following methods have been explored.

Brining

Paneer dipped in 5 per cent brine solution lasts for nearly 20 days as against control that is spoiled after 6 days of storage at 8-10°C. The sensory attributes are rated higher for salted samples. Since paneer is mostly salted and spiced before consumption, the salting at the time of dipping can be advantageously used in extending the shelf life of paneer (Sachdeva 1998).

Use of Chemical Preservatives

A shelf life of 32 days under refrigeration was achieved on treating paneer with a combination of delvocid and hydrogen peroxide (Sachdeva and Singh 1990). Shelf life of 40 days using benzoic acid (1200 ppm) under refrigeration conditions and 20 days at 37°C has been reported (Modi and Jain, 1988). Singh *et al.* (1989) reported a shelf life of 36 days at room temperature by adding sorbic acid to milk (0.15 per cent) and subsequent wrapping of paneer in Sorbic acid coated paper.

Freezing

On storage of paneer at sub zero temperature *i.e.* 13°C and -32°C for 120 days, the flavour and appearance is not affected but its body and texture deteriorates and the product becomes crumbly and fluffy on thawing (Arora and Gupta 1980). Blast freezing has recently been used to enhance the shelf life of paneer. The paneer block is cut into pieces of approx. 1.5x1. 5x1.5 cm size and blast frozen at a temperature below -20°C. The product can be stored under frozen conditions (Below-19°C) for more than one year without any deterioration in its quality (Punjrath *et al.*1997).

Vacuum Packaging

Vacuum packaging of paneer in laminated pouches can help increase the shelf life of paneer to about 30 days at 6+1C (Sachdeva *et al.*, 1991). The body and texture of paneer also improves on vacuum packaging as it becomes more compact and shows better sliceability. Paneer packaged in high film (EVA/EVA/PVDC/EVA) under vacuum and heat treated at 90°C for one minute is reported to have a shelf life of 90 days under refrigeration (Punjrath *et al.*1997).

Heat Sterilisation of Paneer

Although the refrigerated shelf life improves markedly by the various treatments given to raw paneer, the shelf life at room temperature does not improve noticeably. Heat sterilisation of paneer is an effective treatment for improving the shelf life of paneer at room temperature (Sachdeva 1998). Paneer packed in tins along with water/brine and sterilised in an autoclave at 1Kg/cm^2 for 15 min lasts for 4 months. The perception of an oxidised flavour renders the product unacceptable afterwards. A slight amount of cooked flavour accompanied by maillard browning, the intensity of which increases slightly during storage, was noticed.

Paneer Curry

Hurdle technology, a new concept of extending the shelf life of foods, was used by Rao (1993) for preparation of paneer curry that had a shelf life of nearly a month at 30°C and more than three months at 15°C. The paneer curry was prepared through combined effects of reduced water activity (0.95), low pH (5.0), retort packaging and use of Potassium Sorbate.

Convenience Products

Convenience and ready-to-use foods require either no or minimum processing before consumption, hence have attained great significance in the recent past particularly in the developed countries. The technologies of many traditional long-life convenience dairy products namely, khoa powder, gulabjamun mix powder, rasogolla mix powder and dried kulfi mix, have been developed in our country because these offer several benefits; some of these are as given below:

1. Long shelf life at ambient temperature
2. Economic use of seasonal and regional milk surpluses
3. Produce sweets of consistent quality at the convenience of users
4. Adaptable to medium and industrial scale dairy processing operation.
5. Allow product diversification with manageable investments for improved productivity of the dairy industry
6. The products offer good export potential.

Khoa Powder

Under the sub-tropical conditions prevailing in India the shelf life of khoa can be extended only to a limited period. The prevailing methods cannot help preserving khoa produced in flush season, till the onset of lean months, *i.e.* for about 6 months. Manufacture of khoa powder has therefore tremendous scope. Patel and De (1977) were probably the first workers followed by Thomkinson and De (1981) to pursue research efforts for the manufacture of khoa powder. The processes developed by them were, however, technologically cumbersome, expensive, energy intensive, and produced powder of poor acceptability. Rizvi *et al.* (1987) used reverse osmosis process for partial concentration of milk followed by heating in a thermutator and finally spray drying. The product had burnt flavour and caramel colour. The more successful technologies for the manufacture of khoa powder for small scale and industrial applications have been developed by Ranganadham (1988) using tray drying and spray/roller drying techniques. In the tray drying process, khoa prepared by conventional method was grated, spread in trays and dried in commercial hot air ovens at 70°C followed by grinding of dried product in a laboratory grinder. Roller and spray drying processes involved pre-concentration of standardized buffalo milk, its heating in a tubular heater to generate desired heated flavour followed by drying. The shelf life of these products ranged from 5-7 months at room temperature. The technology of manufacturing burfi and gulabjamun from khoa powder was also standardized.

Gulabjamun Mix Powder

Many ready made gulabjamun mixes with different brand names are available in the market. Since the mixes are governed by the patent regulations, general public are not aware about their composition, and proportions of various ingredients. In addition, the cost of these mixes is very high. The basic ingredients of most of these mixes are skim milk powder, maida and/or semolina, vegetable fat, baking powder and flavouring agents. The method of preparation involves: blending of dry ingredients in power driven blender, addition of vegetable fat in molten state in instalments to ensure proper mixing followed by packaging of mix. The method of formulating gulabjamun mix powder from roller as well as spray dried skim milk has been standardized at NDRI (Ghosh *et al.*, 1986; Rajorhia and Pal 1989). The shelf life of the gulabjamun mix powder in metallized laminate pouch of 500 gm capacity is about nine months at room temperature. The method of manufacturing gulabjamun from this mix has also been standardized.

Rasogolla Mix Powder

Manufacture of rasogolla is probably most difficult amongst all the milk based delicacies. It requires lot of art and experience in addition to the right type of raw materials. Cow milk is invariably used in the manufacture of good quality spongy rasogolla. The shelf life of channa is too short, and its cold storage results into production of inferior quality of sweets. The use of ultrafiltration has been made at NDRI to produce base for the rasogolla mix powder (Pal *et al.*1994). Cow skim milk is ultrafiltered to about 3-fold concentration to achieve a product containing all the milk proteins, part of the minerals and lactose as obtained in the chhana. To reduce the mineral and lactose level to almost the same level as in chhana, UF retentate has to be diafiltered. The pasteurized cream is added to diafiltered retentate followed by spray drying adopting standard conditions. The dried retentate is blended with selected additives to produce desired flavour and texture. The conditions of making rasogolla sweets from dried mix have also been optimized. The average yield of mix powder from cow milk is 7 per cent which can produce about 55 kg of sweet on drained weight basis. This yield is almost 20 per cent higher than that obtained by traditional method. Dried rasogolla mix can be stored for about 5 months at 30°C.

Instant Kulfi-Mix-Powder

Kulfi is a popular traditional frozen milk product of India. Traditionally, it is prepared by concentrating whole milk by atmospheric boiling to about two fold level, adding sugar and dry fruits (optional) to the concentrated milk, filling in to aluminium/ plastic cones and freezing in an earthen pot having a mixture of ice and salt. Inferior microbial quality is a serious drawback of market kulfi, in addition to other inherent problems. Ghosh (1991) has developed a technology for the manufacture of kulfi mix powder by spray drying process. It involves formulation of mix from milk fat (11 per cent), MSNF (16 per cent), sucrose (15 per cent) and isabgol husk (0.2 per cent), mixing of only 25 per cent of the total sugar required, homogenization of mix at 6.83/3.43 Mpa, its heating at 100°C for 10 min in a tubular heat exchanger followed by cooling to 4°C, and finally spray drying. The remaining sugar in ground form is dry blended with the powder and packaged in the tin cans. The product has a shelf life of 7 months at 30°C which can be extended upto 10 months with the addition of butylated hydroxy anisole and nitrogen gas flushing. Kulfi mix powder can be instantly reconstituted and frozen to get kulfi of consistently good quality all the year round at an affordable price. It ensures the production of pathogens-free frozen product.

REFERENCES

Aneja, V.P., Rajorhia, G.S. and Makkar, S.K. (1978). Design and development of continuous channa making machine. *J. Inst. Engineers (India)*, 58: 1-4.

Abichandani, H., Bector, B.S. and Sarma, S.C. (1995). Continuous ghee making system-design, operation and performance. *Ind. J. Dairy Sci.*, 48: 646.

Agrawala, S.P., Sawhney, I.K. and Bikram Kumar (1987). Mechanized conical process vat. Patent No. 165440.

Aneja, R.P. (1997). Traditional dairy delicacies-A compendium. In: *Dairy India*, Fifth Edition.

Arora, V.K. and Gupta, S.K. (1980). Effect of low temperature storage on paneer. *Ind. J. Dairy Sci.*, 33: 374.

Banerjee, A.K. (1997). Processes for commercial production. In: *Dairy India*, 5th Edn, pp. 387.

Banerjee, A.K., Verma, I.S. and Bagchi, B. (1968). Pilot plant for continuous manufacture of khoa. *Indian Dairyman*, 20: 81.

Bector, B.S., Abichandani, H. and Sarma, S.C. (1996). Shelf life of ghee manufactured in continuous ghee making syatem. *Ind. J. Dairy Sci.*, 49: 398.

Chriestie, I.S. and Shah, U.S. (1992). Development of three stage khoa making machine. *Indian Dairyman*, 44: 1-4.

De, S., Singh, B.P. (1970). Continuous production of Khoa. *Indian Dairyman*, 22(12): 294.

Dodeja, A.K., Abichandani, H., Sarma, S.C., Pal, D. (1992). Continuous Khoa making machine: design, operation and performance. *Ind. J. Dairy Sci.*, 45: 671.

Ghosh, J. (1991). Process development for the manufacture of instant kulfi mix powder. *Ph.D Thesis*, Kurekshetra University.

Ghosh, B.C., Rajorhia, G.S. and Pal, D. (1986). Complete gulabjamun mix. *Indian Dairyman*, 38: 279.

Mistry, C.D., Singh, S. and Sharma, R.S. (1992). Physico-chemical characteristics of paneer prepared from cow milk by altering its salt balance. *Aust. J. Dairy Technol.*, 47: 23.

More, G.R. (1987). Progress of the project. *NDRI Annual Report*, p. 89-90.

Modi, V.K. and Jain, S.C. (1988). The preservation of paneer by chemical additives. *2nd Intrn Food Conf*, IFCON, Mysore.

Pal, D., Verma, B.B. and Garg, F.C. (2000). Application of membrane systems for improving the quality of traditional dairy products. *NDRI Annual Report.*

Pal, D. and Cheryan, M. (1987). Application of reverse osmosis in the manufacture of khoa: process optimization and product quality. *J. Food Sci. Technol.*, 24: 233.

Pal, D., Rajorhia, G.S., Garg, F.C. and Verma, B.B. (1994). Production of dried rasogolla mix from ultra filtered milk retentate. *24th Int. Dairy Congr.*, Melborne (Australia), Sept 18-20, pp. 424.

Patel, A.A. and De, S. (1977). Production and storage of dried khoa. *Ind. J. Dairy Sci.*, 30: 325.

Punjrath, J.S. (1974). New developments in ghee making. *Indian Dairyman*, 26: 275.

Punjrath, J.S., Veeranjanyalu, B., Mathunni, M.I., Samal, S.K. and Aneja, R.P. (1990). Inclined scraped surface heat exchanger for continuous khoa making. *Indian J. Dairy Sci.*, 43(2): 225-230.

Punjrath, J.S., Kumar, R. and Bandyopadhyay, P. (1997). Tapping the potential of traditional dairy foods. In Souvenir: *28th Dairy Industry Conference*, Bangalore April 27-29.

Rajorhia, G.S. (1993). Ghee. *Encyclopaedia of Food Science, Food Technology and Food Nutrition*, Academic Press, London, pp. 2186.

Rajorhia, G.S. and Sen, D.C. (1988). Technology of chhana: A review. *Ind. J. Dairy Sci.*, 41: 141.

Rajorhia, G.S., Indu Sikka and Srinivasan, M.R. (1984). Use of ionizing radiation for sterilization of packaging materials for dairy products. *Asian J. Dairy Res.*, 3(2): 91.

Rajorhia, G.S. and Pal, D. (1989). Storage quality of gulabjamun mix powder as influenced by flexible packages. *Ind. J. Dairy Sci.*, 42: 326.

Ranganadham, M. (1988). Process innovation in the manufacturing of khoa powder. *Ph.D. Thesis*, Kurukshetra University, Kurukshetra.

Rao, K.J. (1993). Application of hurdle technology in the development of long life paneer based convenience food. *Ph.D. Thesis*, NDRI Deemed Univ.

Rao, K.V.S.S., Zanzad, P.N. and Mathur, B.N. (1992. Paneer technology: A review. *Ind J. Dairy Sci.*, 45: 281.

Rizvi, S.S.H., Mann, R.S. and Ali, S.I. (1987). Case study of appropriate technology transfer: Development of an automated, continuous khoa powder manufacturing process. *Indian Dairyman*, 39: 63.

Sachdeva, S. (1998). Innovation in the manufacture and preservation of paneer. In: *Lecture Compendium on Advances in Traditional Dairy Products*, Dairy Technology Division, NDRI, Karnal.

Sachdeva, S. and Singh, S. (1988). Optimization of processing parameters in the manufacture of paneer. *J. Food Sci. Technol.*, 25: 142.

Sachdeva, S. and Singh, S. (1990). Shelf life of paneer as affected by antimicrobial agents- Part 1. Effect on Sensory Characteristics. *Ind. J. Dairy Sci.*, 43: 60.

Sachdeva, S., Prokopek, D. and Reuter, H. (1991). Technology of paneer from cow milk. *Jap. J. Dairy Food Sci.*, 40: A85.

Sachdeva, S., Patel, R.S., Kanawjia, S.K., Singh, S. and Gupta, V.K. (1993). Paneer manufacture employing ultrafiltration. *3rd Intern. Food Conf.*, IFCON 93, Mysore

Sawhney, I.K., Sarma, S.C., Kumar, B. and Rajdorhia, G.S. (1980). Development of village level khoa pan. *J. Inst. Engrs.* (India), 61: 13.

Sharma, D.K. and Reuter, H. (1991). A method of chhana making by ultrafiltration technique. *Ind. J. Dairy Sci.*, 44: 89.

Singh, L., Mohan, M.S., Puttalingamma, V. and Shankaran, R. (1989). Preservation of paneer by sorbic acid. *J. Food Sci. Technol.*, 26: 129.

Thompkinson, D.K. and De, S. (1981). Utilization of dried khoa for peda making. *J. Food Sci. Technol.*, 18: 259.

10

Application of Reverse Osmosis for Concentration of Milk

Dharam Pal
Dairy Technology Division,
National Dairy Research Institute, Karnal – 132 001

Introduction

Reverse osmosis (RO), also known as Hyperfiltration, is essentially a dewatering technique of fluids. The use of RO was first started in 1960's after the development of synthetic asymmetric membrane by S.Sourirajan. In its initial stage the RO process was mainly used for desalination and water treatment. The process gained entrance in food industry, particularly the dairy, in the seventies and its applications increased dramatically in the 1980's. RO is now considered as the most economical process of removing water from liquid foods including milk (Cheryan 1998).

RO and other membrane processes, *viz.* ultrafiltration (UF), nanofiltration (NF) and microfiltration (MF), are based on the semi-permeable membranes of the appropriate physical and chemical nature. The key to effective separation using RO is the type and structure of the membrane. The active layer of RO membranes is believed to be very thin, approximately 5-30 nm, with a pore size of about 5-20 Å. Over 130 membrane materials have been tried for membranes but all these are not suitable for food applications. Cellulose acetate, polyamide and thin film composite (TFC) membranes have widely been used for processing of milk and other foods. RO membranes are asymmetric in nature in which top layer where separation takes place is very thin which is supported on a more porous backing, for example, fibre glass, filter paper, metals (stainless steel), etc. These membranes are assembled in any one of the following configurations:

1. Tabular
2. Spiral wound
3. Plate and frame
4. Hollow fibre

Presently, most of the RO plants are using TFC membranes for commercial applications because of their superiority over others. Modern RO membranes under optimum applied pressure will retain all the ions and most organics having molecular weight greater than 200 daltons. Their retention co-efficient, that is the fraction of any particular dissolved substance the membrane retain, is 0.999 for sugars and larger molecules (proteins, fat, etc.); 0.90–0.98 for smaller ions such as sodium; and between 0 and 90 for various low molecular weight organic compounds.

Advantages of RO Process

1. RO is a simple process. It consists merely of pumping the feed solutions under pressure over the surface of a suitably supported membrane. The removal of water is accomplished without a change in phase or state of the solvent.
2. The energy requirements of RO are very low compared with other water removal processes, like, evaporation, freeze concentration, freeze drying etc. For example, concentration of milk by RO requires energy in range of 9-19 KWh/1000 kg of water removed in comparison with 37-52 KWh by 5-7 effect evaporator with MVR, and 92 KWh by freezing process (Marshall 1985).

3. Since the RO process can be operated at ambient temperature, the thermal or oxidative degradation problems, common with evaporation processes, are avoided or minimized in RO process. Thus the nutritional and functional properties of milk constituents, like vitamins, proteins, etc. are least affected. There may, however, be occasions when operation at considerably low temperature (for example, to prevent microbial growth problems or denaturation of heat sensitive compounds) or at higher temperature (to minimize microbial growth problems or to lower the viscosity of retentate, thus lowering the pumping costs and improving mass transfer) is necessary.
4. No complicated heat transfer or heat generating equipment is needed, and the membrane operation which requires only electrical energy to drive the pump motor, can be situated far from the prime power generating plant. Also no additional steam capacity need to be installed to handle the RO unit. Since no condensers are needed, problems like thermal pollution and over loading of sewage treatment systems are avoided.

Limitations

1. The efficiency and economy of RO process depend largely on the flux, *i.e.* rate of permeate removed from feed per unit time per unit area (expressed as $L/m^2/h$), and the life of a membrane. In a membrane system the flux start decreasing as concentration of feed increases. The relationship between milk total solids and the flux is shown in Figure 10.1.

   RO process is, therefore, quite limited in its upper solids limits. As an example, current technology permits milk to be concentrated by multiple effect evaporation to about 50 per cent total solids, while the upper limit using RO is about 28 per cent total solids for skim milk and 38 per cent for whole milk (Pal and Cheryan 1987a). It is the osmotic pressure of feed and the concentration polarisation that limit the flux thereby restricting solids level in the RO process.

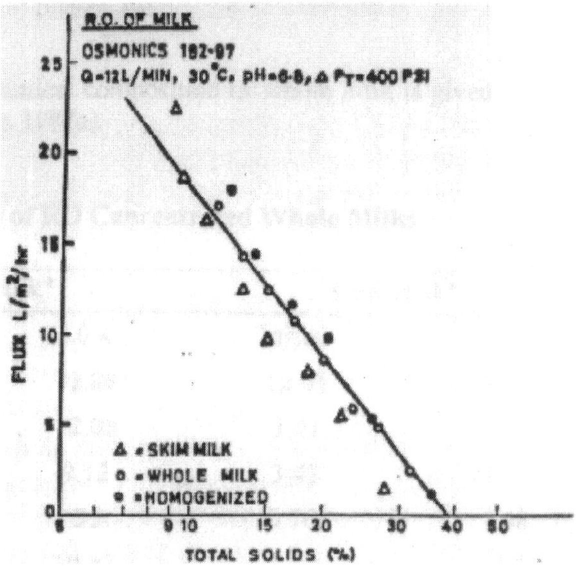

Figure 10.1: Relationship Between Total Solids and Flux During Reverse Osmosis of Milk (Δ: Skim milk; O = Whole milk; – : Homogenised whole milk). The transmembrane pressure was 400 psiga and the flow rate (Q) was 12 L/min.

2. Other problems that plagued early membrane applications were the fouling of membranes, poor cleanability of some early modules, and restricted operating conditions (temperature, pressure, pH and flow rate), although some of these problems have been overcome through the development of superior membrane materials and improved module design. (Pal and Cheryan 1987b; Roberts 1994).

Applications

Since RO is the most energy efficient dewatering process, it can be used for pre-concentration or final concentration of liquid feed for different purposes. RO concentrates can be used in fluid form either as pasteurized or UHT processed milk and for the manufacture of spray/roller powders, yoghurt, ice cream, etc. In India, RO concentration has tremendous scope in the preparation of many concentrated traditional dairy products namely khoa, *rabri, basundi,* and *kheer.* In this write up main focus is given to the following applications:

1. Concentration of milk for utilization in fluid form
2. Concentration of milk for bulk transportation
3. Concentration of whey and UF- permeate

Concentration of Milk

Several factors influence the performance of RO unit and the quality of product. Only those related with quality of retentates are discussed here.

1. The feed in RO process is pumped at very high pressure (30-35 bar) that causes the damage of fat globules. This is probably due to the shearing action of the high pressure pumps causing distinct homogenization action on milk fat, resulting in decrease in the size of fat globules and reduction of the free fat content (Gupta and Pal 1993a). If lipase enzyme is not inactivated before RO process, the hydrolysis of fat take place leading to spontaneous development of rancid flavour in milk and increase in FFAs (Table 10.1). These problems have been solved by modification of RO unit and giving suitable heat treatment, preferably, the thermization or Pasteurization to the milk.

Table 10.1: Effect of RO Proces (30°C, CA membranes) on Bacteriological Quality and Free Fatty Acids Levels of Whole Milk (Abbot *et al.,* 1979)

Concentration Ratio	*Bacterial Count*		*Free fatty acids (milli equiv./100 g fat)*	
	Raw Milk	*Heated Milk**	*Raw Milk*	*Heated Milk**
1.0	*4.1×10^3*	*1.4×10^2*	*1.7*	*1.7*
1.5	*1.8×10^4*	*2.1×10^2*	*9*	*2.3*
2.0	*2.6×10^4*	*4.3×10^2*	*10.2*	*1.7*
2.5	*2.8×10^4*	*5.1×10^2*	*10.9*	*1.8*

* *Milk heated to 80°C for RO*

The use of restriction pipe of smaller diameter instead of pressure control value in the retentate line, and the cooling of retentate below 10°C before pressure release have also helped controlling the release of free fatty acid (Berbano *et al.,* 1983). In addition to the above benefits, heating of milk prior to its RO processing decreases the fouling of membrane and improves the flux.

2. The flux is dependent on the characteristics of membranes and initial feed concentration, and the levels of final concentration, in addition to the operating parameters. The non-cellulosic membrane gives higher flux as well as better rejection properties (Pal and Cheryan 1987a, Cheryan *et al.*1990). Since the system is operating in the mass transfer controlled region, the relationship between flux and solids concentration appears to follow the film theory. The difference between the flux of three milks *viz.*, skim milk, whole milk and homogenized milk is due to slight difference in their initial composition, pre-treatment and fouling of membrane (Figure 10.1). By extrapolation of curve in Figure 10.1, it appears that maximum concentration that can be obtained for whole milk is 38 per cent TS and for skim milk 28 per cent TS. These levels can not be considered as economical because the flux is almost negligible at this stage. The economical levels of RO concentration for skim milk is 22 per cent TS and for whole milk upto 30 per cent TS.

3. The results of a typical survey of RO plant having CA membrane, showed that first portion of product coming out of plant had high microbial load. This problem, however, was less serious in plant having non-cellulosic membrane. The RO plant therefore, requires a proper cleaning and sanitizing cycle not only after use but also immediately before each run for getting a product of satisfactory quality. While deciding on the cleaning schedule for a membrane system, the nature of feed constituents that foul the membrane, the compatibility of the membrane to a particular detergent, and the recommendations of the manufacturer of membranes should be considered (Pal 1993).

Utilization of RO Concentrated Milk in Fluid Form

The RO concentrated whole milk can be used either as such or after dilution in the manufacture of certain fluid products like plain milk, flavoured milk, skim milk and high protein milk. The processes involve concentration of whole milk to about 2-fold, storage for 24 hours at 5°C, dilution with water to the composition of natural milk or to that desired, pasteurization, homogenization, addition of flavours, etc. Taste panels of the various liquid dairy products have shown that the majority (106 out of 111) could detect no significant effect attributable to inclusion of RO processing and later dilution with water on the organoleptic quality of any of the subsequent products. The effect on vitamins levels in milk due to RO process was negligible. Bacterial counts on control milk and reconstituted RO concentrates after pasteurization and during storage showed no significant difference. The shelf life of reconstituted milk was about 9-12 days under refrigerated conditions. (Drew and Manners 1985).

The RO concentrates could also be used as such after pasteurization and packaging. The keeping quality of the pasteurized RO concentrates remained almost similar to other dairy products. In a study where pasteurized RO concentrate (about 23-24 per cent TS) was distributed for household evaluation in Australia, the respondents returned 67 per cent favourable, 22 per cent neutral and 11 per cent negative attitudes to the product (Dixon 1985). The results of the study conducted on buffalo milk RO processing and its utilization in fluid form also did not show any appreciable change in the organoleptic quality during storage (Gupta and Pal 1993b).

Studies conducted on UHT processing of diluted RO milk and 2 fold RO concentrated milk showed very encouraging results (Kocak 1985). These studies indicate that the dairy industry can confidently explore the application of RO technology for concentrating whole milk at the collection/chilling centre followed by its processing in fluid form at the city plant with potential saving in transport cost. In addition, the water obtained during concentration of milk from RO plant is of high purity and can be used for drinking purpose.

Concentration of Milk for Bulk Transportation

The transportation of milk in "single strength" is very expensive. Gupta and Pal (1993b) studied the feasibility of transporting raw milk from rural chilling centres in RO concentrated form to city dairies. There was no losses of milk solids into permeate when buffalo whole milk was concentrated up to 2 fold. The quality of RO concentrated milk during transportation and storage at refrigeration conditions (at 4±1°C) upto 48 hours did not show any appreciable change in the organoleptic and chemical properties. The energy required for concentrating buffalo milk to 1.5 and 2-fold levels were calculated as 369.7 and 470.9 KJ/kg of water removed, respectively with corresponding costs 3.64 and. 4.64 paise. These workers (1993b) also evaluated the economics of bulk transportation of RO concentrated milk under Indian situation.

Economics of Bulk Transportation of RO Concentrated Milk

The following assumptions have been made to evaluate the economic aspects of transporting RO concentrated milk:

1. About 10,000 litres of raw buffalo milk is available at 25°C at a chilling/collection centre located about 300 km away from the main processing plant.
2. Milk is transported by a tanker on hired basis @ Rs. 8/= per km (both sides distance included), and makes only one trip daily.
3. Milk is preheated at 60°C in a plate heat exchanger with 80 per cent regeneration efficiency prior to RO processing.
4. Concentration by RO is done up to 2 fold (50 per cent reduction in volume) in two shifts, each of 8h duration.
5. About 4 h are required for cleaning of RO plant involving approximately 25 per cent of the total operating costs.
6. The present cost of membrane is about Rs.10,000 per m^2, and total membrane area required for 2 fold concentration is $13m^2$.

Cost Calculations

1.	Cost of transportation of fluid buffalo milk	= Rs.17,52,000 per year
2.	Total operating costs	= Rs.2,03,850
3.	Cost of transportation of 2 fold RO milk	= Rs 8,76,000
4.	Total cost of transportation (2+3)	= Rs.10,78,850
5.	Net saving (1-4)	= Rs.6,73,150
6.	Payback period	= 10 months

The application of RO for concentration of milk before bulk transportation curtail the costs to about 38 per cent with a pay back period of less than a year. Other advantages of adopting this technology are: less storage space required for RO milk, very high quality of water obtained in form of permeate, the RO concentrated milk may be directly used for manufacture of *Khoa, Kheer, Basundi,* roller powder etc. thereby providing additional energy savings.

Concentration of Whey

The production of cheese, paneer and casein create large volumes of whey (approximately 80-90 per cent of the initial volume of milk). Whey contains between 5.5 and 6.5 per cent solids that have a proximate composition (dry basis) of 12 per cent protein, 1 per cent fat, 70-75 per cent lactose, 8-10 per cent inorganic and 0.1-1.0 per cent lactic/citric acid. The presence of many valuable milk constituents (protein and lactose), and its huge pollution load (30,000-50,000 ppm B.O.D.) justify the need for economic processing of whey. However, the large proportion of water in the whey makes its processing energy intensive, per unit of solids, compared with other dairy processes. Operations such as pumping, transportation, concentration, fractionation and drying, all use more energy per unit of product than the dairy industry can afford. RO has the potential for removing a major part of the water from whey more economically than other evaporative processes (Table 10.2).

Table 10.2: Energy Cost of Efficient Evaporation vs. Reverse Osmosis for Whey (Pepper and Orchard, 1982)

Energy per Tonne of Whey	*Efficient Evaporation*				*RO Conc. to:*		
	1 Stage	*2 Stage*	*7 Stage*	*MVR*	*12% TS*	*18% TS*	*28% TS*
Steam (Tonnes)	0.62	0.14	0.09	0.03	0	0	0
Electricity (KWh)	0	3	3	25	3.6	8.8	9.6
Total equivalent (KWh)	387	90	60	44	3.6	8.8	9.6
Equivalent energy	192	46	31	35	3.6	8.8	9.6

Initial total solid in whey = 6 per cent

MVR = Mechanical vapour recompression

TS = Total Solids.

RO may be used (a) as a pre-concentration step prior to transportation of whey and (b) for its concentration for using as such or for making whey powder and other products (Jayaprakash *et al.*, 1995). Whey can be easily concentrated to about 18 per cent total solids on a batch type RO plant, but with the use of multistage recycle designs (MSR), concentration up to 26-28 per cent total solids is now technically and economically feasible. Flux for RO of whey is normally higher than that for milk mainly because of lower total solids in the former.

Concentration of UF Permeate

Applications of ultra-filtration (UF) process in concentration and fractionation of whey, skim milk and whole milk are increasing rapidly for various purposes, particularly for cheese making. Consequently, a large quantity of permeate is produced as a by-product. The UF permeate contain 5.5-6.5 per cent solids of which approximately 90 per cent is lactose alone and 8-9 per cent are minerals. Such a high concentration of lactose obviously favours the UF permeate as a potential source for manufacture of lactose.

Conventially lactose is prepared by concentration, fractionation, crystallization and centrifugation of cheese or casein whey. Whey and UF permeate are dilute solutions of lactose (5.0 to 5.5 per cent) and for getting higher yield and better purification their concentration up to 50 per cent total solids before crystallization is desirable. Thus, RO can play an important role in pre-concentrating UF permeate with minimum energy requirements and least thermal degradation. It is possible to concentrate UF milk permeates from initial total solids of 5.8 per cent to about 25 per cent by RO and maintain a reasonable flux (Pal and Cheryan 1987b). Higher concentration than this may not be economically feasible with the present generation of membranes and current energy costs.

The quality of lactose can be improved by demineralization of permeate using nanofiltration process before the RO processing. About 98-99 per cent of COD of the UF-permeate is retained in the concentrated product. Therefore, the permeate from the RO processing is free from any organic matter and thus a permissible waste.

REFERENCES

Abbot, J., Glover, F.A., Nair, D.D. and Skudder, P.T. (1979). Application of reverse osmosis to the manufacture of dried whole milk and skim milk. *J. Dairy Res.*, 46: 663.

Barbano, D.M., Bynum, D.G. and Senyk, G.F. (1983). Influence of reverse osmosis on milk lipolysis. *J. Dairy Sci.*, 66: 2447.

Cheryan, M. (1998). *Ultrafiltration and Microfiltration Handbook.* Technomic Publication Co., US.

Cheryam, M., Veeranjaneyulu, B. and Schlicher, L.R. (1990). Reverse osmosis of milk with thin-film composite membranes. *J. Membrane Sci.*, 48: 103.

Dixon, B.D. (1985). Dairy Products prepared from reverse osmosis concentrated milk. *Aust. J. Dairy Technol.*, 40: 91.

Draw, P.G. and Manners, J.G. (1985). Microbiological aspects of reverse osmosis concentrated milks. *Aust. J. Dairy Tech.*, 40: 10.

Gupta, S. and Pal, D. (1993a). Application of reverse osmosis for concentration of buffalo milk. *J. Food Sci. Technol.*, 30: 344.

Gupta, S. and Pal, D. (1993b). Studies on concentration of buffalo whole milk for bulk transportation employing reverse osmosis. *Ind. J. Dairy Sci.*, 46: 444.

Jayaprakash, H.M., Patel, R.S. and Renner, E. (1995). Optimization of pre-crystallization process for non hygroscopic whey powder by reverse osmosis concentration. *Jap. J. Dairy Food Sci.*, 44:A 113-117.

Kocak, H.R. (1985). Dairy products manufactured from whole milk concentrated by reverse osmosis Part 1. UHT products. *Aust. J. Dairy Technol.*, 40: 96.

Marshall, S.C. (1985). The reverse osmosis process. *Aust. J. Dairy Technol.*, 40: 86.

Pal, D. (1993). Membrane techniques: Principle of ultrafiltration. *Encyclopaedia of Food Science, Food Technolgy and Food Nutrition*, Academic Press, London, pp. 2975-2980.

Pal, D. and Cheryan, M. (1987a). Application of reverse osmosis in the manufacture of Khoa. Process optimization and product quality. *J. Food Sci. Technol.*, 24: 233.

Pal, D. and Cheryan, M. (1987b). Membrane technology in dairy processing Part 1. Reverse Osmosis. *Indian Dairyman,* 39: 251.

Pepper, D. and Orchard, A.C.J. (1982). Improvement in concentration of whey and milk by reverse osmosis. *J. Soc. Dairy Technol.,* 35: 49.

11

Trends in Technology of Drying

Y.K. Yadav
Department of Agriculture Processing and Energy, CCSHAU, Hisar

Introduction

With the growing world population, the provision of adequate food supply has assumed vital importance. Currently, about one-fourth of the world's population is undernourished, and this fraction is likely to increase with an increase in population unless urgent steps to avoid the situation are initiated. To cope with current and future demands two types of action traditionally have been emphasized: 1) reducing future demand by reducing population growth, and 2) augmenting food supply by expanding production. A third vital complimentary measure *i.e.* reducing the food losses between harvesting and consumption, has not been emphasized adequately. These postharvest food losses have become increasingly important for developing countries as the substantial amounts of food are being lost annually in the postharvest period. The staples, major vegetables and fruits in developing countries, comprise over 39 per cent of the food crops consumed. The amount of moisture present in food product is the most important factor in determining the extent of deterioration during storage.

Importance of Drying

Drying of the product, therefore, is the most vital operation in the chain of food handling. Proper drying inhibit the germination of seeds and prevents the growth of fungi and bacteria. It also retards the attacks of insects and mites on grains. Traditional drying techniques are specific to each locality and commodity, however, these may be categorized as open air drying or shade drying. In open air drying the crop is spread on a hard floor and rotated occasionally to ensure uniform drying. In shade drying the crop is covered to protect it from weather. Both these methods are unsatisfactory for a number of seasons. There is little or no control over the crop drying rate and drying in non-uniform. Over-excessive exposure to sun or heat, can damage the seed coat, cause bleaching, scorching, discoloration, loss of germination and nutritional changes. Very rapid drying of crops with high moisture content also can cause damage *e.g.* case hardening and bursting. Under-drying or slow drying results in deterioration of food due to fungi and bacteria. Therefore, drying under controlled conditions of temperature and humidity will help the crop dry seasonably rapidly to a safe moisture level and also ensure a superior quality of the dry product.

The amount of moisture present in food product is the most important factor in determining the extent of deterioration during storage. The moisture content of the product is reduced by various drying techniques. The principal objective in a drying operation is the supply of the required heat in an optimum manner to yield best quality of product with minimum overall expenditure of energy.

The maximum allowable temperature for drying of some selected products alongwith the final moisture content for safe storage are present in Table 11.1.

Process of Drying

In drying, heat is transferred to the surface of the product, part of this heat propagates to the interior of the product and the remaining amount is utilized in evaporation of moisture from the surface. The moisture from the interior diffuses to the surface to replenish the evaporated surface moisture; water vapour in the air surrounding the product also gets removed continuously by diffusion and flow of air. Drying can be accelerated by increasing

Table 11.1: Maximum Temperature Allowable for Drying and the Initial and Final Moisture Contents of Various Products

Product	*Moisture Content*		*Maximum Temperature Allowable for Drying (°C)*
	Initial (%)	*Final (%)*	
Paddy, raw	22–24	11	50
Paddy, parboiled	30–35	13	50
Maize	35	15	60
Wheat	20	16	45
Millet	21	4	–
Corn	24	14	50
Rice	24	11	50
Green peas	80	5	65
Cauliflower	80	6	65
Carrots	70	5	75
Green beans	70	5	75
Onions	80	4	55
Garlic	80	4	55
Cabbage	80	4	55
Sweet potato	75	7	75
Red lauan	90	20	–
Potatoes	75	13	75
Spinach	80	10	–
Cassava	62	17	–
Cassava leaves	80	10	–
Chillies	80	5	65
Fish, raw	75	15	30
Fish, water	75	15	50
Onion rings	80	10	55
Prunes	85	15	55
Apples	80	24	70
Apricots	85	18	65
Peaches	85	18	65
Grapes	80	15–20	70

Contd..

Table 11.1–Contd..

Product	Moisture Content		Maximum Temperature Allowable for Drying (°C)
	Initial (%)	*Final (%)*	
Bananas	80	15	70
Guavas	80	7	65
Mulberries	80	10	65
Fig	80	24	–
Okra	80	20	65
Pineapple	80	10	65
Yams	80	10	65
Nutmeg	80	20	65
Sorrel	80	20	65
Coffee	50	11	–
Coffee beans	55	12	–
Cocoa beans	50	7	–
Cotton	50	9	75
Cotton seed	50	8	75

the flow rates of air and/or by increasing the temperature of the drying air. The initial process of drying is the surface drying after that the rate of drying depends upon the rate of moisture migration from the interior of the product to the surface; this naturally depends on the type of material to be dried. For nonhygroscopic materials, drying can be carried to a level of zero moisture content. Hygroscopic material such as grains, fruits and food stuffs in general will have residual bound moisture, such as water trapped in closed capillaries, the water component of juices, water absorbed by surface forces and unbound water held within the material by the surface tension of water itself. In such a situation the escalation of product temperature is more effective in the removal of moisture than the increase in the flow rate of the drying air. The process of drying is therefore controlled by the properties of the drying, often known as external parameters, and the properties of the product *i.e.* internal parameters.

Drying Parameters

The external parameters of importance are the dry and wet-bulb temperatures and hence relative humidity, humidity ratio, temperature and enthalpy. In the drying process, evaporation of water is caused by the heated or unheated air from the wet product. The dry-bulb temperature decreases approaching dew point, while the wet-bulb temperature remains constant. If the moist air is heated without addition or removal of water, its humidity ratio remains constant, whereas the relative humidity decreases with an increase in temperature. This dry-bulb temperature is used in removing the moisture as well as raising the temperature of the product.

The internal parameters affecting the process of drying are moisture content, drying ratio, exposure time and the latent heat. The magnitude of the moisture content at the stage of equilibrium *i.e.* when the rate of loss of moisture from the product to the surrounding environment equals the rate of moisture absorption by the product, is termed as the equilibrium moisture content (EMC). The associated value of the relative humidity of the surrounding air is called the equilibrium relative humidity. The relationship between equilibrium relative humidity and equilibrium moisture content is devoted for a particular temperature on an equilibrium moisture curve known as sorption isotherm.

Effect of Parameters on Drying Process

The drying rate is dependent upon many parameters, namely air temperature, air flow rate, relative humidity, exposure time, types, variety and size of grains, initial moisture content, depth of drying material etc.

Among various parameters, air flow rate and the temperature of flowing air are the two main parameters that determine the quality of the dried product. Too rapid drying may result in the formation of hard drying surfaces that will inhibit the migration of moisture from the interior of the material to the surface.

The temperature of the drying air must be kept below some maximum value depending on the intended use of the product. The recommendations for temperature limits to avoid deleterious effects usually are stated on the basis of drying air temperatures, whereas the important consideration is the temperature of the grain. In fact, the deleterious effect on grain is determined by a combination of temperature and humidity of drying air and the duration of exposure and not by temperature alone. Thus, a high temperature can be used for a short time. In the beginning of the drying process, the grain kernel is at a lower temperature than the air used for drying. However, in the later stages of drying when the kernels are nearly dry, the grain kernel temperature approaches the air temperature. Therefore, different maximum temperatures are recommended for different food products depending on initial moisture content and desirable final moisture contents of the products.

Alongwith the temperature of the drying air, flow velocity of air also determines the drying time. Although the evaporation of water requires heat, the driving force for drying is provided by the gradient of water vapour pressure between the ambient air and the evaporating surface. The best indication of the drying potential of the ambient air is relative humidity, measured as the wet-bulb depression. Therefore, temperature and humidity of air play an important role in the drying process.

Different Methods of Drying

Drying methods can be broadly classified on the basis of either the mode of heat transfer to wet solid handling characteristics and physical properties of the wet material. According to the mode of heat transfer, the drying method can be divided into:

1. Conduction drying
2. Convection drying
3. Radiation drying
4. Dielectric drying
5. Chemical drying
6. Sack drying
7. Spray drying
8. Vacuum concentrator drying
9. Fluidized bed drying
10. Freeze drying and
11. Desiccant air drying.

Conduction Drying

In this method of drying, the heat of drying is transferred to the wet product mainly by conduction through a solid metallic surface. This method is also known as contact drying and the vaporized moisture is removed independently of heating media.

Conduction drying is the most efficient method of drying and can be carried out either continuously or batchwise. Cylinder dryers, drum dryers, steam tube rotary dryers are some of the continuous conduction dryers whereas vacuum tray dryers, freeze dryers agitated pan dryers are examples of batch conduction dryers.

Convective Drying

In this method of drying, the drying agent (hot air or gases) in contact with the wet solid is used to supply heat and carry away the vaporized moisture and heat is transferred to the wet solid by convection. The energy used in this method is higher than that of conduction drying.

Convection drying is most popular in grain drying and can be carried out either continuously or batchwise. Continuous tray dryers, continuous sheeting dryers, pneumatic convecting dryers, rotary dryers, tunnel dryers are of continuous types whereas tray and compartment dryers, batch circulation dryers are of batch type. The convective dryers may also be of natural air drying and heated air drying type depending on the type of the material and market requirement of the product.

Radiation Drying

Radiation drying is based on the absorption of radient energy of the sun and its transformation into heat energy by the wet product. Sun drying can also be accomplished with the aid of special infra-red lamps. Moisture movement and evaporation is caused by the difference in temperature and partial pressure of water between product and surrounding air.

Dielectric Drying

In dielectric drying, heat is generated within the solid by placing it in a fixed high frequency current. In this method, the substance is heated at the expense of the dielectric loss factor. The molecules of the substance, placed in a field of high frequency current are polarised and begin to oscillate in accordance with the frequency. The oscillations are accompanied by friction, and thus a part of the electric energy is transferred into heat. The main advantage of this method is that the substance is heated with extraordinary rapidity. The dielectric heating is in use in timber, plastics and cereal processing industries.

Chemical Drying

Various chemicals such as sodium chloride, calcium propionate, copper sulphate, ferrous sulphate, urea etc. have been tried for preservation of wet products. Common salt has been proved to be very effective for paddy preservation as when wet paddy is treated with common salt, water is removed from rice kernel by osmosis. The care should be taken that in no case the salt should penetrate into the wet product.

Sack Drying

This method is particularly suitable for drying of small quantity of seed to prevent mixing of varieties and conserve strain purity and viability. In this method, the bags are laid flat over holes cut on the floor of a tunnel system so that heated air is forced up through the grains from an air chamber underneath. The sacks are turned once during drying operation for uniform drying. The sack drying process involves higher labour cost.

Spray Drying

In this method water from the material is removed by evaporation and at that moisture content, powder is obtained which reaches equilibrium with exit drying air. Spray dryer is used intensively in pharmaceutical, chemical and food industries. Design changes from right angle chamber to large volume system which is fitted with spray jet through which drying air passes continuously. The three methods named high pcrystallization, centrifuge and double fluid crystallization are suited for bringing fine droplets of fluid. This method of drying is faster because of small diameter drops and the material dries at the bottom layer or at the walls of the material holder.

Vacuum Concentrator Drying

This method is used for concentrating milk, fruits and vegetable juices, jams and jellies etc. by boiling a small amount of water. The boiling point of concentrated fluid is controlled by the temperature of condensed water mixture leaving the condensor. Condensed water mixture is often separated by a pump and vacuum is often maintained by vapor extractor through a vacuum pump. In this method, high rate of heat is possible with minimum heat exchange surface.

Fluidized bed drying

In this method of drying, products are being dried under fluidized condition in a dryer. The material are fluidized by drying air with sufficiently high velocity to cause suspension. In this process, higher rate of moisture migration and uniform drying of products takes place. This method is preferred for vegetable seeds.

Freeze Drying

This method of drying is based on the sublimation of frozen moisture from the wet product placed in a drying chamber. The pressure in the drying chamber is low. Heat is supplied by radiation or conduction from heated trays and the temperature of the product is not raised above 0°C. The sublimated moisture is condensed on refrigerated plates in the drying chamber but these plates do not come in contact with products.

Osmotic Drying

This method of drying can be successfully applied to dewatering of cellular products such as fruits, vegetables, meat etc. For the moisture removal, osmo-active substances like, 60 per cent aqueous solution of saccharose or 25 per cent aqueous solution of sodium chloride is used. The difference in osmotic pressure on both sides of the cell wall having properties of a semi-permeable membrane causes moisture removal from the products.

Desiccant Air Drying

In this method of drying, the ambient air is passed through a desicant medium *e.g.* silica gel. The desiccant absorbs moisture from air, as a result, the relative humidity of air is reduced and at the same time there is an increase in its temperature. When such air comes in contact with wet product, transfer of moisture to drying air takes place. The drying action occurs due to convective heat transfer. The method is preferred for thermosensitive food material or maintaining very low moisture content of the product.

12

Quality Management System in Dairy Industry

C.M. Kapoor and R.S. Dabur
Department of Animal Products Technology, Haryana Agricultural University, Hisar

"QUALITY NEVER COMES ACCIDENTALLY. IT IS ALWAYS THE RESULT OF HIGH ATTITUDE, SINCERE EFFORTS, INTELLIGENT DIRECTIONS AND SKILFUL EXECUTION"

The Indian dairy industry is confronted with Herculean task of meeting the ever-increasing demand of milk and milk products. About 70 Millions of small milk producers, holding 2 or 3 cows/ buffaloes, scattered over an area of 3.29 million km^2, inhabited in about 5.75 lakh villages contribute nearly 78 MMT of milk/annum (Chawla, 2000 and Rajorhia, 2001). Indian milk production system is a sub system of agriculture mainly depending on the crop residue resulting in lesser farm gate price.

Milk Processing and Utilization Pattern

Indian milk utilization pattern is too skewed towards liquid milk (45 per cent) and rest is converted into milk products such as *Ghee* (28 per cent), *Dahi*/curd (7 per cent), *Khoa* (6.5 per cent), butter (6.5 per cent), Milk powder (2.6 per cent), *Chhana, cheese* and *paneer* (2.0 per cent), cream (0.5 per cent), ice-cream (0.2 per cent)and other products (1.7 per cent).

There is a wide variation in the share of milk processed in dairy industry. In USA, New Zealand and Japan, share of milk processed is as high as 98 per cent, while in India, organized dairy sector handles about 12 per cent of the total milk produced through about 575 dairy factories in the cooperative, public and private sector. Out of this 9-10 per cent of milk is processed in to products by cooperatives and public sector dairies while the rest of the milk is distributed as liquid milk. Private sector dairies process about 85 per cent of milk into products and rest is distributed as liquid milk.

Although India has achieved the laudable position of being top milk producer in the world, its share in the international trade of dairy products is almost negligible (Table 12.1). Over the recent years, however, India's export increased steadily- it was worth Rs. 11.94 crore during 1996-97, Rs. 12.54 crore during 1997-98 and Rs. 16.13 crore in 1998-99.

Table 12.1: Share of Major Exporting Countries and India in the Export of Milk Products

	Share in Export %							
	Butter/Butter oil		*SMP*		*WMP*		*Cheese*	
	1998	*1999*	*1998*	*1999*	*1998*	*1999*	*1998*	*1999*
EU	20.5	21.8	18.0	24.0	42.2	40.2	36.3	32.2
USA	1.4	0.8	11.4	15.8	1.4	1.2	3.0	3.1
Australia	13.3	16.1	20.4	21.0	7.9	9.8	13.5	15.8
New Zealand	39.6	38.2	17.0	15.4	25.8	25.5	18.8	19.6
India	0.1	*	0.07	*	0.01	*	0	*

* India's share in export during 1999 was similar to that during 1998

(Varshney, 2000)

Present Scenario of Quality Control (QC) of Milk in India

As major part of milk is being produced by landless labourers, small and marginal farmers, their poor socio-economic conditions is one of the serious constraint in the poor quality of raw milk production. Only 25-35 per cent of the milk produced is traded and most of this is handled by vendors/contractors. The quality of milk is further deteriorated by adding poor quality water, urea/synthetic milk, H_2O_2 and some other additives to extend the quantity and shelf life of raw milk. Most of the dairy plants follow chemical quality control with the intention of acceptance of milk for processing, payment to producers/contractors and end-product analysis for meeting mandatory requirements of PFA-Act.

Microbiological Quality of Milk is Far from Satisfactory Right from Production of Raw Milk to the End Product

QC on the basis of final product analysis may lead to product failure. By the time results are known, it may be too late to take remedial measures. In the newer context on Quality, concept and organizational set up has moved from *Inspection* to *Quality Control* to *Quality Assurance* to *Continuous Improvement* and finally to *Total Quality Management.*

Quality Management Systems

The global economy today centers around technology, transnational production systems and international competitiveness. The emergence of global market of unprecedented magnitude has created a demand for globally standardized products. Implementing international standards and third party certificate provides confidence to the consumers, thus facilitates trade.

Legislative Controls

PFA Act (1954) and Rules (1955)

PFA Act and Rules come under the Ministry of Health and Family Welfare. This Act prescribes the minimum mandatory standards and other provisions to prevent adulteration of foods (including milk and milk products). The adulteration has been defined in wide perspective in terms of purity, nutrition, wholesomeness, safety against contaminants, toxicants, bacterial toxins, preservatives, colourants, insecticides and pest infection, misbranding or mislabeling and anything else, which may be injurious to health.

Table 12.2: Evolution of Quality Concept

< 1900	Craftsman (Single worker)
1900-1910	Operator (Single or group of workers)
1910-1920	Foreman
1920-1940	Inspection
1940-1955	Statistical Quality Control
1955-1960	Quality Control
1960-1965	Total Quality Control (In line testing)
1965-1975	Quality Assurance
1975-1990	Longitudinal Integrated Safety Approach (LISA) HACCP, GMP
>1990	TQM, TQC, HACCP, ISO-9000

Definitions and standards of quality of different classes and designations of milk (such as: buffalo milk, cow milk, sheep and goat milk, mixed milk, standardized milk, recombined milk, toned milk, double toned milk and skimmed milk and full cream milk) and milk products are mentioned under Appendix B-A.11 of PFA Rules.

Standards of Weight and Measures (SWM) Act 1976 and SWM (Packaged Commodities) Rules 1977

Ministry of Law, Justice and Company Affairs govern the Act and Rules, which describe provisions for labeling, declaration and quantity to be packed. Packing and labeling of foods (including that of milk and milk products) has also been covered under Part VII of PFA Rules (1955)

Export (Quality Control and Inspection) Act 1963

The Export Inspection Council operates this Act. Recently Government of India has exempted Agriculture and Food products, Fruit products, Fish and Fish products from compulsory pre-shipment inspection provided exporter has firm letter from overseas buyers.

Pollution Control Board

No objection certificate from respective State Pollution Control Board is required for every dairy plant.

Consumer Protection Act (1986)

This Act provides six consumer's rights *viz.* right to safety, right to be informed, right to choose, right to be heard, right to seek redressal and right to consumer education. This Act provides a simple, speedy and inexpensive redressal by District Forum/State/National Commission.

Bureau of Indian Standards (BIS-1986)

BIS earlier known as ISI, was established under Indian Standards Act-1987. It took over the functions, assets, liabilities of ISI which was functioning since 1947 under the Ministry of Civil Supplies, Consumer's Affairs and Public Distribution. Functions of BIS are:

1. Standards formulation
2. Certification mark scheme
3. Recognition of laboratories
4. Publications
5. National and International cooperation and coordination
6. Promotional and Extension activities

BIS has recently revised and formulated new standards on pasteurization, pasteurized milk, pasteurized butter, SMP and low fat infant formula with more emphasis on microbial requirements.

Agricultural Produce (Grading and Marketing) Act 1937 (Amended-1986)

The Directorate of Marketing and Inspection under the Ministry of Agriculture is responsible to implement this Act. Three dairy products *viz.* ghee, butter and fat spreads are graded under "AGMARK".

Milk and Milk Products Quality Standards/Testing at International Level

Following quality control standards/methods of analysis of milk and milk products have been acceptable at international level:

ADPI (ADMI-1929)

American Dry Products Institute (ADPI), earlier known as ADMI formulated standards for different grades of whole milk powder and skim-milk powders. These standards and methodology are still acceptable and in use in most of the research and quality control laboratories. However, these standards are not recognized/ approved under WTO.

AOAC (Association of Official Analytical Chemists)

AOAC was established in 1884 and its methodology is acceptable/recognized for analysis of food and other products all over the world.

IDF (International Dairy Federation)

IDF with its headquarter at Brussels was established in 1903. Thirty-six countries are its members. IDF has made significant technical and scientific contributions. It has six specialized commissions (A to F). Commission "D" deals with food legislation and Commission "E" deals with analytical techniques and laboratory methods. IDF has formulated 180 standards.

ISO (International Organization of Standards)

ISO consists of 120 members (one from each country). It has formulated more than 10,000 standards, out of which 55 dairy standards are dealing with sampling, methodology, pesticides residues, sensory evaluation, quality control. These standards are voluntary in nature.

ISO-9000 is a Quality System and ISO-9001, ISO-9002 and ISO-9003 are the quality assurance models, the components of which are:

ISO-9001: It assures quality in design, development, production, installation and services.

ISO-9002: It assures production, installation and services.

ISO-9003: It assures final inspection and testing. (Annexure-I).

CAC (Codex Alimentarious Commission) and Codex Regulations

The Codex Alimentarious Commission is an inter-government body, which came into existence in 1962,under the joint sponsorship of FAO and WHO. This body is responsible for:

1. Formulation of International Standards for the safety and quality of food.
2. Good Manufacturing Practices (GMP).
3. Methods of analysis and other guidelines.
4. To remove unfair trade practices in international trade.

One hundred and sixty-three countries are members of CAC. In the pre-WTO era, the CAC had no authority over its members for the implementation of Codex standards. It used to be a recommending body to the governments. But since the constitution of WTO and signing of two vital agreements, namely agreement for Sanitary and Phytosanitary (SPS) measures and agreement for Technical Barriers to Trade (TBT), the CAC standards have become the basis for *international trade*. CAC is inter-Government in nature and not open to NGO's. Seventeen *Commodity Committees* and eight *Horizontal Committees* undertake the work of Codex standards, guidelines and recommendations. The Commodity Committees work in a vertical manner on a specific food allotted to it.

The Horizontal Committees deals with general subject matters such as food additives and contaminants, pesticides residues, veterinary drugs residues, heavy metals and aflatoxins, food hygiene, labeling, method of sampling and testing etc. Table 12.3 provides Codex work completed. Codex maximum pesticide residue limits (MRL-mg/kg or ppm) and maximum Veterinary Drugs residues limits (MRL-ppb) in milk are presented in Annexure-II. and Annexure-III, respectively.

Table 12.3: Codex Work Completed

Food Standards for commodities	~	240
Codex of hygienic/technological practices	~	40
Limits for pesticides residues	~	3280
Guidelines for contaminants	~	25
Food additives evaluated	~	1000
Veterinary drugs evaluated		55

Food Safety Management Systems

In the quest of quality, the dairy industry is passing through revolutionary changes.

The holistic approach to Total Quality Management/Quality Assurance is a must for entering in the Global Market. Any food/ dairy industry, which is quality oriented and customer focused, is bound to become the market leader.

Total Quality Management (TQM)/Quality Assurance (QA)

TQM is an integrated approach towards the customers by meeting their expectations on a continuous basis through every one involved in the organization working on continuous improvement in all spheres. Three major elements of TQM are: a) Total employees involvement, b) Total waste elimination and c) Total quality control. Principles of TQM are:

1. Customer focused
2. Do it right, first time, every time, quality first and always
3. Communication and education
4. Quality measures and records
5. Do it together

Integrated TQM Model

Total Employees Involvement

1. *Kaizen*-a Japanese word, which means "continuous improvement" through internal consultation development.
2. *Hoshin Kanri*- a Japanese concept of policy deployment.

Total Waste Elimination

House- keeping-JIT (Just- in-time). The basic principle of HK is "A place for everything and everything in its place" Five steps (Five S's) for better HK are:

1. Segregation *(Seiri)*
2. Arrangement *(Seiton)*
3. Cleanliness *(Seiso)*
4. Maintenance of standards *(Seiketsu)*
5. Discipline *(Shitsuke)*

Total Quality Control

1. Problem solving tools

2. PDCA (Plan, Do, Check and Act)
3. SPS (Statistical Process Control)
4. HACCP/ISO 9000 (Vyas, 1999)

HACCP (Hazard Analysis of Critical Control Point) System

HACCP was introduced in the food industry as a spin of the NASA-USA space programme in 1959. M/S Pillsbury Co., a major food supplier to NASA hit upon the concept of food safety by exercising control over the raw material, process, environment and personnel.

Amongst various quality control/assurance systems, CAC recognizes only a HACCP based approach for food safety. A number of countries have made HACCP system mandatory for production of various foods including dairy products. The most important element of a HACCP based system is its preventive nature and control throughout the food chain *i.e.* from primary production to consumption. HACCP is a systematic method using the seven principles for analyzing a food process:

1. Conduct a Hazard Analysis.
2. Identify the Critical Control Points for each step.
3. Establish Critical Limit(s).
4. Establish a system to monitor control of the CCP(s).
5. Establish the corrective action to be taken when monitoring indicates that a particular CCP is not under control.
6. Establish procedures for verification to confirm that the HACCP system is working effectively.
7. Establish documentation concerning all procedures and records appropriate to these principles and their application.

Area of concern for adoption of HACCP system in Milk and Milk products:

1. Good hygienic practices for milk production (GHP)
2. Good manufacturing practices (GMP)
3. Inadequate testing facilities
4. Level of contaminants

5. Lack of concept/attitude of pest or disease free area
6. Insufficient training and education
7. Top management commitment

Good Manufacturing Practices (GMP)

GMP standards define requirements for the management, control of activities and operations involved in the manufacture, storage and distribution of foods. Elements of activities and operation are identified as determinants of product quality and food safety. These are:

1. Employees working in a plant are required to wash their hands with a sanitizing soap prior to beginning or returning to handle food.
2. Loose hair poses a serious sanitation problem in food plant. All workers or visitors must wear head covering.
3. Sanitary precautions while sneezing or coughing. Workers with infectious skin and communicable diseases not to be allowed to handle food processing.
4. Workers in production area should wear clean uniforms. Shirts should be buttoned and tucked into the trousers.
5. Workers in the production area not to wear watches, rings, neck and ear jewellery.
6. Smoking, spitting or chewing of tobacco be prohibited in the production and storage area.
7. Consumption of beverage or food is allowed in the designated area only.
8. Glass containers and equipment made of glass to be avoided in production area. Good housekeeping in the production area is necessary for work efficiency and workers safety.

Sanitary and Phytosanitary (SPS) Measures

Under the Codex regulations, the agreement for SPS recognizes only Office International Epizootics (OIE) for animal health concerns and CAC for food safety concerns. The Indian dairy industry has not so far taken OIE and CAC seriously. These vital issues remained unattended due to financial constraints or India is represented by a non-technical spokesperson, which is unable to argue India's case forcefully. India is unable to provide the certified evidence because it has not as yet promulgated the Animal Diseases Act. This is likely to become a major stumbling block in the way of India's dairy export. SPS measures are on:

1. Food additives
2. Contaminants
3. Toxins
4. Veterinary drugs and pesticides residues
5. Plant/animal quarantine
6. Procedures related to prevention of spread, control and/or eradication of pests/diseases
7. Processing methods with implication for food safety
8. Labeling/packaging requirements directly related to food safety
9. Sanitary requirements on imported food/feed products
10. Certificate of food safety/animal or plant health certificate

International Office of Epizootics (OIE)

OIE, created in 1924, is the world organization for animal health. In April 2000, the number of Member Countries was 155. OIE Standards, Guidelines and Recommendations concerning animal health are:

1. General Provisions: Notification of animal diseases, Veterinary Ethics, Certification of international trade, import risk analysis and import/export procedures etc.
2. Recommendations Applicable to Specific Diseases: List A Diseases, Multiple Species and Bovine diseases in List B Diseases.
3. Diagnostic Tests for International Trade: Prescribed and Alternative Diagnostic tests for List A and B diseases.
4. Health Control and Hygiene in Establishments
5. Quarantine Recommendations
6. Inactivation of Pathogens and Vectors.
7. Transport of Animals: Principles applicable to all forms, specific form and air transport.

8. Epidemiological Surveillance System.
9. Model International Certificates

India is yet to comply with the important requirement of enactment of a central Act on "Prevention of Infectious and Contagious Diseases in Animals Act " (Chawla, 2001).

Microbiological Concern

Codex has been seriously concerned on the microbiological quality of milk and milk products and suggests measures to minimize them. Pasteurization, as per its definition kills all pathogens but initial high bacterial count in milk is undesirable for several reasons:

1. Pasteurization influenced by initial count.
2. Activated/germinated spores spoil the milk.
3. Bacterial inhibitors inactivated by high temperature. Growth of surviving bacteria/germinated spores is stimulated.
4. Toxins produced by some bacteria are stable to temperature of pasteurization.
5. Some bacteria produce enzymes like proteases and lipases, which are, heat stable.
6. Psychrotrophs proliferate even at 4°C.
7. Dead cells remain in the milk.

Under Codex Standards," Recommended International Code of Practice: General Principles of Food Hygiene" measures have been suggested to produce milk of good microbiological quality.

For implementation of Good Hygienic Practices at primary milk production level, chilling centers and dairy plant level, a model (Varshney,2000) is suggested in Figure 12.1.

Conclusion

1. Make available good quality feed in sufficient quantity at reasonable rates.
2. Enactment of "Animal Disease Act" for providing the certifying evidence.
3. Clean milk production, both in terms of microbiological quality and toxic residual level by educating milk producers through extension education programmes.

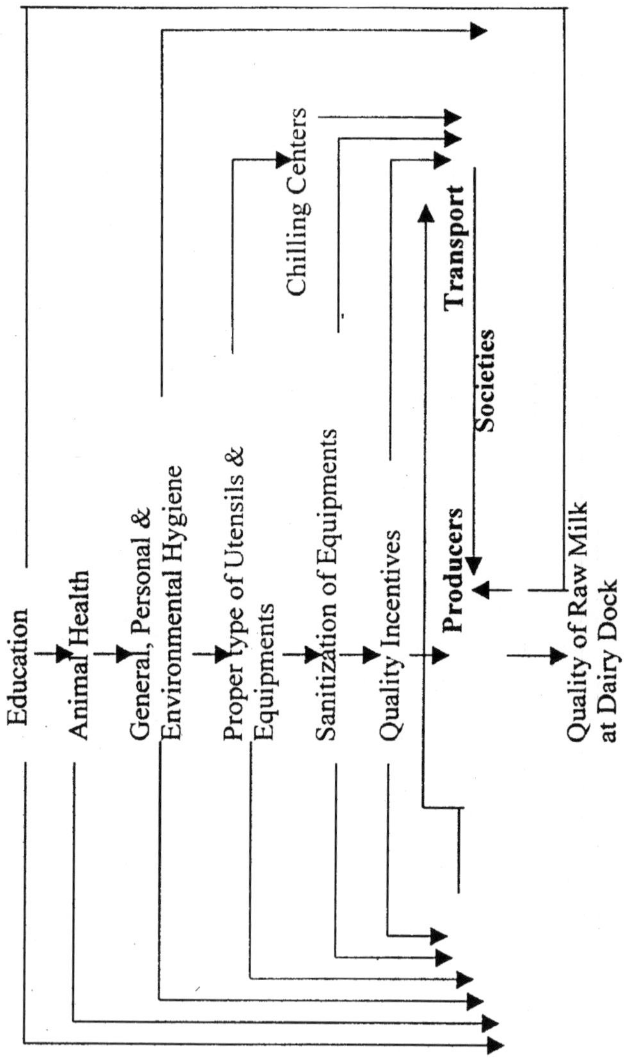

Figure 12.1: Strategy to Meet Quality Challenges

4. Incentives to producers for producing high quality milk.
5. Develop chilling facilities.
6. For producing quality and safe milk products, Dairy industry should adopt ISO 9000-HACCP Quality Management System not only for international market but also for internal market.
7. WTO recognizes Codex Standards, which are more stringent than PFA -Standards. There is urgent need to harmonize PFA–Standards with those of Codex Standards.

REFERENCES

Chawla, N.K. (2000). Effect of WTO and Codex regulations on Indian Dairy industry. *Indian Dairyman*, 52: 11–27.

Rajorhia, G.S. (2001). Opportunities for dimensional growth of Indian dairy processing industry. *Indian Dairyman*, 53: 9–35.

Varshney, N.N. (2000). Codex Standards and Quality needs. *Indian Dairyman*, 52: 11–33.

Vyas, B.M. (1997). A policy paper on Indian dairy. *Indian Dairyman*, 49: 5–125.

Annexure-I
Elements of ISO–9000 Quality Systems

Sl.No.	*Elements*	*ISO-9001*	*ISO-9002*	*ISO-9003*
1.	Management responsibility	+	+	+,–
2.	Quality system	+	+	+,–
3.	Contract review	+	+	+
4.	Design control	+	–	–
5.	Document control	+	+	+
6.	Purchasing	+	+	+,–
7.	Purchases supplied products	+	+	+,–
8.	Product identification and trace ability	+	+	+,–
9.	Process control	+	+	–
10.	Inspection and testing	+	+	+,–
11.	Inspection, measuring and testing equipment	+	+	+
12.	Inspection and test status	+	+	+
13.	Control of non–conforming products	+	+	+,–
14.	Corrective action	+	+	+,–
15.	Handling, storage, packaging and delivering	+	+	+
16.	Quality record	+	+	+,–
17.	Internal quality audits	+	+	+,–
18.	Training	+	+	+,–
19.	Servicing	+	+	–
20.	Statistical techniques	+	+	+,–

+: Must be included in the system;
–: Not present in the system;
+,–: May or may not be included in the system.

(Nandi and Gokulakrishnan, 1999).

Annexure-II
Codex Maximum Pesticide Residue Limits in Milk

Contaminant	*Recommended MRL, mg/kg (ppm)*	*PFA-MRL, mg/kg (ppm)*	*Remarks*
Abamectine	0.005	*	
Acephate	0.1	*	
Aldicarb	0.01	*	
Aldrin	0.006	0.15	F
Amitraz	0.01	*	V
Anilazine	0.01	*	
Azocyclotin	0.05	*	V
Bendiocrab	0.05	*	V
Bentazone	0.05	*	
Bifenthrin	0.05	*	
Cabaryl	0.1	*	MP
Carbendazim	0.1	0.1*	F*
Carbofuran	0.05	0.05*	F*
Chinomethionate	0.01	*	
Chlordane	0.002	0.05	F
Chlorfenvinphos	0.008	0.2	F,V
Chlormequat	0.1	*	
Chlorpyriphos-ethyl	0.01	0.01	F,V
Chlorpyriphos-methyl	0.01	*	F
Clofentezin	0.01	*	
Cyfluthrin	0.01	*	F,V
Cyhexatin	0.05	*	V
Cypermethrin	0.05	0.01*	F,V
Cyromazin	0.01	*	V
2,4,Dichlorophenoxy ascetic acid	0.05	0.05	
DDT	0.05	1.25*	*F
Deltamethrin	0.02	*	F,V
Diazinon	0.02	*	F,V

Contd..

Annexure-II–Contd..

Contaminant	*Recommended MRL, mg/kg (ppm)*	*PFA-MRL, mg/kg (ppm)*	*Remarks*
Dichlorvos	0.02	*	
Dicofol	0.05	*	F
Diflubenzuron	0.05	*	
Dimethipin	0.02	*	
Diquat-bromide	0.01	*	
Disfulfoton	0.02	*	
Dithiocarbamate	0.05	*	
Edifenfos	0.01	0.01*	F*
Endosulfan	0.004	*	
Endrin	0.0008	*	F,MP
Ethiofencarb	0.02	*	
Ethion	0.02	0.5	F,V
Etrimfos	0.01	*	F
Fenbutatin-oxide	0.02	*	
Fenitrothion	0.002	0.05	F
Fenpropathrin	0.1	*	F
Fenthion	0.05	0.05	F,V
Fenvalerate(cis,trans)	0.1	0.01	F
Flusilazol	0.01	*	
Glyphosate	0.1		
Heptachlor	0.006	0.15	F
Propioconazole	0.01	*	
Isofenphos	0.01	*	F
Lindane	0.01	0.2	F,V
Mecarbam	0.01	*	
Methacrifos	0.01	*	F
Methamidophos	0.01	*	
Methidathion	0.0008	*	F
Methiocarb	0.05	*	
Methomyl	0.02	*	
Methoprene	0.05	*	F,V

Contd..

Annexure-II–Contd..

Contaminant	*Recommended MRL, mg/kg (ppm)*	*PFA-MRL, mg/kg (ppm)*	*Remarks*
Monocrotophos	0.002, (0.02)*	0.02	MP*
Myclobutanil	0.01	*	
Paraoxon-methyl	0.01	*	
Paraqoat-dichloride	0.01	0.01	
Penconazole	0.01	*	
Permethrin	0.1	*	F
Phenthoate	0.01	0.01	F
Phorate	o.05	0.05*	F*
Phosmet	0.02	*	F,V
Phoxim	0.05	*	F,V
Pirimicarb	0.05	*	
Pirimiphos-methyl	0.05	0.05	F
Prochloraz	0.1	*	
Profenofos	0.01	*	
Propargite	0.1	*	F
Propoxur	0.05	*	
Terbuphos	0.01	*	
Thiabendazole	0.05	*	
Triadimefon	0.1	*	F
Triadimenol	0.01	*	
Triazophos	0.01	*	
Trichlorfon	0.05	0.05	V
Vinclozdin	0.05	*	

Note: F: Residue as fat soluble; V: Additional veterinary use; MP: Milk products.

Varshney (2000).

Annexure-III

The Codex Maximum Residues Limits (MRLs) of Veterinary Drugs in Milk

Veterinary Drug	*Recommended MRL,ppb*
2,4,5-Trichlorophenoxy acetic acid	0.05
Benzylopencillin	4
Ceftiofur	100
Dihydrostreptomycin	200
Gentamycin	100
Neomycin	500
Oxytetracyclin	100
Spectinomycin	200
Sulfadimine	25
Thiabendazole	0.1

Several other drugs are under consideration of Codex.

(Varshney, 2000).

13

Quality Assessment of Dairy Products

D.K. Thompkinson

Dairy Technology Division, National Dairy Research Institute, Karnal – 132 001

In view of the highly perishable nature of Dairy products which are extremely vulnerable to chemical and microbiological contaminants, dairy industry must have a strong network of efficient quality assurance programme to monitor the quality and safety of these products before reaching the consumers. Ensuring acceptable level of food quality can be achieved by implementing effective quality control programme along the entire food chain where ever appropriate and possible. This can be largely possible with the application of recent developments in methodologies that are sensitive, accurate and rapid as tools in quality assurance programme.

Quality of any processed food is the combination of characteristics (physical, chemical, nutritional, microbiological and aesthetic) that have significance in determining the degree of acceptability of the product to the consumers. Quality is judged both by subjective as well as objective tests. The purpose of quality control is to ensure production and distribution of food products within the well defined and accepted standards, thereby protecting health and legal rights of consumers and financial interest of manufacturer. Quality control of foods run parallel to the production of food and should adhere strictly to the prescribed standard.

The consumers today are increasingly quality conscious and have started exercising their rights to expect that the food they eat is safe and suitable for consumption. The quality environment in the Indian food processing sector basically revolves around compulsory legislation, which specifies minimum standard and certification system for both processed and agricultural products.

With the progress in globalization it appears that the World would be, in due course of time, become a single market place with all kinds of produce crossing international borders. Such changes will lead to fierce competition both in domestic and international markets. The new international food laws have come into force emphasizing the shift towards quality, suitability and safety of the products.

Quality Control

All the changes to which milk is susceptible are mostly post secretory events. Milk is a produce of biological origin and microbiologically fragile. It is easily attacked by microbes as soon as it is secreted by the cow. It also accommodate additives and adulterants without any apparent change in its look. Quality control therefore begins when mammary functions ends. It has to be excersied at different points of processing in order to ensure good quality of finished product. Various stages of checking the quality of milk are:

Initial Stage

The quality of raw material and other ingredients used in manufacture of milk products to assess their suitability for use. Raw milk for processing can be assessed by various plat form tests.

Intermediate Stage

Quality should be checked after standardization and before the product is ready for packaging, to assess if it confirms to the prescribed standards. The product is generally tested for various sensory and chemical parameters to assess quality.

Final Stage

The quality of finished product is checked after packaging and before it is ready for marketing. Various chemical, sensory, nutritional and microbiological parameters are tested in detail to ensure its fitness for human consumption and confirmation to standards.

Raw material controls are difficult without reference to process control and vice versa. Once the product has been made, little can be done to alter its quality. In order to assure good quality of dairy product for market supply everyone involved in milk production, processing, handling and distribution should be quality conscious and maintain strict hygienic conditions. Besides, lab quality control checks should be in no way strict than the police command.

Sensory Analysis

In a modern food company, sensory evaluation plays a critical role. It is one of the simplest analytical tools for monitoring quality control at all stages of food product processing. Foods are submitted to sensory evaluation to provide information on product improvement and quality maintenance. Tests may be conducted to evaluate raw material selection, to study processing effects, to correlate sensory and chemical and physical measurements and to determine consumer reaction. The success and validity of sensory evaluation programme depends largely on trained panelists and adoption of precise sensory tests. Careful selection of judges is essential in order to achieve maximum discriminability. Sensory methods have been classified into following three categories:

Discriminative Tests

It is on the basis of perceived difference between two products and is a most useful analytical tool available to the sensory professionals. There are variety of specific methods namely paired comparison test, duo-trio test, triangle test, multiple sample test etc. It is important that each test is conducted with proper consideration for all aspects of test designed, product preparation, implementation and data analysis and interpretation.

Descriptive Analysis

It is defined as a sensory methodology that provides quantitative description of products based on the perception of a group of qualified panelists. It takes into account all sensations that are perceived–visual, auditory, olfactory and so on, when the product is evaluated. It provides basis for determining sensory attributes that are important for acceptance of the product.

Affective/Acceptance Testing

It is a valuable and necessary component of every sensory programme and is performed at consumer's level. It refers to measuring preference or liking for the product which can be measured directly by comparison of two or more products with each other. Indirect measurement can be achieved by determining which product has scored significantly higher rating than the other. The two methods frequently used are paired comparison test and nine point hedonic scale.

Chemical Analysis

This requires determination of product composition to establish their acceptability on part of quality assurance programme. Varieties of methods are available to assay a particular component. Speed, accuracy and precision are the key factors that determine the choice of methods. Process control samples are usually analyzed by rapid methods whereas legal requirements generally require use of official reference methods. The traditional Gerber method for fat estimation and Kjeldhal for protein estimation are being replaced with more rapid instrumental methods which are less costly and fairly accurate and comparable with reference methods. There is now a growing range of relatively simple-to-use and sophisticated instruments for quick analysis of major milk constituents in the quality assurance programme.

Milko Tester

It is a photometric method for fat determination based on measurement of scattered light. A chelating agent eliminates turbidity caused by casein micelles making light scattering solely due to the fat. The sample is homogenised for uniform distribution of fat globule size. The amount of light passing through the sample and reaching the photocell is proportional to the fat content which is displayed on a digital readout. It is a quick and accurate method.

Pro-milk MKII

This instrument is used for the determination of protein content of milk by dye binding method. The protein dye mixture is filtered and optical density of the filtrate is measured in a colorimeter for direct readout by a suitably calibrated galvanometer.

Infra Red Milk Analyser

This instrumental method measures absorption of infra red energy by the carbonyl group in the ester linkages of fat molecules, peptide linkages of amino acids in protein molecule and hydroxyl group of lactose molecule. It uses optical filters to isolate specific wavelengths absorbed by fat, protein, lactose and reference wavelengths not absorbed by these components. Several improved models are now available with modifications based on experiences gained. The method is rapid and accurate for estimation of fat, protein and lactose content of the milk samples.

Milko Scan-133B

This is the latest microprocessor controlled instrument for the determination of fat, protein, lactose, solid-not-fat and total solids contents of milk samples. It has a special B filter designed to measure the fat content with high accuracy in milk samples around the year regardless of breed, feed and seasonal variations.

Lactostar Automatic Milk Analyser

It is a new, versatile, easy to operate microprocessor controlled fully automatic analyser for determination of fat, protein, lactose, solids-not-fat and freezing point of milk. It applies a combination of thermo-optical method to analyse the constituents of milk. The devise measures both thermal and optical properties of milk and milk products.

Microbiological Analysis

To ensure and protect public health and reduce economic losses various microbiological analytical methods are necessary for maintaining the quality of dairy products. Micro-organisms pose a serious challenge to the dairy industry. Many of them produce toxins and cause serious infections. It is for these reasons microbiological quality of dairy foods is scrutinized closely. These tests are necessary and performed at a number of points along the chain from producers to consumers. The specific test must be chosen as per the application and cost. Rapid tests are preferred for taking timely action in any untoward conditions. Some of the important rapid methods are outlined hereunder.

Bacto-scan

This is based on direct microscopic counts of the microbial cells. It is an automated counting of bacteria in milk using fluorescent microscopy. The somatic cells, fat globules and sediments are removed by centrifugation. Bacteria obtained are stained with a fluorescent dye like "Alcidin orange" and applied as layer on a disc and passed under microscope. Fluorescent impulses are converted into electric impulse and the count is recorded. The method is quick (68 samples per hour) and has good correlation with SPC (r=0.88).

Epifluorescent Filter Technique

This is a new rapid method for bacterial count, which uses membrane filtration and epifluorescent microscopy. Milk is treated with trypsin and non-ionic detergent to disperse fat and somatic cells, passed through membrane filter (0.6 um), stained with fluorescent dye containing fluorescent brightener and then passed under the microscope for estimation of bacterial count. It is a rapid, inexpensive method suitable for milk containing 6000–10 million bacteria/ml. The method has a correlation (r=0.92) with SPC.

Bioluminescence

It is an indirect method based on changes produced by bacteria due to metabolic activity. The function and significance of ATP in metabolism of living cells suggest that its assay could be an excellent monitor of biological activity in the sample. The light produced by enzymic reaction is used as an assay as it is specific for ATP. The light is recorded by photometer and is proportional to the concentration of ATP. Accurate estimation of bacterial ATP can be made using this rapid method suitable for detecting bacteriological quality of raw milk. It has correlation of 93 per cent for range of 10^4–10^7 cells/ml with SPC.

Pyruvate Test

This is a key intermediate of metabolism as protein, lipids and polysaccharides degradation leads to pyruvate formation. In the presence of reduced form of co-enzyme $NADPH_2$ pyruvate is converted to lactate by the enzyme lactate dehydrogenate. $NADPH_2$ is oxidized to NAD. Decrease in $NADPH_2$ is proportional to amount of pyruvate which is determined spectrophotometrically at 340 nm.

Application of Biotechnology

Quality assurance programme in dairy industry can be effectively implemented adopting emerging biotechnological innovations from time to time, with overall emphasis on improving the quality of such products. To monitor critical control points within HACCP system, reliable and sensitive methods are required. Development of innovative methods for rapid detection of emerging high risk food pathogens is extremely important. The use of modern biotechnology has proved important in the development of rapid, sensitive and accurate methods for the detection and analysis of bacterial contaminants and pathogens or their toxins in foods. Biotechnology has led to the development of tests that can be completed in a fifth of the time required by conventional methods.

Immunological Assays

The immunoassays currently being used for the detection of food pathogens are marketed in a variety of formats such as immunofluorescent assay, immunomagnetic assay, ELISA and agglutinin assays. Immunomagnetic separation technique has been successfully used as a rapid alternative for 24 hours culture enrichment step for determination of *Salmonella* and *Listeria* in foods. The main advantage is the quantitative enumeration that can be made in the food samples. ELISA remains the most widely used format for immunoassays. This format has been used to develop methods to detect pathogenic bacteria and bacterial toxins in foods. It has been successfully used for detection of *Salmonella, Listeria* and other pathogens in foods. A recent arrival in the market is a dipstick of immunoassy of *Salmonella* with a sensitivity of 10^6 cfu/ml and the assay time of 20 min.

DNA Probe Based Assay

Nucleic acid based assays are supposed to be exquisitively sensitive and reliable as the genome of the organism is responsible for the expression of biochemical and serological characteristics of the target organism. Rapid test kits have been developed which rely on the interaction of nucleic acid probes with marker sequence of the target nucleic acids. Two types of nucleic acid probe based tests *viz.* colorimetric and chemitiminescent are available for *Salmonella* and *Listeria.*

PCR Based Detection

Polymer chain reaction has recently emerged as a most powerful molecular technique in the area of diagnostics. This technique is now being increasingly used in dairy and food industry for monitoring of food pathogens. PCR based detection of food pathogens has revolutionised the area of diagnostics as the results are available in less than 24 hours for appropriate follow up action to avoid any risk. By using different formats of PCR it is now possible to simultaneously detect more than one food pathogen in samples of food. The recent development in PCR is the use of nested primers and magnetic separation of the PCR generated fragments along with different signals transducing systems. Development of PCR based methods in conjunction with immunological assays would prove to be extremely valuable for proper monitoring of the safety of milk and milk products produced in India.

Biosensors

These represent new generation of powerful analytical tools incorporating biologically derived material with a physiochemical transducer microsystem. The techniques are being developed for rapid time direct or indirect detection of foodborne microorganisms, toxins or undesirable metabolites. All sensors comprise an energy transfer mechanism, which may be of biological, chemical or physical in nature. These are self contained analytical system that respond directly or indirectly to biologically important species. Biosensors use antibodies or enzymes to react selectively with targeted toxins or pathogens and then one of the several possible transduction mechanisms to detect that interaction. Several types of biosensors based on fluorescence, surface enhanced infra red absorption and quartz crystal microbalance have been studied. New technologies such as acoustic wave biosensors and radio frequency identification sensor tags promise to greatly improve food safety.

REFERENCES

Ahmad, N., Grover, S., Patnaik, P. and Batish, V.K. (1998). Biotechnology ushers in a new era in waste handling and environmental pollution control. *Indian Dairyman*, 49: 25-34.

Biggs, D.A., Jhonson, G. and Sjaunja, L.O.(1987). Analysis of fat, protein, lactose and total solids by infra-red absorption. *IDF Bull.*, 208: 21-29.

Brooks, S.I., Higgins, I.J., Newman, J.D. and Turner, A.P.F. (1991). Biosensors for process control. *Enzymic Microbial Technol.*, 13: 946-955.

Clout, P.R. (1998). Enzyme biosensors. In: *Biosensors Food Analysis.* A.O.Scott (Ed.). The Royal Society of Chem., Cambridge, UK. p. 28-36.

Deshpande, S.S. and Rocco, R.M. (1994). Biosensors and their potential use in food quality control. *Food Technol.*, 76: 146-150.

Electronic milk tester. Instructional manual, Rajasthan Electronics and Instruments, Jaipur.

Fung, D.Y.C. (1995). Rapid methods and automation in food microbiology-an update. In: *Microbiology Control in Foods and Agricultural Products*. (Ed.) Fuy, D.Y.U. VCH Publ. Inc.

Hoban, T.J.(1998). Improving the success of new product development. *Food Technol.*, 52: 46-49.

Karel, M. (2000). Tasks of food technology in 21st century. *Food Technol.*, 54: 56-64.

Kumar, Sanjeev (1999). Rapid detection of *E.coli* in milk by RNA targetted PCR. *M.Sc.Thesis,* National Dairy Research Institute (Deemed University), Karnal.

Meilgaurd, M., Civilli, G.V. and Carr, B.T. (1991). *Sensory Evaluation Technique*. CRC Press, Florida, USA.

Murthy, I. (2000). The portable nose. The Times of India (Sunday Review), Feb. 6, p. 6.

Pal, D., Sachdeva, S. and Singh, S. (1995). Methods for determination of sensory quality of foods: A critical appraisal. *J. Food Sci. Technol.*, 32: 357-367.

Rasooly, L. and Rasooly, A. (1999). Real time biosensor analysis of Staphylococcal enterotoxin in food. *Intern. J. Food Microbiol.*, 49: 119-127.

14

Advances in Milk Processing

R.S. Dabur and C.M. Kapoor

Department of Animal Products Technology, Chaudhary Charan Singh Haryana Agricultural University, Hisar – 125 004

The world scenario is fast changing during the post-GATT period. India being the member of World Trade Organization shall have to abide by the guidelines formulated regarding international trade in milk and milk products. Major challenge that emerged out of globalization and liberalization of Indian economy with reference to dairy co-operative sector in meeting the International Standards related to milk and milk products and positioning it as a market leader. The standard guideline formulated by the Codex Alimentarious Commission (A body jointly formed by the FAO and the WHO) has now acquired a mandatory status for international trade in food. In order to comply with these requirements and to become competitive in the international trade, it is extremely necessary to improve the quality of raw milk, which in turn will lead to a better quality product in a cost effective manner. Apart from the standards and conditions prescribed by various national and international regulatory authorities, there are internal market forces also. The consumers today are more enlightened and conscious about quality. In this era of globalization, a country like

India with highest milk production figure and a growth rate of 4.5 per cent cannot remain insulated from the rest of the world.

Dairy industry is also fast diversifying to develop special food ingredients derived from milk, which can be deployed advantageously to improve the sensory quality attributes. This includes fractionation of individual milk constituents such as fractionated caseinates, whey protein concentrates, and enzymatic hydrolysates prepared by employing innovative energy efficient techniques such as membrane technology. Membrane processing also permits the isolation and preparation of various bioactive substances that permit tremendous value addition for application in the pharmaceutical, cosmetic and functional foods industry. The main thrust in research priorities for next decades is, therefore, likely to include:

1. Raw milk quality improvement.
2. Increased processing efficiency with a reduced environmental effect.
3. Utilization of milk components for value addition.
4. Development of health oriented foods
5. Rapid detection of high risk microorganism

Raw Milk Quality Improvement

If India wants to compete in the world market, our milk and milk products should meet International Standards. Moreover, production of high quality milk products would require high quality raw milk. At times, when initial quality of raw milk is bacteriologically poor, the physical properties of the final products gets disturbed and it runs the risk of being rejected by the discerning consumers. In the west, processed milk has a shelf life of 14 days under refrigeration as compared to that of 2/3 days in India. Whereas, there is practically no difference between the quality of milk from udder in the East and the West, the east surpasses the west by leaps and bounds as far as bacterial count is concerned in each subsequent stages starting from the producers' pail.

Because of our indifference to the quality of the bulk processing and handling of food products, the Indian products have become the victim of the common consumer perception of having an inferior quality. Until this stigma is removed, Indian food products would continue to be browbeaten by the multinational/foreign products, both in the national and international markets. Hence, there is a need to produce clean milk/good quality milk or milk should be cleaned during processing. It is necessary to look at the bacteriological and hygienic quality of raw milk and eliminating pesticide residues, antibiotics, drug, hormone, heavy metals, adulterants, etc. While it is necessary to give emphasis on clean milk production to produce high bacteriological and hygienic quality of raw milk, it is equally important to preserve this quality until milk is processed in the dairy.

Lactoperoxidase System

The lactoperoxidase enzyme catalyses oxidation of thiocyanate (SCN) yielding intermediate oxidation products *i.e.* $OSCN^-$ and O_2SCN^- that has antibacterial effect. The lactoperoxidase-H_2O_2-thiocynate system (LP system) has been exploited in preservation of raw milk. Several studies have demonstrated that a substantial improvement in the bacterial quality of milk was achieved during collection and transportation after activation of LP system at collection centers. Thus, LP-system has emerged as an attractive alternative for milk preservation especially in developing countries located in tropical zones where application of refrigeration for preservation of milk presents insurmountable techno-economic problems.

Bactofugation

Bactofugation is a special form of separation in which specific type of microorganisms are removed from milk and inactivated thereafter. Bactofuges are special nozzle clarifying separators with high separation precision that can remove microorganisms from milk on the principle of their higher density. This is applicable especially for spore-forming cells or spores, as the concentrated cell plasma of the spore enhances the density difference from milk, and it permits complete (100 per cent) separation of this group of microorganisms by a bactofugation.

Increased Processing Efficiency with a Reduced Environmental Effect

Newer technologies are available which can be used to increase the processing efficiencies and to reduce environmental impact.

There has been considerable interest in utilizing the potential Ultra High Temperature processing and membrane processing:

UHT Processing

Insurmountable difficulties have been encountered in the Indian context for maintenance of cold chain due to several techno-economic factors under the tropical climatic conditions. Furthermore, there is world wide concern for conservation of energy in processing and distribution of foods. All these factor have led the planners to opt for UHT processing as national policy for market milk distribution. Unfortunately attempts to introduce UHT processing technology have met with limited success owing to high cost of highly automated plant and high cost of packaging material. Recently, scaled down versions of UHT plants have been developed that employs lesser degree of automation and use low cost laminated material to reduce capital and running costs in UHT processing of market milk. These modified versions of UHT processing have been successfully installed at Pune, Hyderabad, Baramati and some more projects are in pipeline. Future developments in market milk industry will involve a range of flavoured and refreshing beverages from milk.

Membrane Processing

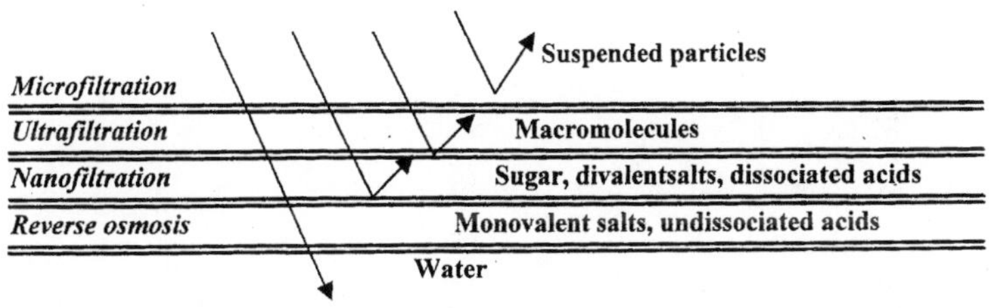

Microfiltration

Microfiltration process can be effectively employed for reducing the bacterial counts in skim milk by more than 99 per cent. Subsequently treatment of skimmed milk and its ready mixture with cream permits production of superior quality of milk with extended shelf life. M/S Alfa-Laval has patented a process called 'Bacto catch' for removal of bacteria from skim milk. With skim milk at 50°C the operational flux have been 500 litres/hr m^2. The bacterial count was reduced to 99.91 per cent. The retentate (having bacteria and some

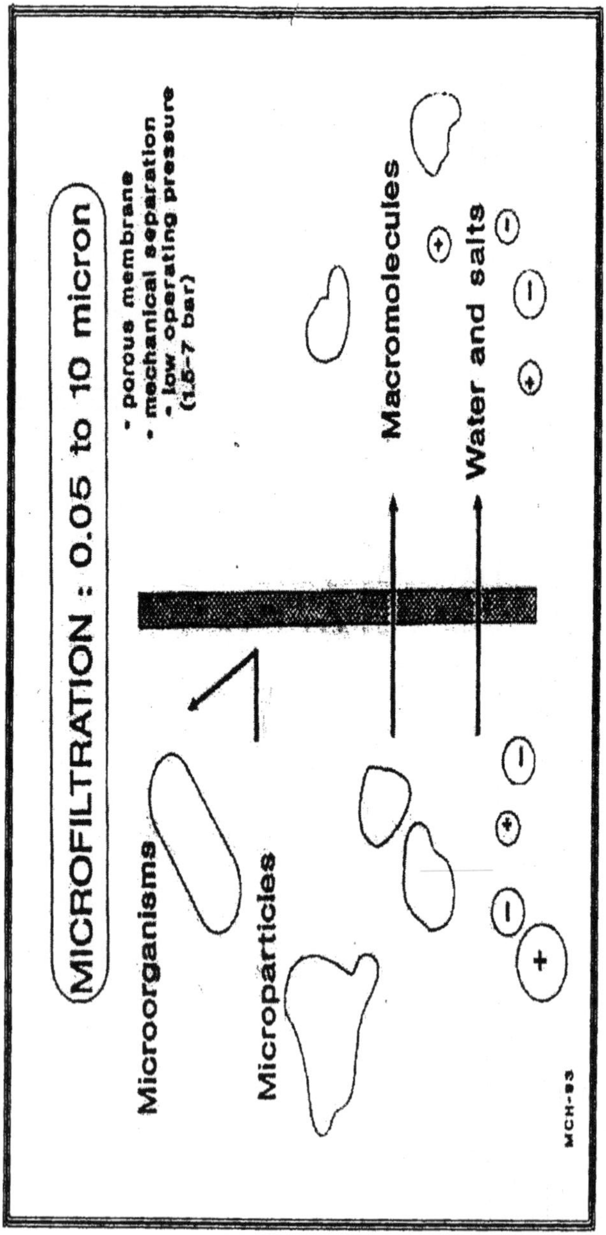
MICROFILTRATION : 0.05 to 10 micron
- porous membrane
- mechanical separation
- low operating pressure (1.5-7 bar)
Microorganisms
Microparticles
Macromolecules
Water and salts
MCH-93

milk solids) is heat treated (130°C) and recombined with filtrate and whole lot is pasteurized. The keeping quality of milk is enhanced by 3–4 days. Removing of *Clostridia* spores using MF was 10 times better compared to bactofugation, regardless of initial count.

Reverse Osmosis

Reverse osmosis is well-established process now for concentration or preconcentration of raw milk. It is beyond doubts that this process consumes 10 times less energy for removing water as compared to conventional evaporators, with however, limitation of concentration levels. Whole milk and skim milk powders can be manufactured using RO as a first concentration step followed by either evaporation and spray drying or spray drying alone. Since milk is normally concentrated to 45–50 per cent total solids before spray drying, RO cannot be used as a complete replacement of conventional evaporation. The reason for such a limitation of the process is that, with the current practice of RO, the permeation of water through the membrane practically ceases beyond 36–38 per cent TS in case of whole milk and 27 per cent TS in case of skim milk. Nevertheless, major saving in energy can be achieved by concentrating whole milk to about 30 per cent TS and skim milk to 20–22 per cent followed by concentrating to about 48 per cent TS using multistage evaporator, and finally spray drying. The flavour and solubility of powders made by using RO concentrated milk were found superior than that of conventionally prepared powder. RO processed milk has also been successfully used in the manufacture of yoghurt, dahi, ice cream and several others.

1. Reducing the bulk of raw milk to efficiently transport RO concentrated milks.
2. Utilization of pasteurized RO concentrate in place of market milk.
3. Preconcentration of milk for khoa making and spray drying.
4. Removal of major part of water from whey.

Ultrafiltration

In contrast to reverse osmosis, which is diffusive process, UF is essentially a sieving process in which the constituents of milk are fractionated according to their molecular size. Ultrafiltration is now

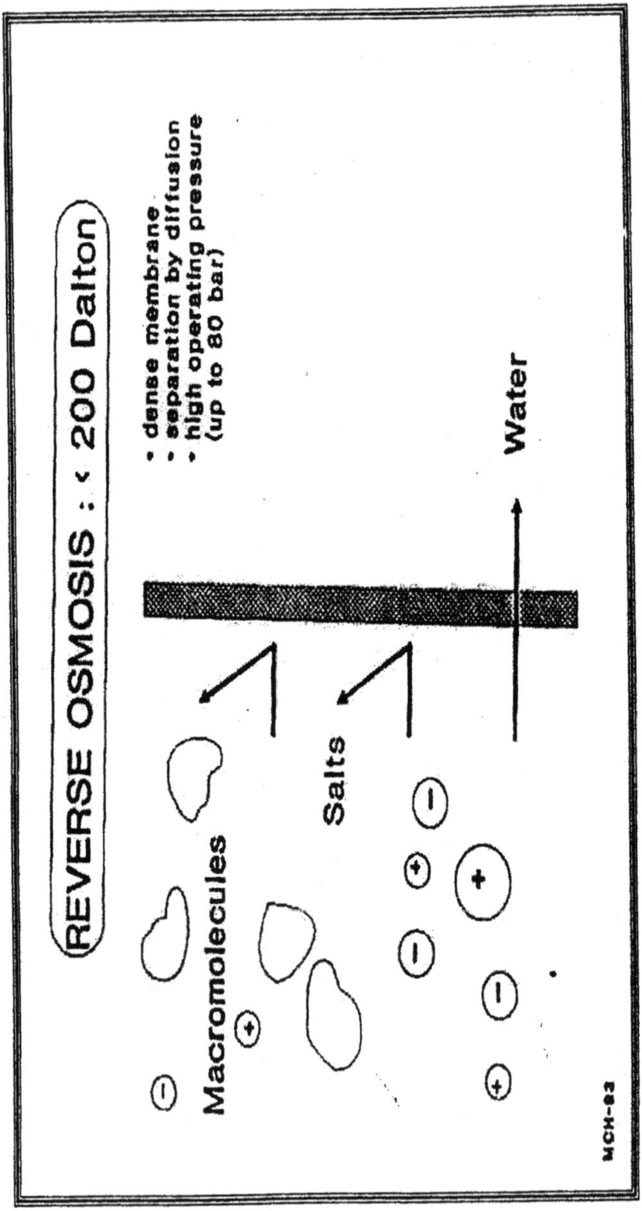
REVERSE OSMOSIS : < 200 Dalton
• dense membrane
• separation by diffusion
• high operating pressure (up to 80 bar)
Macromolecules
Salts
Water

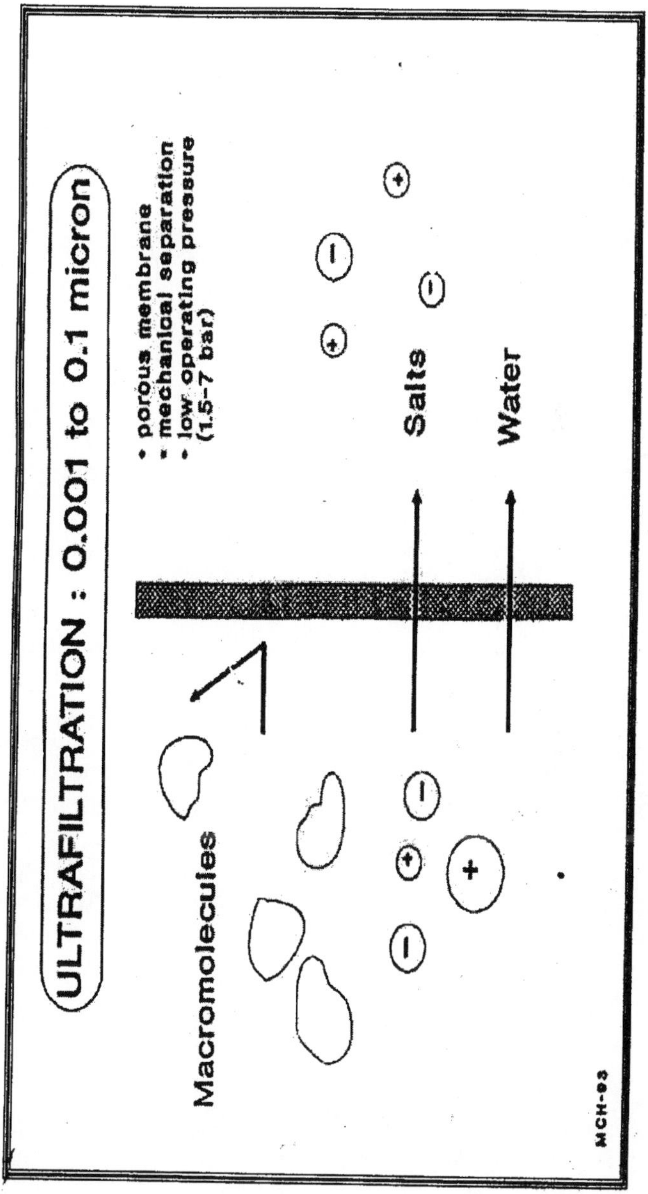
ULTRAFILTRATION : 0.001 to 0.1 micron
• porous membrane
• mechanical separation
• low operating pressure (1.5-7 bar)
Macromolecules
Salts
Water
MCH-93

considered to be a need based processing operation in dairy industry. The two largest applications are preconcentration of liquid (milk and whey) and fractionation of whey. In the dairy industry, traditional application of UF is in the separation and fractionation of individual milk proteins from lactose and minerals. Significantly, the permeate stream is sterile due to the retention of microorganisms. Other industrial applications include enzyme recovery in various operations, for example, lactose hydrolysis by soluble lactase preparations. The application of UF in making the various types of milk products, which may have certain advantages over conventional process such as:

1. Higher yield
2. Reduced energy requirement.
3. Whey disposal problems are reduced.
4. Reduces the space requirement.

Nanofiltration

The principal application of this membrane process is for separation of mineral ions in the 10.9 m size exclusion range. Nanofiltration (NF) has recently emerged as a membrane processing technique offering the possibilities of concentrating dairy fluids such as whey and milk while at the same time effecting partial demineralization. It has been established that this technique can be profitably used for demineralization of whey to the extent of 30–32 per cent for various purposes such as lactose manufacturing. NF processing of whey and UF permeate offers an opportunity to mitigate the problem of their disposal as it greatly reduces the BOD requirement of these byproducts. By using selective membranes for partial mineral removal aimed at improving the quality of certain indigenous milk products such as chhana from buffalo milk, and paneer and khoa from cow milk. Hence research attempts need to be made in this direction.

1. Selectively removal of ionic particles
2. Demineralization of UF permeates whey prior to manufacture the lactose
3. Deacidification of sour whey
4. Desalting of salt whey
5. Mechanization and automation in product manufacturing.

Utilization of Milk Components for Value Addition

A number of membrane-oriented processes has been developed for the preparation and purification of drug principles from milk having pharmaceutical significance. A process has been developed at the university of Hamburg, Germany which allows the separation of milk into its insoluble (fat and casein) and soluble (whey protein, lactose, peptide and NPN components. This separation is advantageous in the sense that the casein and fat can be processed separately, avoiding the inevitable interaction between the casein and whey protein fractions under the influence of heat. This approach allows reconstitution of milk with 'bio-protective factors' intact. Alternatively casein and fat rich retentate can be used for cheese production and the whey protein permeate can be ultrafiltered to obtain undenatured whey protein isolates of extremely high purity (pharmaceutical grade) which display prophylactic quality. Various prophylactic biological preparations have been found tested under medical supervision to treat or prevent a range of human ailments such as arthritis, toothache, allergies, various kinds of viral infections such as common cold etc.

Newer R and D efforts are being made for the utilization of whey with refined whey protein components intended for unique functional and pharmaceutical application, which may permit greater value addition. Lactoferrin (LF) and lactoperoxidase (LP) have a very high market value as fine chemicals functional foods as well as pharmaceutical products. Recent developments in membrane processing have paved the way for the industrial preparation of nutraceuticals such as LF and LP and high purity, more than 95 per cent and of pharmaceutical quality.

The bio-protective nature of LF is of great significance in products such as infant formulas, health foods, skin creams, and as antibacterial products. Another significant development in field of cancer is the potential beneficial application of LF for treatment of leukemia. It has been postulated that in association with the human interferon, LF suppresses early human monocyte or mono-blastoid cell line. LF also enhances the action of certain antibiotics, which enhances the efficiency of pharmaceutical preparations.

Development of Health Oriented Foods

The exploring area of functional foods and probiotics shows considerable promise to expand the dairy industry into new arenas.

Functional foods contribute bioactive components to human diet. They have a significant role in the prevention and treatment of diseases and in improving body functions. Milk and its products are an excellent source of vital nutrients and functional constituents.

Probiotics

Milk is an excellent medium to carry or generate live and active cultured dairy products. Probiotics may be defined as a food or supplement containing concentrates or defined strains of living microorganisms that on ingestion in certain doses exert health benefits beyond inherent basic nutrition. They can be incorporated in fermented dairy products either as culturing agents or fermented milk may be used as a carrier for them. Now capsules/tablets are available containing active probiotics.

Potential mechanisms by which probiotics may exert their beneficial effects are:

1. Competition with other microflora for nutrition
2. Production of acids inhibitors for certain enterpathogens
3. Production of bacteriocins or inhibitory metabolites
4. Immunomodulation
5. Competition for adhesion to intestinal mucosa.

Beneficial effects of consuming certain strains of cultures and/or their metabolites are:

1. Enhanced immune system
2. Balancing of colonic micro flora
3. Anticarcinogenic or antimutagenic activities
4. Alleviation of symptoms of lactose intolerance
5. Reduction in serum cholesterols
6. Reduction in blood pressure
7. Decreased incidence and duration of diarrhoea
8. Prevention of vaginitis
9. Maintenance of mucosal integrity.

Fat Replacement in Dairy Products

High fat consumption has been linked to several chronical diseases including cardiovascular diseases and obesity. Nutritional experts recommend a total fat intake of less than 30 per cent of total daily calories. These dietary recommendations are one reason of the increasing demand for lower fat food products in the world market, which has been flooded with the food products carrying the labels "low fat" "no fat" or "reduced fat". Fat mimics or fat substitutes are normally used to produce low fat foods, fat mimics are substances that help replace the mouth feel of fat but not substitute for fat on a gram for gram basis and cannot be used for applications involving frying. Substances whose physical or thermal properties resembles fat are termed as fat substitutes and can replace fat on gram for gram basis and can also be used for frying applications. Low fat cheese, processed cheese, cultured products, frozen desserts, butters and spreads have been successfully developed using commercially available fat mimics/replacers.

Table 14.1: Specific Probiotics Strains in Commercial Use and their Reported Health Benefits

Strain	*Reported Benefits*
Lactobacillus acidophilus DDS1	Serum cholesterol reduction, immune enhancement, prevention or cancer initiation
Lactobacillus acidophilus L1	Serum cholesterol reduction
Lactobacillus LA1	Immune enhancement, adjuvant, gastritis treatment
Lactobacillus LC1	Immune enhancement, vaccine adjuvant, balancing oral tract microbiota
Lactobacillus acidophilus NCFM	Fecal enzymes reduction
Lactobacillus casei shirota	Immune enhancement, balancing of GI tract microbiota, Fecal enzymes reduction, treatment of superficial bladder cancer
Lactobacillus gasseri ADH	Fecal enzymes reduction
Lactobacillus reuterii	Immune enhancement
Lactobacillus rhamnosus GG	Prevention of treatment of diarrhoea, crohn's disease, dental caries, rheumatoid arthritis, Immune enhancement, vaccine adjuvant
Enterobacter faecium GAIO	Serum cholesterol reduction
Bifidobacterium bifidum	Treatment of rotavirus diarrhoea, balancing of GI tract microbiota
Yogurt culture	Alleviation of lactose maldigestion symptoms, immune system modulation

Source: Chandan, 2001.

Milk Beverages by Direct Acidification

The technological process developed for the manufacture of long-life directly acidified milk based soft drink is simple and can easily be adopted by dairy industry for commercial production. It is more nutritious and energy giving as compared to other carbonated soft drinks and may bring a great revolution in the "beverage world" once it will be introduced by the beverage or food industry for commercial production (Pagote, 2001). These milk drinks have the following attributes:

1. Very pleasant sweet-sour taste.
2. More nutritious than colas and other carbonated soft drinks.
3. No bacterial culture is required.
4. Can be stored at ambient temperature for more than three months.
5. Good for health.

Milk as Tablets and Compressed Milk Products

Now, we are in the era of global competitions and once focused in the food sector, competition is intense and failure is unforgiving. Therefore, it is necessary to focus on areas, which will become predominant like neutraceutical, convenience, hygiene, safety and price, etc. Milk in tablets and compressed form offers certain advantages in terms of their portability, convenience to use, easy handling, requires no refrigeration, etc. These products may find good market in places of scarcity. They can also serve as an alternative to UHT milk. However, more R and D is needed both at the basic and applied level to understand compression and tabletting aspects of milk components that will solve the needs of tomorrow (Verma and Kanawjia, 2001).

Rapid Detection of High Risk Pathogen

Microbiological safety of milk and milk products is now the foremost issue confronting Indian dairy industry to enable it to maintain high quality standards of these foods produced therein. However, to make any worthwhile impact both in the internal and International market, the industry has to make sure that the dairy products manufactured within the country are absolutely safe and free of high risk pathogens particularly *Salmonella* and *L. monocytogenes*. Hence, there is a need to develop Cost effective, reliable and simple ready to use kits for rapid detection of these pathogens at field level. Developments of such ready to use kits and their subsequent introduction in dairy industry deserve immediate attention and top priority so that dairy products could be quickly monitored on routine basis right from production of milk to consumption of the end products.

Conclusion

During WTO regimen, major transformations in dairy industry are anticipated to impart greater competitiveness and opportunities for value addition. It will be necessary to pay greater attention on improving the quality of raw milk, increasing the efficiency of milk processing, development of value added products and control on food born disease.

In order to improve the quality standards, it will be necessary to increase the quality consciousness amongst milk producers as well as milk processors through proper training. It is necessary to improve the processing efficiencies through use of modern processing techniques and quality management.

References

Chandan, R.C. (2001). Functional foods and bioactive dairy ingredients. *Indian Dairyman*, 53(9): 43.

Pagote, C.N. (2001). Milk beverages by direct acidification: A new approach. *Indian Dairyman*, 53(9): 59.

Verma, R.B. (2001). New forms of milk: Milk tablet and compressed milk products. *Indian Dairyman*, 53(9): 37.

Part III

FRUITS AND VEGETABLE PROCESSING

15

Free Radical Scavenging Enzymes and Shelflife of Fruits and Vegetables

Dharam Singh and Santosh Dhillon
Department of Biochemistry, CCS Haryana Agricultural University, Hisar – 125 004, India

Ripening and postharvest storage/shelf life of fruits and vegetables, like senescence is accompanied by loss of chlorophyll and proteins and increase in lipid peroxidation and membrane permeability (Thompson *et al.*, 1987; Halliwell and Gutteridge, 1989). The presence of oxygen in cellular environment poses a constant oxidative threat to cellular structure and processes. The reactive oxygen species (ROS) such as superoxide radical ($\cdot O_2^-$), hydrogen peroxide (H_2O_2), hydroxyl radical ($\cdot OH$), singlet oxygen ($^1O_2$), etc., are normally produced in plant cells as a result of normal metabolism. Superoxide radical is produced by electron transport systems in chloroplast and mitochondria while H_2O_2 is produced by β-oxidation in glyoxysomes (Fridovich, 1995; Thompson *et al.*, 1987). Since their presence causes oxidative damage to cellular constituents such as nucleic acids, proteins and lipids, they are considered detrimental to cellular functions and affect storage/shelf life adversely. Besides ROS, other important contributing factors in fruit ripening and storage are increased lipoxygenase activity and several ageing promoting substances such as, ethylene and jasmonic acid (Leschem, 1998). Lipoxygenase is an enzyme that catalyzes the reaction of polyunsaturated fatty acids (PUFAs) with O_2 to give hydroproxides (ROOH) that has an important role in the breakdown of membrane lipids.

Mediators of Oxidative Damage

Activated oxygen species are the primary mediators of oxidative damage both during ripening and storage. They mainly include superoxide radicals, hydroxyl radicals, hydogen peroxide, singlet oxygen, alkoxyl radical (RO·) peroxyl radicals (ROO·), PUFAs and semi-quinone radicals. Some of these reactive oxygen species, particularly superoxide radicals are strong oxidizing agents that can rapidly attack all kinds of biomolecules, especially DNA leading to irreparable metabolic dysfunction and eventually cell death (Halliwell and Gutteridge, 1989).

$$O_2 \xrightarrow{e^-} \cdot O2^- \xrightarrow[2H^+]{e^-} H_2O_2 \xrightarrow[H^+]{e^-} H_2O + \cdot OH \xrightarrow[H^+]{e^-} H_2O$$

Superoxide Radical ($\cdot O_2^-$)

The superoxide radical is unique among ROS; favourable thermodynamics enable it to act both as an oxidant and reductant. The superoxide is a charged molecule and therefore cannot cross biological membranes. Subcellular compartmentation of defence mechanism is therefore crucial for its efficient removal at their site of production throughout the cell (Alscher, 1997). Superoxide radicals which are generated in chloroplasts, mitochondria, cytosol and microbodies (peroxisomes) are one of the major sources for the production of other ROS in cells.

Hydrogen Peroxide (H_2O_2)

Furthermore, the superoxide radical will accept one electron and two protons to form H_2O_2. Unlike oxyradicals, H_2O_2 can readily diffuse across biomembranes and cause oxidative stress far from the site of generation. Various flavoprotein oxidases are also capable of catalyzing two electron transfer reactions that produce H_2O_2. The dismutation of superoxide radical to H_2O_2 and O_2 is another major source of H_2O_2 in cells.

$$\cdot O_2^- + \cdot O_2^- + 2H+ \xrightarrow{\text{SOD}} H_2O_2 + O_2$$

Hydroxyl Radical (·OH)

The hydrogen peroxide can accept an electron and proton to form a molecule of water and a hydroxyl radical which is the most reactive of all oxyradicals. The greatest potential for hydroxyl radical formation *in vivo* is from H_2O_2 through the Fe^{2+} catalyzed Haber-Weiss reaction. Superoxide reduces Fe^{3+} to Fe^{2+} which then reacts with H_2O_2 as follows:

$$Fe^{3+} + \cdot O_2^- \longrightarrow Fe^{2+} + O_2$$

$$Fe^{2+} + H_2O_2 \longrightarrow Fe^{3+} + \cdot OH + OH^-$$

Singlet Oxygen ($^1O_2$)

One of the unpaired electron in ground state oxygen shift to an orbital of higher energy with an inversion of spin. Singlet oxygen is not therefore a free radical but is reactive by virtue of having a vacant molecular orbital. A major source of singlet oxygen in biochemical systems appears to be its formation from O_2^-. This can occur by spontaneous disproportionation of O_2^-

$\cdot O_2^- + . O_2^- \xrightarrow{2H^+} H_2O_2 + {}^1O_2$, by a direct transfer of electron,

$\cdot O_2^- \xrightarrow{\text{spontaneous}} e^- + {}^1O_2$, through a reaction with H_2O_2,

$\cdot O_2^- + H_2O_2 \longrightarrow \cdot OH + OH^- + {}^1O_2$, through a reaction involving hydroxyl radical,

$$\cdot O_2^- +. OH \longrightarrow {}^1O_2 + OH^-$$

Thus enzymes and electron transport systems that generate superoxide, presumably serve as indirect source of singlet oxygen. In plants, chlorophyll may also be a source of singlet oxygen. More recently, lipoxygenase have also been shown to produce it.

Alkoxy (RO·) and peroxy radicals (ROO·)

A number of reactive oxygen species are known to initiate peroxidation of lipids. It is likely that the physical structure of biological membranes in which fatty acid side chains are placed in close proximity, facilitates the autocatalytic propagation of lipid peroxidation following initiation,

$$\cdot OH + RH \longrightarrow R\cdot + OH^- + H^+, \text{ initiation}$$

$$\left.\begin{array}{l} R\cdot + O_2 \longrightarrow ROO\cdot \\ ROO\cdot + RH \longrightarrow ROOH + R\cdot \end{array}\right\} \text{Propagation}$$

RH = Lipid; R. = lipid radical; ROO. = proxy radical; ROOH = lipid hydroperoxides. Lipoxygenase can also initiate the formation of fatty acid hydroperoxides following lipid peroxidation.

In fruits and vegetables, ageing and ripening induced membrane deterioration, lipoxygenase (LOX; EC 1.13, 11.12) plays a central role by peroxidizing free PUFAs and generating free radicals (del Rio *et al.*, 1998). Increase in lipid peroxidation is associated with concomitant production of melonaldehyde (MDA), a secondary end–product of PUFAs oxidation. Lipoxygenase, specifically initiates oxidation of linoleic and linolenic acids. Both soluble and membrane bound lipoxygenases capable of oxygenation of PUFAs esterified to phospholipids in the membrane bilayer have been identified in tomato fruits.

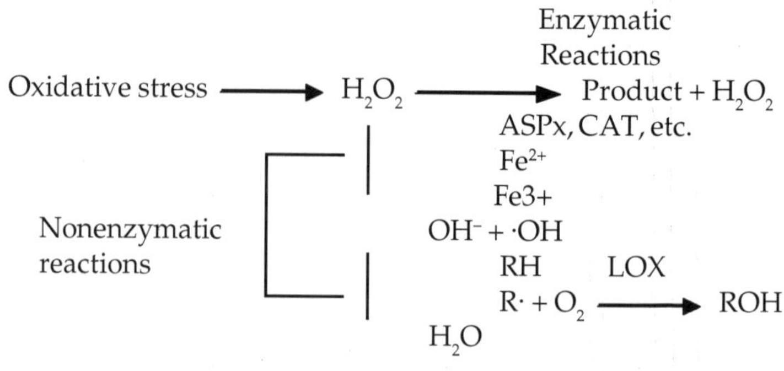

Figure 15.1: A Proposed Scheme for Enzymatic and Nonenzymatic Pathways Activated Under Stress Inducing Excess H_2O_2 Production.

RH: Fatty acid; R·: Fatty acid radicals; ROOH: Lipid hydroperoxide; ROH: Reduced lipid hydroperoxides; LOX: Lipoxygenase.

Antioxidant System in Plants

The plant cells (leaves, fruits and vegetables) are known to possess antioxidant defence systems to provide protection against harmful effects of reactive oxygen species. The ROS are scavenged in a sequence of reactions involving enzymic and non-enzymic oxidants which play an important role in regulating the levels of ROS.

Antioxidant Systems	
Enzymatic	
Superoxide dismutase	(SOD; EC 1.15.1.1)
Catalase	(CAT; EC 1.11.1.6)
Ascorbate peroxidase	(ASPx); EC 1.11.1.11)
Dehydroascorbate reductase	(DHAR; EC 1.8.5.4)
Glutathione reductase	(GR; EC 1.6.4.2)
Monodehydroascorbate reductase	(MDHAR; EC 1.6.5.4)
Peroxidase	(PER; EC 1.11.1.7)
Nonenzymatic	
Reduced glutathione (GSH)	
Ascorbic acid	
Carotenoids	
α-Tocopherol	
Flavonoids	

Superoxide dismutases are a family of metalloenzymes that play an important role in protecting cells against toxic effects of superoxide radicals. In higher plants three major types of SODs have been described, each defined by the use of metal as their prosthetic group (Mn, Fe and Cu/Zn). They are distributed in different cellular compartments, mainly chloroplasts, mitochondria and peroxisomes. Superoxide dismutase is a major scavenger of superoxide radical and is the first line of defence against free radical injury. The enzymatic action results in the formation of H_2O_2 and O_2.

$$\cdot O_2^- + \cdot O_2^- + 2H^+ \xrightarrow{\text{SOD}} H_2O_2 + O_2$$

Cytoplasmic superoxide and possibly that associated with cellular membranes can be removed in this manner. The capacity of plant tissues to scavenge superoxide radical decrease with age.

The oxidative damage to different cellular components by H_2O_2 could be minimized mainly by catalase in peroxisomes and ascorbate peroxidase in cytosol and chloroplast through concomitant oxidation of ascorbate. The principal means of removing H_2O_2 from plant cells is through two enzymes, catalase and peroxidase. Catalase is mainly localized in the peroxisomes and not in chloroplast and mitochondria and it is unique among H_2O_2 degrading enzymes as it degrades H_2O_2 without consuming cellular reductant. Hence, catalase provides the cell with energy efficient mechanism to remove H_2O_2.

$$2\,H_2O_2 \xrightarrow{\text{Catalase}} 2\,H_2O + O_2$$

Alternatively, peroxidases reduce H_2O_2 to H_2O and O_2 in the presence of appropriate oxidazable substrates:

$$SH_2 + H_2O_2 \xrightarrow{\text{Peroxidase}} 2\,H_2O + S$$

where, SH_2 is a reduced substrate.

Ascorbate-Glutathione Cycle in Chloroplast

The ascorbate-glutathione cycle is an efficient way for plant cells to dispose of H_2O_2 in certain cellular compartments where there is no catalase (chloroplast, mitochondria and cytosol). This cycle makes use of nonenzymic-antioxidants, ascorbate and glutathione in a series of reactions catalysed by four enzymes, ascorbate peroxidase (ASPx), monodehydroascorbate reductase (MDHAR), dehydroascorbate reductase (DHAR) and glutathione reductase (GR).

Stromal Pathway

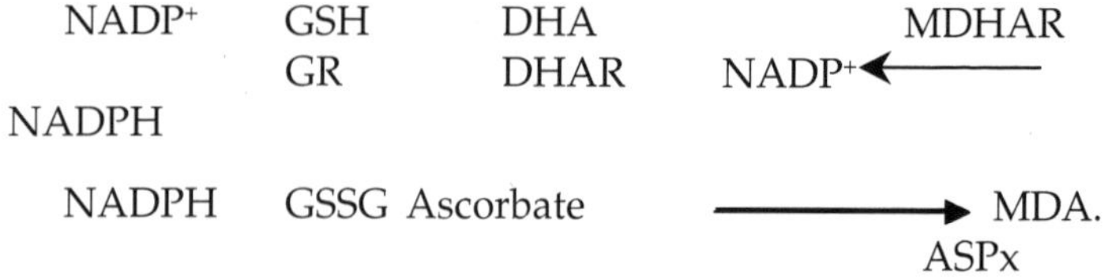

Thylakoid pathway

$$\cdot O_2^- \xrightarrow[\text{Thylakoid membrane}]{} H_2O_2 + H_2O$$

Figure 15.2: Scavenging of ROS in Chloroplast Involving SOD. ROS that escape the thylakoid pathway are dismutated by Fe-SOD and stromal ASPx. Monodehydroascorbate radicals (MDA.) produced by ASPx are converted to ascorbate through MDHAR reaction. Reduction of DHA to ascorbate is catalyzed by DHAR via stromal ascorbate-glutathione pathway, which is also referred to as "Halliwell-Asada pathway'.

In this cycle DHA is reduced to ascorbate by DHAR and GSH with the resultant production of GSSG. Finally GSSG is reduced to GSH by NADPH in a reaction catalyzed by GR. Thus, this cycle involves the successive oxidations and reductions of ascorbate, glutathione and NADPH by the enzymes ASPx, DHAR, and GR, respectively.

Antioxidants

α-Tocophenol and Ascorbic Acid

Two antioxidants,α-tocophenol (vit E) which is hydrophobic and associated with membranes and ascorbic acid (vit C) which is water soluble, are important components of cellular defence against oxidaive injury.Vitamin E reacts with superoxide radical and also can quench singlet oxygen. However, primarily it functions as a source of protection against lipid peroxidation because of its lipid solubility. It reacts with lipid peroxyradicals to form tocophenol radicals (tocophenol semiquinone) and thus serves to terminate the chain reaction of lipid peroxidation. The tocophenol radical formed can be reduced by ascorbate.

$$\alpha\text{-Tocopherol} + ROO\cdot \longrightarrow \text{Tocopherol-semiquinone (radical)} + ROH$$

$$\text{Tocopherol – semiquinone} + \text{Ascorbate} \longrightarrow \alpha\text{-Tocopherol} + \text{Dehydroascorbate}$$

The dehydroascorbate so produced can be reduced back to ascorbate by dehydro ascorbate reductase in the presence of reduced glutathione. Ascorbic acid occurs naturally in fruits and vegetables, almost exclusively in reduced form (L-ascorbic acid). Ascorbic acid is widely used as food ingredient because of its reducing and antioxidant properties. The antioxidant role of ascorbic acid is multifunctional with ascorbate inhibiting lipid peroxidation by several mechanisms. These include, scavenging of singlet oxygen, reduction of oxyradicals and regeneration of other antioxidants such as reduction of tocopherol radicals.

Carotenoids

They are Nature's most widespread pigments. A large majority of these pigments are synthesized by the ocean's algae population. In higher plants, carotenoids in chloroplasts are often masked by the more dominant chlorophyll pigments. The yellow-orange colour of carotenoids becomes evident only after decomposition of chloroplasts at maturity and ripening of fruits. Carotenoids provide photoprotection by quenching and inactivating ROS formed by exposure to light and air. Most prominent role of carotenoids in the diets of human beings and other animals is their ability to serve as precursors of vitamin A. The most common carotenoid found in plant tissues is β-carotene, which is also used as colorant in foods. Carotenoids are comprised of two structural groups: the hydrocarbon carotenes and the oxygenated xanthophylls. In addition to cellular and *in vitro* protection against singlet oxygen, superoxide radical, hydrogen peroxide and carotenoids at low partial pressures inhibit lipid peroxidation. Not all carotenoids are equally effective as photochemical protectors. For instance, lycopene is known to especially efficient in quenching singlet oxygen relative to other carotenoid pigments.

Flavonoids

Phenolic compounds comprise a large group of organic substances and flavonoids are an important subgroup. This subgroup contains the anthocyanins, one of the most broadly distributed pigment groups in plant world. Anthocyanins are responsible for a wide range of colours in plants including blue, purple, violet, magenta and can scavenge free radicals, particularly singlet oxygen due to their reversible oxidation-reduction properties.

All antioxidants and antioxidative enzymes described thus far are known to scavenge free radicals and minimize their cellular levels in all plant cells and thereby may enhance the keeping quality of fruits and vegetables. In spite of such efficient anti-oxidative systems, oxidative damages may still occur in plant cells either due to uncontrolled production or inefficient removal of ROS after harvest of fruits and vegetables. A gradual loss in the ability of antioxidative system to neutralize free radicals may be responsible for many of the metabolic changes associated with ripening.

The literature on this subject is conflicting. For instance, SOD activity has been reported to decrease during senescence of tomato but increase during senescence of apple fruit. In tomato fruit, SOD and peroxidase activities declined steadily from immature green to red ripe stage (Rabinowitch *et al.*, 1982). Superoxide dismutase activity decreased in ageing detached pea leaves (Pastori and del Rio, 1997). Rogiers *et al.* (1998) found that ripening of saskatoon fruit was associated with reduced ability to catabolize ROS. An increase in fruit peroxidase activity at the time of ripening has been documented in blue berries (Miesle *et al.*, 1991), papaya (Silva *et al.*, 1990), peaches (Quesada *et al.*, 1992), tomatoes and pears (Frenkel, 1972) and grapes (Caldron *et al.*, 1993). Similarly, ripening tomato fruit displayed increases in peroxidase activity (Andrews *et al.* 2000) and ion leakage and saturation index of membrane lipids (Palma *et al.*, 1995). The latter was attributed to loss of lenoleneate. Fatty acid unsaturation levels also fell in apple fruits during post climacteric stage of ripening. A loss of microsomal membrane integrity has been osberved during senescence of bell pepper and maturation and ripening of muskmelon and apple fruits (Lurie and Ben-Arie, 1983).

Conclusions

From these studies, it can be concldued that under normal conditions of growth and devleopment, the enzymic and nonenzymic neutralizaiton of ROS in plants provide adequate cellular protection against them. However, during senescence, ripening and storage, when activity of various antioxidative enzymes decline and levels of antioxidants also fall but contrarily continued generation and build up of ROS overcomes the defence system. The levels of ROS so increased cause disruption of cell structure and organization which ultimately affect shelf life of fruits and vegetables adversely. Conversely, prolonged activity of free radical scavenging enzymes along with sustained level of antioxidants during ripening and storage can increase the shelf life of fruits and vegetables significantly.

Molecular biology techniques has provided a desirable tool to investigate, understand and manipulate the biochemistry of fruit ripening and ageing. These techniques are being used to workout the role of hydrolases, ethylene synthesis, pigment production and starch degradation, etc., during fruit ripening, ageing and softening. Consequently, now it appears feasible to manipulate the process of ripening in fruits because scientists have already influenced fruit ripening in tomato by down regulation of polygalacturonase gene. Besides this, ethylene synthesis have also been manipulated to increase shelf life of fruits such as tomato and apples. The ability to apply this technology to various fruits is expanding as the number of fruit bearing plants that can be transformed and regenerated is increasing steadily (Grierson, 1998).

REFERENCES

Alscher, R.G., Donahue, J.L. and Cramer, C.L. (1997). Reactive oxygen species and antioxidants: relationship in green cells. *Physiol. Plant.*, 100: 224-233.

Andrews, M., Malone, D.S., Thompson, L.C.H.O. and Burton, K.S. (2000). Peroxidase isozyme patterns in the skin of maturing tomato fruit. *Plant Cell Environ.*, 23: 415-422.

Caldron, A.A., Zapata, J.M., Munoz, R. and Barcelo, A.R. (1993). Localization of peroxidase in grapes. *Hort. Sci.*, 28: 38-40.

del Rio, L.A., Sandalio, L.M., Palma, J.M., Pastori, G.M., Bueno, P. and Lopez, E. (1998). Metabolism of oxyradicals in peroxisomes and cellular implications. *Biochem. Soc. Trans.*, 24: 434-438.

Frenkel (1972). Involvement of peroxidase and IAA oxidase from pear, tomato and blue berry fruit in ripening. *Plant Physiol.*, 49: 757-763.

Fridovich, I. (1995). Suiperoxide radical and superoxide dismutases. *Ann. Rev. Biochem.*, 64: 97-112.

Grierson, D. (1998). Manipulation of fruit ripening by genetic modification In: *Transgenic Plant Research,* (Ed.) K. Lindsey. Harwood Academic Publishers, pp 109-124.

Halliwell, B. and Gutteridge, J.M. (1989). *Free Radical in Biology and Medicine*. Cladron Press, Oxford.

Leschem, Y.Y. (1998). Plant senescence processes and free radicals. *Free Radical Biology and Medicine*, 5: 39-49.

Miesle, T.J., Proctor, A. and Lagrimini, L.M. (1992). Peroxidase activity, isoenzymes and tissue localization in developing blue berry fruit. *Phytochem.*, 29: 1051-1056.

Palma, T., Marangoni, A.G. and Stanley, D.W. (1995). Environmental stresses affect tomato microsomal membrane functions differently than natural ripening and sensescence. *Postharvest Biology and Technology*, 6: 257-273.

Pastori, G.M. and del RIO, L.A. (1997). Natural senescence of pea leaves: function of ROS for peroxisomes. *Plant Physiol.*, 113: 411-418.

Quesada, M.A., Sanchez-Roldan, C., Heredia, A. and Berkovac, M.J. (1992). Peroxidase and IAA oxidase activities in seeded and seedless peach fruits. *J. Plant Growth Regul.*, 11: 1-6.

Rabinovich, H.D., Sklan, D. and Budowski, P. (1982). Photoxidative damage in the ripening tomato fruit: protective role of superoxide dismutase. *Physiol. Plant*, 54: 369-374.

Rogiers, S.Y., Mohan Kumar, G.N. and Knowles, N.R. (1998). Maturation and ripening of fruit of saskatoon are accompanied by increasing oxidative stress. *Annals Bot.*, 81: 203-211.

Silva, E., Lourenco, E.J. and Neves, V.A. (1990). Soluble and bound peroxidases from papaya fruit. *Phytochem.*, 29: 1051-1056.

Thompson, J.E., Legge, R.L. and Barber, R.E. (1987). The role of free radicals in senescence and wounding. *New Phytol.*, 105: 317-344

16

Packaging of Fresh Fruits

J.K. Sandooja

Department of Horticulture, CCS Haryana Agricultural University, Hisar (Haryana)

"Packaging is to substance that body is to soul". Packaging has to protect beautify and make substance durable.

Packaging is the use of containers and components to add decoration, labeling, protection, transportation and marketing of produce it contains. It demands specialized skills, advanced machinery and well equipped facilities to produce packages of demand. During pre historic days primary intention of packaging was just to transport agricultural goods from field to Bazaar *i.e.* Basket era.

India produces 30 million tones of fruits as against world production of 300 Mt. However, its share in global market is just 0.4 per cent. Hence, there is potential for India to emerge as major exporter of fruits and processed products. Future of fruit export potential is very bright in view of liberalization policies of Govt. of India. In order to remove export major emphasis has to be laid on export quality and proper packaging of fruits as per demand of exporting country. 25-40 per cent of fruits get spoiled or became substandard during storage and transportation. Increase in production can make an impact on the availability of fruits to consumer only when it is packed in unadulterated, hygienic conditions at an economic price.

In a country like India which has long distance with varying climatic conditions ranging from sub zero to 45°C and humidity varying from 20-100 per cent packaging has another important role to play. Problem of food packaging is further aggregated by diversified food habits, low purchasing power, non-availability of packaging material of specific properties at economic price, non-availability of suitable packaging machinery and paucity of technology in packaging.

Main aim of packaging is to protect commodity during storage, transportation and distribution against deterioration which may be physical, chemical or biological. Packaging is generally done at point of production, processing or at distribution centers. So it is last link in the chain of production and first in chain in storage marketing and distribution.

Fruits being highly perishable loose their firmness in a short period under prevailing retail marketing condition due to high rate of physiological activities like respiration and transpiration. Studies have shown that proper packaging in transportation and consumer packaging during retail marketing can reduce losses to 50 per cent. This reduces physiological changes and also prevents mechanical damages during sale and ensures hygienic distribution.

An ideal package is one which fulfills the following conditions and objectives:

1. It serves as an efficient handling unit for customers and dealers.
2. It serves as convenient ware house or storage unit.
3. It should protect fruit from mechanical injury.
4. It should reduce the wastage of moisture loss.
5. It should keep fruits in clean and hygienic manner.
6. It should be convenient for transportation and service.
7. It should be attractive to customers in different modes.
8. It should be economical in every respect.

9. It should have height adjustment in such a way that fruits of lower stack do not get pressed by upper stack.
10. It should provide adequate ventilation to fruits to avoid storage and transportation loss.
11. It should have retailer acceptance to increase turnover and sales at profit.
12. It should be acceptable in all marketing areas.
13. It should fulfill Government compliance problem, manufactures responsibility for safety and regard of consumer health, welfare and essential interest.
14. It should specify company, its product and imagine.

Type of Packaging

A wide range of packaging are used for fresh fruits of their processed products. In most of cases more than one type of packaging is also used for individual commodity. Different size and shape of these packaging are used by various commodities depending upon variety of fruits, their intrinsic value and also depending on market, some of material used for packaging are (*i*) Paper, (*ii*) Plastic for fresh fruits. Aluminum, steel, plastics, glass are mainly used for packaging of processed foods. Paper can be used for fresh fruits in the form of paper cutting as cushioning material, paper lining (news paper lining in boxes), as corrugated card board boxes. Card board boxes as of different strength 3ply, 5ply, 7ply.

Table 16.1: Extent of Losses in Different Types of Packaging

1.	Conventional wooden boxes	25.03 per cent	For same fruits and same distance of transport
2.	Folding box	20.24 per cent	
3.	Card board boxes	12.20 per cent	
4.	Corrugated plastic cartoon	10.63 per cent	

Wood is mainly used as wooden boxes traditional or modified wore bound folding wooden boxes. These are mainly used for long distance transport and have dimension of 48.5 × 20.5 × 360 cms and is commonly, called as petti or 48.5 × 20.5 × 22.5 and commonly called as Dabba. Positive points of three boxes are good stacking strength, weather resistant and ease of assembling. These are constantly using our renewable resources to great extent and will trigger ecological changes. Moreover extent of losses in these boxes is higher than that of card board boxes. So these are being replaced by card boxes of different strength in majority of fruits.

In the present age called as 'plastic era' is replacing almost every other packaging because of its durability. So use of corrugated plastic cartoon is becoming popular which is more durable and can be reused and has less losses. However, it is costlier than cardboard boxes. Now a days plastic basket of 20-25 kg capacity have became more popular. These can be easily stacked one alone other during transportation and generally avoids transportation jerks. Only problem in these basket is that initial cost is high and these are to be returned back to production care for processed products glass containers is handling in transportation specially due to breakage. So these are being replaced by steel, plastics and aluminum. However aluminum is costly steel has problem of rusting. So plastic is again emerging as an most important packaging in present era. They are mainly used in form of cans, bottles, basket, drums and barrels etc.

Packaging Technology for Different Fruits

Apple

It is one of important fruit of India and is grown in hilly areas. These are harvested in pedded basket and transferred to bigger basket (kilt) which is also pedded from inside to avoid injury to fruits. Then they are transported to tractor loaded trailor, brought under shed, graded, sorted and packed made possible for HPMC and transported to different mandies.

For retail packaging plastic bags, trays and sleeves are generally used. Plastic bags are most common in almost all parts of country because of its low cost and needs no special technique for its packaging. In some super stores of metro cities they are also packed in rigid trays and wrapped with plastic film or stretch film. This technique is costly and requires special packaging machine for stretch film. Sleeve is kind of net bag to make packaging attractive and unit pack. However this is not very popular because it has no extra advantage.

Mango

Mango at time of harvesting are generally hard. For their transportation to distance less than 1000 km there are heaped in lorries without any packaging. However mats made of wild date leaves are spread on floor and sides of lorry. For long distance transport *i.e.* more than 1000 km there are either packed in Bamboo basket with paddy straw or wooden boxes of size of 21" × 21" × 40" with loading capacity of 10 kg mango with paper lining as cushioning material in Milhabad areas and wooden boxes of 45" × 30" × 30" with capacity of 15-20 kg with dry grass and newspaper cushioning material are used in Ratnagiri areas. For long distance special train starts from Andhra to Delhi (Mango special). For small distance road, transport is preferred.

Stone fruits (Pulm, cherry, almond): All these fruits coming from Himachal are being packed in card board boxes of different sizes. Whereas fruits coming from Kashmir are still being packed in wooden boxes. Almonds are generally packed in 4 kg capacity card board packs. Whereas cherry is generally packed in 1 kg capacity card board box. Peach coming from Kashmir is still packed in wooden boxes 20 × 28 × 42 cm. There are also lined by paper cutting, paper lining or pine needless as cushioning material.

Kinnow

It is generally grown in Haryana and Punjab areas. These are generally transported in Bamboo basket, gunny bags and to small extent in card board boxes. The leads to great loss in them during transportation. These are also to be changed to card board boxes just like apples which are most successful one.

Grapes

After harvest trimming is done carefully to remove defective sunburnt, shriveled diseased berries. In Haryana they packed in bamboo basket and transported, but, in Andhra and Maharashtra packaging has been shifted to card board boxes with paper lining and paper cutting as cushioning material. Sulphurdioxide releasing papers that kills germs one also used in packaging material.

Guava

Guava is also generally packed in Bamboo basket or wooden boxes which causes great loss. These are mostly preferred one due to economic reasons as guava fruits are cheapest one.

From the above discussion it can be concluded that future of packaging of fresh fruits lies with card board boxes which has all advantages of ideal package.

17

Postharvest Management of Horticultural Crops

J.K. Sandooja

Department of Horticulture, CCS Haryana Agricultural University, Hisar (Haryana)

India having varied agro-climatic conditions produces wide range of tropical, subtropical temperate and zone fruits in all seasons in one or other parts of the country. During last few years considerable emphasis has been given to production of horticultural crops in our country. Accordingly areas under fruit production has risen from 1.22 million ha in 1961 to 2.32 million ha in 1993. Productivity/ha has nearly doubled from 5.52 to 10.28 tonnes/ha and total production has increased by 32 per cent. Today India has emerged as world leader in fruit and vegetable production. There is considerable gap between production and net availability of fruits and vegetables. This is mainly due to postharvest losses in our country due to lack of elaborate harvesting equipment, collection centers in major producing areas, commercial storage plants and lack of cold chain. Total losses in fruits ranges from 20.30 per cent amounting to nearly 3000 crore annually. These losses vary from fruit to fruit, varieties, storage conditions and quality.

Causes of Losses

Fruits are highly perishable commodities as it contains 80-95 per cent water. Even 5 per cent loss of water in fruits due to transportation leads to shrvilling and fruits became unmarketable. Further large unit size, high respiration rate, soft texture, susceptible to bruising rotting and senescence limit shelf life of these commodities. Transpiration and respiration are major cause of deterioration in the fruits. These losses can be reduced to great extent by harvesting fruits at proper stage, proper method of harvesting, breeding, sorting, application of pre and postharvest treatments, Packaging Transportation and appropriate marketing. Important sites where postharvest losses are noticed in India are farmers' field 15-20 per cent, packaging 15-20 per cent transportation 30-40 per cent marketing 30-40 per cent. Reduction of these losses will have following advantages:

1. More availability of fruits without bringing additional land into production
2. Reduction in pollution arising from rotting of fruits
3. Saving of various inputs which will be required for additional production
4. Better nutrient will be made available with more satisfaction of consumers with same renewable resources.

Control of Postharvest Losses

In order to minimize these losses efforts are being done in two direction throughout country:

1. To develop technological needs (Research activities infrastructure for PHT)
2. To strengthen postharvest facilities of the country (Developmental programmes).

To Develop Technological Needs (Research Activities and Infrastructure)

Infrastructure

Considering large amount of postharvest losses in fruits All India Coordinated research project on PHT was started by ICAR in 1978 through PL-480 grant with 4 centres in beginning and 11 centres in second phase.

Research Activities

It includes works on these lines:

1. Development of cultivars with larger shelf life
2. Cultural practices for better quality of fruits
3. Proper maturity indices
4. Method and time of harvesting
5. Grading and sorting
6. Use of pre and postharvest treatments for better shelf life
7. Packaging
8. Transportation and storage.

Development of cvs with Larger Shelf Life

Fruits cvs. with larger shelf life are best suited for long distance transport and storage. This can be achieved through traditional breeding and gene transfer technology. Hybrids which are having genes of higher shelf life have been prepared in many fruits such as tomato, citrus etc. by inhibiting certain genes which are responsible for softening of fruits.

Cultural Practices for Better Quality of Fruits

Use of various cultural practices specially fertilizers (N, P, K) have their role to play in improving quality of fruits. Improved quality of fruits leads to better shelf life. Irrigation at appropriate time has also an important role to play to effect quality of fruits.

Proper Maturity Indices

Harvesting fruits at proper maturity effects shelf life of fruits. Harvesting fruits before proper maturity or after maturity may decrease shelf life of fruits. So maturity indices of each fruit should be standardized.

Method and Time of Harvesting

Harvesting of fruits by faulty means leads to decay of fruits because it creates suitable medium for growth of micro flora at point of injury of fruits which has been created its surface by faulty harvesting technique.

Time of harvesting has also an important role. Harvesting fruits in early morning hour helps to enhance shelf life of fruits. Harvesting in noon hour is most harmful due to maximum field heat present in fruit which reduces their shelf life. In some vegetables during is also required to enhance shelf life of fruits.

Grading and Sorting

Systematic grading is pre-requisite for efficient marketing system. This is done manually or by mechanical graders. This is done either on weight or dimension basis. Mechanical graders has been developed by apple and citrus (Nagpur mandarin). This helps producer for better prices. Sorting not only helps to remove diseased fruits and prevent decay loss but also helps to differentiate fruits for differentiate fruits for different purposes.

Use of Pre and Postharvest Treatments

Use of various chemical, waxes, oil emulsions growth regulators applied as pre and postharvest treatment are helpful to enhance shelf life of many fruits for example Pre-harvest spray of 10-15 ppm of GA proved useful for a tree storage of mango. Use of GA at 200 ppm as postharvest dip are helpful to enhance shelf life of lemon fruits.

Packaging

Present packaging system for fresh fruits in our country is unsuitable and unscientific. Use of traditional forms of packaging like bamboo basket gunny bags are still prevalent. These packages being unhygienic and do not allow adequate aeration, proper handling and ticking. Keeping in view long terms needs of ecosystem and to achieve overall economy other alternative packages such as fibre board boxes, plastic trays, moulded pulp, card board boxes shrink wrapping and stretch film have to be evaluated for different fruits.

Marketing system in India involves chain of commission agents from producers to consumers. These people rotate market as per their desire. In this chain producers have to sell their produce at throw away prices and consumer has to purchase it at high prices. In order to reduce this gap use of cooperative societies is most successful one. NAFED (National Agril. Cooperative Marketing Fed.) at National level in collaboration with state Federation and Primary marketing societies has taken up procurement and marketing of potato, onion, ginger, apple, etc. HPMC (Hort. Produce Marketing and Processing Cooperative Ltd.) in Himachal is another attempt which has become landmark for apple industry in India and helped fruit growers to improve their income and availability of apple all over the country. NDDB (National Dairy Dev. Board) has also established outlet for fresh fruits and vegetables in Delhi. Concept of *Apni Mandi* in Chandigarh is another good example and where producer sell their produces directly to consumers without chain of commission agents.

To Strengthen Postharvest Facilities of Country (Development Programmes)

Same attempts have been made in this direction in following areas:

1. Packaging stations
2. On farm storage
3. Pre-cooling chambers
4. Cool store and cold chain
5. Irradiation
6. CA and MA storage
7. Rural and urban markets
8. Support price for perishables
9. Export promotion
10. Primary processing

Packing Stations

Concept of packing station is completely lacking in India and packing is done mainly on farm in unhygienic manner. Mostly fruits are transported as such to *mandies* even without packing. Packing stations should have facility for grading, sorting, pre-cooling and packaging in hygienic manner. This helps to reduce postharvest losses and better prices to farmers. Little attempts has been initiated by Government in this direction in 8th Five Year Plan by establishing NHB (National Horticulture Board). NHB has set up 127 packing sheds, 64 pre-cooling units, 39 cold storage and 76 special transport vehicles and many retail outlet in different parts of the country. Grading and postharvest treatment plant for Nagpur oranges has been installed at National Research Centre for Citrus, Nagpur. Mechanical washing, waxing and grading unit has been installed at Bangalore.

On Farm Storage

In tropical country like India deterioration in quality of fruits takes place immediately after harvest due to lack of on farm storage. To overcome this problem low cost, environmental friendly zero energy chambers have been developed by IARI. These can be easily constructed at small and marginal farmers field and are helpful to keep temp. 15-30 low than atmospheric temp. These can be used for short-term storage of products at the farmers field itself.

Packaging and Precooling Chambers

Unhygienic and traditional packaging are major cause of postharvest loses specially during transportation. Use of modern packaging such as ventilated corrugated fiberboard boxes (CFB), plastic trays, HD film are mainly helpful to reduce these losses and enhance shelf life during storage. Pre-cooling chambers are best to reduce field heat and enhance shelf life of fruits. Provision of such chamber at doorstep of farmers field can help to great extent. These techniques will also help to boost export potential of fresh fruits.

Establishment of Cold Chain and Cold Storage

In all developed countries adoption of cold chain is most successful for handling of fruits. Maintenance of low temp. during transit at different stages of handling help in reduction of losses and retention of quality. APEDA (Agri. and Processed Food Products Export Dev. Authority) has initiated some efforts in this direction near metro cities like Delhi and Bombay. This should be extended to some other cities to network and to boost export of fruits and vegetables. Construction of more cold storage near production centers can help to store produce for off season and regulate export and domestic market.

Irradiation

Small doses of irradiation can help to kill germs of fresh fruits and inhibit process of sprouting in onion, potato, garlic. It also helps to delay ripening of fruits like mango and banana. Although research on irradiation is being done at BARC and other station since long but now these facilities are being made available on commercial scale. First facility of this kind was made available at VASHI on out skirt of Bombay. Another such plant is being set up near Nasik for Irradiation of potato and onion and is named as proton.

CA and MA Storage

Main purpose of controlled atmosphere or modified atmosphere is to adjust composition of gases in storage chamber different from that of surrounding atmosphere so that respiration rate can be reduced. This helps to store fruits for longer time. CA storage has fixed composition of various gases in chamber *i.e.* increased CO_2, N_2 and less O_2 concentration from that of atmosphere but it is not maintained precisely. These chambers should be made available commercially for storage of fruits in order to regulate market and avoid glut in on season. Satisfactory attempt has not been done in this direction by any agency.

Rural and Urban Markets

As already discussed middle man has main role to play in regulation of market for fresh fruits and vegetables. More efforts are needed to construct more *mandies* at rural and urban level with efforts of cooperative and on concept of *Apni Mandies*. Different agencies involved in this process should pay more emphasis in this direction.

Support Price

Like cereals, pulses and other agricultural commodities Government should give some support price for all fruits and vegetables. If price of these commodities fall below this level then these commodities can be procured by some agencies at certain minimum support price and can be used for processing, storage etc. This will also the farmers but also help to reduce postharvestlosses of these commodities, because sometimes these commodities are such low priced in market that farmers are unable to get production and transportation cost. So he has compelled to dumpit and all resources utilized by him in production are wasted.

Export Promotion

Export of horticultural produce in India is yet to make a dent in International market except few commodities like cashew, spices, mango and onion etc. Even in these commodities our export, market is not regular. In spite of largest producing country, our export potential is negligible. Main strategy of concern to increase export potential includes (i) Production of export quality material, (ii)

Improving postharvest handling and (iii) Strengthening of infrastructure support.

For making export a sustained activity substantial investment are essential to establish auction houses specially for flowers, separate terminal under contoured environment and cool chain facility at airports and seaports. Although little attempt has been made in this direction by APEDA by setting perishable cargo center at international air port in New Delhi for fresh fruits vegetables and flowers. Such facilities are needed at many airports in India.

Primary Processing

Fresh fruits and vegetables are marketed in our country immediately after harvest without primary or minimal processing. So large amount of inedible parts are transported for marketing. These parts generate waste at that place and undue transportation and packing cost. Main advantages of primary processing are:

1. Reduction of bulk
2. Longer shelf life
3. Less transportation cost
4. Conventional handling
5. Employment generation

Conclusion

Postharvest scenario of fruit crop is not very encouraging in our country. There is an urgent need to look more closely at some basic aspects of postharvest management of fruits with following thrust areas:

1. Screening and breeding of hybrids of fruits for better shelf life.
2. Standardization of storage technique including CA and MA storage to enhance shelf life.
3. Development of indigenous fruit products
4. Utilization of fruit waste for viable by products.
5. Development of infrastructure facilities such as grading, packing, transportation and marketing

6. To develop cheaper and better packaging for domestic and export purposes.
7. Mechanical loading and unloading of packages to avoid damage due to mishandling.
8. To cut transport charges by removal of inedible parts of fruits.

18

Recent Trends in the Processing of Fruits and Vegetables in India

Susanta K. Roy

Division of Fruits and Horticultural Technology,
Indian Agricultural Research Institute, New Delhi – 110 012

The diverse agro-climatic zones of India ranging from tropical to temperate, allows us to produce a wide spectrum of fruits and vegetables. India has made significant progress in production of fruit and vegetables since Independence. India produces about 132 million tonnes fruits and vegetables and is considered to be the second largest producer in the world. Unfortunately, unlike other horticulturally rich countries average Indians do not get the basic daily requirement of fruits and vegetables since a considerable amount is lost due to improper post harvest operations. As a result there is a huge gap between the gross production and net availability. According to a joint study conducted by the management consultancy firm, McKinsey and Co and the Confederation of Indian Industry (CII), at least 50 per cent of the production of fruits and vegetables in the country is lost due to wastage and value destruction. The wastage cost is estimated to be Rs. 23,000 crores each year. The high rate of population growth in India offsets the higher production. It is a matter of great regret that in spite of becoming one of the world leaders in the production of fruits and vegetables of Human Development Index (HDI) is very low. Furthermore, only a small fraction of fruits and vegetables is utilised for processing and export compared to many countries in the world whose position is much below India's production level. Hardly 2 per cent of the production of fruits and vegetables go for commercial processing whereas this figure is more than 50 per cent in developed countries. It is often found that there is glut of a particular fruit or a vegetable in one part of the country when it is scarce in another part. If the fruits and vegetables are processed adequately and marketed from the places of abundance to places of scarcity not only will the consumers get the produce at a reasonable price but also the producers will not be forced to sell at throw-away prices. Therefore, in order to achieve our target of tiding over the glut, feeding the population as well as meeting the requirements of the processing industry and export trade, only increasing the production and productivity of fruits and vegetables will not be enough. A lot more emphasis needs to be given to postharvest management in general and processing of fruits and vegetables in particular. This article discusses the importance of fruits and vegetables in our thet, primary processing and how best India can process them for the benefit of the producers and the consumers. A few fruit and vegetable processing techniques are mentioned that can add value and find good business opportunities in India.

Fruit and Vegetables in Human Nutrition

Fruit and Vegetables in general, except for a few, are not considered to be the primary source of carbohydrate, protein and fat. However, some of them with storage roots and tubers are rich in carbohydrate, particularly starch, in amounts comparable to the cereal crops, and the leguminous vegetables supply as much as 14 per cent protein, dry seeds supplying still more. The lipid content in most vegetables is less than 0.1 per cent. Most fruit, vegetables and root crops are rich in minerals, carotene (Pro-vitamin A) and vitamin C. Besides, there are some trace elements required by the body like copper, manganese and zinc, which act as enzyme cofactor. These are found in appreciable quantity in fruit and vegetables. The amount of nutrient can vary with fruit and vegetables, cultural practices, stage of maturity, post-harvest handling and storage conditions. Once they arc harvested, their composition goes on changing as a result of physiological and biochemical activities, which are natural processes. Apart from nutritive value of fruit and vegetables, some other constituents, such as antioxidants, bioflavonoids, flavour compounds and dietary fibres deserve attention. It is necessary to point out that certain fruits or vegetables which contain a modest amount of a particular nutrient could contribute significantly because of relatively high level of consumption *e.g.* Potato and Banana. Similarly certain commodities containing high amount of a particular nutrient may not make any contribution because of low intake such as green chilly, coriander leaves, curry leaves etc. It will not be out of place to highlight that the leafy portion of some of the important vegetables are rejected and fleshy portions are consumed regularly without knowing that a rich source of nutrients such as calcium, iron, vitamin-C and carotene are being discarded. In the case of fruits, it is worth while to mention that apart from common and generally expensive fruits, there are a large number of indigenous fruits growing in plenty are also rich sources of nutrients.

Primary Processing

Unlike durable crops like cereals, pulses and oilseeds perishable fresh fruits and vegetables in our country are marketed immediately after harvesting without primary processing. Horticultural produce in general and fruits and vegetables in particular generate a large amount of valuable waste *viz.*, leaves, stalk and other inedible plant parts which end up as garbage. However, if they are gainfully utilised at the proper time they can become value added products *viz.* cattle feed, β-carotene pro-vitamin from cauliflower leaves etc. Vegetables such as cauliflower, peas, leafy vegetables, etc. can be primarily processed at packing stations so that inedible parts are removed before being sent to the metro-city markets in unit packs.

Perishable Nature of Fruits and Vegetables

Fruit and vegetables are living beings and they carry out transpiration, respiration, ripening and other biochemical activities, which adversely affect the quality of the produce resulting in loss of nutrients. In addition, because of their high moisture content fruit and vegetables are inherently liable to deteriorate rapidly. The major causes of deterioration of fruit and vegetables are microbial growth, enzyme action and chemical changes. These actions and reactions take place fastest at high water activities as well as favourable temperature, pH and other environmental factors. The principles of food preservation are based on the manipulation of their environmental factors in order to retain the nutritive value.

Processing of Fruits and Vegetables

The consumption of processed fruits and vegetables is still low all over the world compared to fresh ones but it is progressively increasing and will continue to do so in view of the seasonal availability of fresh commodities. In most of the developed countries consumption of processed products is relatively high. Processing industries are also beginning to emerge and grow in the less developed parts of the world.

There are six basic principles of food processing. These are:

1. Moisture removal–drying, dehydration and concentration
2. Heat treatment–pasteurization and sterilisation
3. Low temperature treatment–refrigeration and freezing
4. Acidity control–fermentation and acidic addition
5. Various chemical–additives and
6. Irradiation.

Low Cost Appropriate Techniques of Processing

Most of the commercial methods of preservation are expensive and beyond the means of the common people. Therefore, it is necessary to adopt low cost methods of processing so that it can reach the grass root level. In this respect the following techniques are suggested:

Solar Drying

Vegetables

Most of the leafy vegetables rich in nutrients can be preserved successfully by solar energy. A low cost solar drying chamber made of black polyethylene has been found effective in drying most of the leafy vegetables. Even though the drying process causes loss of considerable amount of nutrients a substantial amount is still left in the dried vegetables to meet nutritional requirements. The storage of dried vegetables is of great importance and should be done with great care. Dried vegetables, particularly the leafy ones in the powdered form can be easily mixed with wheat flour in order to enhance its nutritional value.

Fruits

It has often been found that during the season of glut a huge quantity of fruit is piled up in the production centre due lack of transportation and marketing facility. In order to avoid wastage the fruits can be solar dried or the fruit pulp can be dried in the form of fruit leather or slab/bar and can be utilised gainfully for the benefit of local people by packaging and storing them appropriately. A substantial amount of nutrition can be obtained by consuming them during the off season.

Home Scale Processing

Most of the nutritious fruit and vegetables can be preserved successfully in the form of juice or pulp by bottling and pasteurising them at home. Once it is bottled successfully, it can be stored for a long time.

Whole Tomato Concentrate

It is a simple and low cost technique of processing of tomato that utilises the surplus produce, and is an easily adaptable technology for the benefit of growers and consumers. The whole tomato is crushed in a container and the volume is reduced to one third by boiling. The final product is filled in bottles after addition of permitted preservatives.

Lactic Fermentation

This is another low cost technique of preservation that can be adopted easily. Most of the winter vegetables can be preserved by this method, simply by applying salt and allowing the vegetables to ferment naturally. It is advisable to mix cabbage with other vegetables during lactic fermentation.

Minimal Processing

Fruits and vegetables are minimally processed in response to the consumer's demand for convenience foods. Minimally processed fruits and vegetables are cleaned, peeled, cut, sliced, packaged and/or lightly processed. These foods are in great demand because of their convenience. Not all fruits and vegetables need to be minimally processed. However several tropical fruits and vegetables are quite large such as jackfruit, papaya, watermelon, pumpkin; ash gourd, yam, etc. It is very often not convenient for the consumer with a small family to purchase the whole fruit, therefore if it is suitably sliced and sold he MU be more inclined to buy it. Pineapple is one such fruit that is very difficult to eat out of hand. Throughout the world only 5 per cent of pineapple produced is marketed in the fresh form.

Recent studies show that pineapple slices and cut pieces can be minimally processed and marketed

Processing of Underutilised Fruits

India has a wide range of tropical fruits that are underutilised. Most of these fruits can grow even under adverse agroclimatic conditions. A large number of these fruits are known for their therapeutic/medicinal and nutritive value and have excellent flavour and very attractive colour. Consumers today are becoming increasingly conscious of the health and nutritional aspects of their food. The tendency is to avoid chemicals and synthetic foods and choose therapy and nutrition through natural resources. The indigenous fruits of India have an important role to play in satisfying the demand for nutritious, delicately flavored and attractive natural foods of high therapeutic value. Some of these fruits are not easy to eat out of hand *e.g.* bael fruit that has a hard shell, mucilaginous texture and numerous seeds; as a result it is not popular as a dessert fruit. Kokum is not acceptable as a fresh fruit because of its high acidity. Similarly fresh aonla is not liked by many people because of its strong astringent taste. It is urged that indigenous/underutilized fruits which are not easily marketed in the fresh form should be processed into acceptable products so that the growers get a remunerative price and consumers all over the world get the opportunity to enjoy the fruit in the form of its processed products.

Bulk Aseptic Packaging

Aseptic packaged food can be defined as specially packaged foods in which the products are sterilised and are then packaged in sterile packaging materials under an aseptic environment without re-eating for sterilisation. From the stand point of processors, aseptically processed food show no juice separation phenomena caused by the re-heating operation, have improved taste, and reduce energy consumption. In addition they can greatly saved on both packaging material and shipping cost because they are mostly packaged in water-tight paper board cartons replacing conventional glass containers. Of all packaged foods, aseptic packaged foods are still lowest in market share. However, they are gradually increase in percentage year by year. Aseptic packaged food will present much demand not only in the consumers market but also in the institutional market in the near future. Bulk aseptic packaging is one of the most recent development in the fruit and vegetable processing industries. Aseptic packaging *i.e.* the bag-in box packaging system was originated in the USA. In most of the developing countries this technology is widely used. There is a tremendous potentiality of using bulk packaging in our internal and export trade. In fruit and vegetable processing industry, fresh whole fruit and vegetables are not always required excepting where fruit and vegetable slices/ pieces are canned, frozen or dehydrated. Most of the fruit products *viz.* beverages, ketchups, sauce etc. are prepared from the fruit pulp or juice. One of the greatest advantages of using bulk aseptically packed pulp is to avoid handling fruits and vegetables as they are highly perishable in nature. In addition, a considerable amount of waste is generated in terms of peel, seed, stone, pomace etc. which could be utilised for value added products. It is advisable to establish bulk aseptic packaging unit in close proximity to production centre.

Quick Frozen Fruits and Vegetables

Freezing is rated as the best technique available for food preservation since it maintains the natural properties by reducing post harvest changes and microbial deterioration to the barest minimum without any influence on the original qualities. The rate of freezing plays a great role on the quality of frozen fruits and vegetables; faster freezing rate is required to obtain better quality. Liquid nitrogen is the most common cryogenic substance used in food freezing. Ultra quick freezing rate, minimum dehydration loss, freedom from oxidative changes, minimum freezing damage of freeze sensitive products, maximum quality retention of texture, colour and flavour of sensitive fruits and vegetables during freezing and the inert nature of the freezant are the advantages of liquid nitrogen freezing. A systematic establishment of quick freezing industry can boost export trade of country. Methods have been standardized under laboratory conditions at IARI for the manufacture of cryogenically frozen, crack free, peeled ripe mango slices having excellent retention of quality attributes, well comparable with those of fresh mangoes in ready-to-serve form and cent-per-cent edible portion. This has a great potentiality in export trade. The problems generally facing the export of fresh mangoes, like short storage life, added bulk of stone and peel, hidden disorders like spongy tissue and stone weevil can be successfully overcome by producing the cryogenically frozen mango slices.

Packaging and Storage of Processed Fruits and Vegetables

The ultimate success of processing of fruits and vegetables depends largely on the packaging and storage conditions. Food packaging has become an increasingly important area. The purpose of processed fruit and vegetable packaging is to safely preserve and distribute them to the consumer, preventing deterioration under various environmental conditions. Excepting frozen products other processed fruits and vegetables are stored wider normal room temperature. India is a tropical country and the prevailing temperature of a large part of this country remains around 30°C or more excepting winter month, this condition is highly detrimental to processed fruit and vegetable storage. In North India the temperature during the summer months is usually more than 40° during summer months. This condition can inflict lots of damage to the fruit and vegetable products in terms of loss of nutrients, change in colour, flavour etc. Therefore it is necessary to store processed products below 30°C otherwise it will lose its market value.

Processing of Price Stability

Prices of perishable fruits and vegetables fluctuate greatly unlike durable crops like cereals, pulses, oilseeds etc. During the period of maximum availability of fruits and vegetables in peak season the prices are not remunerative to the farmer. At other times, these commodities are so highly priced that the ordinary consumer finds it beyond his purchasing power. Another problem is that fruits and vegetables are not uniformly available throughout the country and some areas suffer from inadequate supply even when there is a glut in other parts. Large scale processing of fruits and vegetables and their distribution play a vital role in price stability of perishable fruits and vegetables.

Conclusion

India provides great opportunities for investment and growth in fruit and vegetable processing industry. This sector is recognized as having an important role in improving horticultural productivity, reducing wastage of fruits and vegetables, providing better nutrition and improving fruit and vegetable availability for the domestic market in addition to promoting export. The demand for processed fruits and vegetables in India is likely to multiply significantly in the coming years. Rapid urbanisation, breakdown of joint family system, increase in the cost of household labour, increasing number of working women, rise in per capita income etc. have all contributed to the rapid growth and change in demand patterns. India is also one of the biggest emerging markets with a population of over one billion and with a strong middle class of over 250 million. A massive thrust to food processing and other agro-based industries will add value to the product thereby increasing the income of farmers, create employment opportunities, diversify the rural economy and foster rural industrialisation.

19

Postharvest Handling of Fruits and Vegetables

S.S. Dhawan

Department of Horticulture, CCS Haryana Agricultural University, Hisar (Haryana)

Horticulture plays a significant role in Indian agriculture. The contribution of these horticultural crops towards agricultural output and export earning of total agricultural produce is very considering the percent area covered by these crops. The production figure of fruits is estimated to be about 41.5 million tonnes and vegetables about 80.8 million tonnes including potato and onion. Unfortunately unlike other horticultural advanced countries, average Indians do not get the basic daily requirements of fruits and vegetables. Despite tremendous growth in production of fruits and vegetables during the last 50 years Indian's human development index (HDI) rank for nutrition is 132. This reflects a major deficiency in the quality of life of a vast majority of population. A considerable amount of fruits and vegetables produced in India is lost due to improper postharvest operations; as a result there is a considerable gap between the gross production and net availability. It is estimated that at least 30-50 per cent of total production of fruits and vegetables in the country is lost due to wastage and value destruction which amounts to about Rs.23,000 crores each year.

In India, it is often found that there is a glut of a particular horticultural commonality in one part of the country when it is scarce in the other parts. The unnecessary wastage of valuable commodities can be checked if they are processed into value added products or adequately distributed in different parts of the country. There is a great scope of both domestic and export trade in our country by improving the post-handling and distribution procedures. There is a great scope of improving the system by adopting proper and modern methods of postharvest handling.

Many pre-harvest factors (variety, growing conditions, cultural practices, pest management etc.) influence the postharvest behaviour of horticultural crops. However, by the time of harvest these factors are fixed and cannot be manipulated. From harvest to market usually involves much less time than the production phase but may be more crucial from the stand point of both economic gain and food supply. Therefore, the basic steps involved in postharvest handling of fruits and vegetables are discussed here.

Harvest

Harvest is the process of separating the desired portion of the plant from the undesired. Although it is usually viewed as a simple process, it does require skill and actually consist of 5 steps.

Detecting

Actually seeing the fruit or that portion of the plant to be harvested. The picker must see the fruit or vegetable before he can harvest it. Therefore, judging the proper maturity stage is the most important factor determining the storage life.

Selecting

Making a decision as what to harvest and what to leave. This is the first decision and may be based on one or more of the following. Maturity, ripeness, colour, firmness, location of the plant, size, shape, defects, disease etc.

Detaching

Separating from the rest of the plant. Many fruits and vegetables form an abscission layer as they mature which makes harvesting easier and less damaging to the crop. Depending on the crop, the proper harvesting procedure may be a simple twisting and/or snapping motion, a thumb push on the stem, snip with shears, cut with knife. Tree fruits are sometimes knocked off with stick and (hopefully) caught. Improper harvesting can result in damage to the harvested product and to the plant.

Collecting

Placing the harvested commodity in the picking containers. The container may be a picking bucket, bag, box or crate with someone else carrying it. The primary problem here is impact injury due to tossing rather than placing. This should also be a sorting step in that overripe (or over-mature), decaying or obviously out of grade produce should not be included.

Handling

Employing the harvesting containers into field containers for transport to the packing house. This step is sometimes avoided by transporting the harvest containers directly. This avoids the potential damage (both impact and compression) due to dumping operation and the weight associated with using large bins. However, for large packing operations that may be some distance from the field, it is not practical to transport the smaller picking containers.

Cleaning

Often fruits and vegetables are cleaned prior to marketing. In some cases this may consist of wiping off the fruit during harvest. However, the most common method of cleaning is to wash the commodity. In addition to improving the appearance, this can help to cool either by water temperature and/or by evaporative cooling. If the water is going to be reused, it should be treated (chlorinated). If it is fresh well water that is not recirculated, treatment is not usually necessary. Washing (wetting) also reduces transpiration of water loss which is particularly important on leafy vegetables. It should be noted that warm, wet horticultural crops are very susceptible to decay.

For some fruits such as kiwi fruits and avocados, dry brushing may be sufficient to clean the produce other commodities, such as bananas and carrots require washing. The choice of bruising and/or washing will depend upon both the type of commodity and the type of contamination. Certain produce should not be washed because washing will remove the natural waxes that protect produce from water loss and disease.

Trimming/Topping

Certain crops should be trimmed or topped to remove inedible plant material or excess outer leaves before packing. This simple practice also helps to reduce the high rate of water loss, which occurs through the leaves of root and tuber crops.

Cooling

Extremely important for highly perishable commodities. Preventing temperature rise is equally crucial. If temperature is increased by 100°C the product will deteriorate 2 to 3 times faster. In the field keeping produce (especially dark coloured ones) out of the Sun (in the shade or covered with a light coloured material), will help to keep the temperature at a more reasonable level. Allowing air to move through the produce to dissipate the heat of respiration is also helpful. The higher the temperature, the faster respiration, the more heat produced and the more important it is to able to remove it.

Evaporative cooling is also an inexpensive and simple way to help reduce temperatures. The simplest method is to wet a porous material that will hold water and move air through the wetted material (like desert cooler). Air may be moved by fans or by using prevailing winds. Since cool air is heavier than hot air it is also effective to place the wetted material over the top of the container. Other methods of cooling are vacuum cooling, room cooling, forced air-cooking and package icing.

Sizing/Grading

Grading may vary from the very simple action of sorting and discarding, undesirable or unmarketable product to grading into different quality classes which may be sized into 2 or more sizes; which may be divided into 2 or more colours. Usually the more sophisticated the market, the more precise and detailed the grading procedure. Horticultural Crops that have been fully graded should present a vary uniform package. Discarding of unmarketable product is often done during the harvest operation. Additional grading–colour sorting and/or sizing may also be performed at the field level. Each time a product is handled there is potential bruising. Bruises that are incurred during harvesting, grading etc. often do not become visible until the fruit or vegetable is in the market.

Waxing

Waxing of immature vegetable such as cucumber and summer squash mature fruits and vegetables such as eggplant peppers and tomatoes and fruits such as apples and peaches, citrus fruits is a common postharvest practice. Food grade waxes are used to replace some of the natural waxes removed during harvesting and sorting operations, and can help to reduce water loss during handling and marketing. If the produce is waxed, the wax coating must be allowed to dry thoroughly before further handling.

Bunching/Wrapping

Some commodities are easier to handle and pack if they are bunched and/or wrapped. Wrapping cauliflower in vented polythene film can reduce damage to the delicate heads by protecting against abrasions during postharvest handling. Small items like radishes can be bagged and packed as a unit 500g or 1 kg, while green onions, broccoli and asparagus spears are often bunched with a wide rubber band before packing.

Postharvest Treatments

There are some chemicals used in the packing house that may generally recognized as safe (GRAS) which can help to control physiological disorders and a variety of molds and fungi on fruit crops. Some of the common treatments are mentioned here:

1. A solution of calcium chloride ($CaCl_2$, 2-3 per cent in clean water) can be prepared as a dip and acts to control bitter pit in easily harvested lower maturity apples.
2. Sulphur dioxide (SO_2) is used as a fumigant or a water spray (0.5 per cent) for 20 min. for initial treatments, then 0.2 per cent for 20 min. at 7 days intervals) on grapes to control Botrytis, *Rhizopus* and *Aspergillus* fungi.
3. Sulphur is used on banana as a paste (0.1 per cent) to control crown not fungi.
4. Sodium or potassium bisulphites are used in a sawdust mixture (usually contained within a pad that can be placed

inside a carton) to release SO_4 for control of molds on grapes (5 g for a 11 or 13 kg box).

5. Bacterial soft rot (*Erwinia*) of cabbage can be controlled by using lime powder or a 15 per cent solution of alum (aluminum potassium sulphate) in water. After treatment of the butt-end of the cabbage heads, the produce should be allowed to dry for 20 to 30 minutes before packing.
6. Postharvest heating using hot water or hot forced air to kill or weaken pathogens can be used as a method for decay control in fresh fruits and vegetables. Immediately after the treatment cool water showers should be provided.

Packing

Horticultural produce are packed for to facilitate handling and to protect them from injury during transport and marketing. Large fruits or vegetables may be handled individually and transported in bulk. Very tender, high value products may be warped or padded for additional protection. Containers are supported to product the crop inside. Often they do not offer protection from impact or compression damage due to:

Inadequate Strength

Bags (paper, plastic, mesh, etc.). Fiberboard-may be ok if properly used-but (unless treated) loses most of its strength if it becomes wet.

Improper Use

One of the biggest problems with containers is overfilling. The product may actually be damaged by force applied during filling. In addition, when a container overfilled to the point of bulging is stacked all the weight from the above is on the contents instead of on the corners and walls of the containers. Containers that are too heavier also prone to dropping or rolling which also injures the contents. Most commonly package materials are wood bins/boxes, jute sacks, plastic crates and corrugated fibre board (CFB) boxes.

Transportation

Transport is costly and in many developing countries is unreliable and may not be readily available. When a vehicle is available it is often overloaded and occasionally misloaded with tender soft commodities on the bottom and heavier less fragile ones on top.

These problems, coupled with rough roads, no refrigeration and covered or enclosed trucks take a tremendous toll on the quality of fruits and vegetables. Other ways of getting crops to market include buses (on top), mini trucks, and village cart or cycle rickshaw.

Table 19.1: Appropriate Storage Conditions for Fruits and Vegetables

Product	*(ºC)*	*RH (%)*	*Approx.Storage Life*
Fruits			
Apples	0-2	90-95	1-12 months
Grapes	0-2	90-95	1 month
Kinnow	0-3	80-90	3 months
Peaches	0-2	90-95	4-6 weeks
Mango	13-15	85-90	2-3 weeks
Vegetables			
Cabbage (late)	0	98-100	5-6 months
Cabbage (early)	0	98-100	3-6 months
Carrots (mature)	0	98-100	7-9 months
Cauliflower	0	95-98	4-6 weeks
Cucumber	10-13	95	10-14 days
Brinjal	12	90-95	1 week
Garlic	0	65-70	6 months
Ginger	13	65	6 months
Okra	7-10	90-95	7-10 days
Onions (green)	0	95-100	3-4 weeks
Onions (dry)	0	65-70	1-8 months
Peas	0	95-98	1-2 weeks
Potatoes (late)	4-13	90-95	5-10 months
Tomatoes (ripe)	13-15	90-95	4-7 days
Tomatoes (mature Green)	18-22	90-95	1-3 weeks

Storage

In modern times, the storage of perishables has become more sophisticated because of the demand of fresh fruits and vegetables at all the times of the year and because of need to spread peaks of production over long periods to maximize profit and reduce waste. Certain crops like potatoes and apples can be stored for long period whereas the crops like tomatoes and peaches the storage period is very short. There are a few important factors, which needed to be taken into account before storage of the produce.

1. Suitability of crops for storage
2. Knowledge of appropriate storage conditions
3. Compatibility with other crops for storage.

The use of modified atmosphere packaging (MAP) and controlled atmosphere storage (CAS) for extending the postharvest life of fruits and vegetables has increased steadily world wide during the past 50 years. The trend is expected to continue with the advancement of technologies in attaining and maintaining CA and MA during transport, storage and marketing of fresh produce.

20

Freezing of Fruits and Vegetables

S.S. Dhawan

Department of Horticulture, CCS Haryana Agricultural University, Hisar (Haryana)

In today's world of population explosion the importance of food preservation hardly needs any emphasis. Food preservation has a special relevance to India, a heavily populated country. With its vast resources of varied tropical and marine foods, there is a wide scope for export and internal markets if these can be properly preserved. Foods are being preserved by various methods such as sundrying, pickling, canning, dehydration, pasteurization, fermentation, chemical preservatives, refrigeration and freezing etc. consideration of factors such as (1) product quality (2) process feasibility and (3) economics, shows that freezing preservation of foods is one of the most attractive and practical methods, since freezing is a quick, convenient and an easy method of preserving foods.

In actual practice the freezing of foods slows down, but it does not stop the physical and biochemical reactions that govern the deterioration of foods. When properly handled and processed frozen foods are often perceived by other methods these qualities depend upon (a) the use of raw material, (b) control of the freezing process, (c) careful pre-freezing preparation and (d) post freezing storage of the product.

Superiority of Frozen Foods over the Preserved Foods

It is not far to seek the reasons for the growing popularity of frozen foods; their quality is considered to be far superior to that of similar foods preserved by other conventional methods of preservation like canning, dehydration and chemical preservation etc. Table 20.1 shows the comparative sensory quality of foods preserved by well known methods.

Table 20.1: Comparative Merits of Foods Preserved by Different Methods

Sl.No.	*Quality Attributes*	*Frozen Foods (% Score)*	*Canned Foods (% Score)*	*Dried Foods (% Score)*	*Others Foods (% Score)*
1.	Freshness	71	14	5	10
2.	Taste	56	30	7	7
3.	Appearance	62	28	3	7
4.	Nutritive value	51	20	11	18
5.	Value for money	47	29	10	13
6.	Ease of preparation	40	52	3	3

Source: Paruthi (1999).

Purpose of Refrigeration or Freezing of Foods

The purpose of refrigeration are three fold:

1. To reduce the temperature of a substance (act of cooling)
2. To transform a substance from one state to another (as water into ice)
3. To maintain the substances in a desired healthy state (as in the storage of ice for the preservation of food).

Water can be transformed into ice if its temperature and heat content are changed a gas into liquid, or a liquid into solid. When refrigeration is applied to water, the solid ice will be produced. Production of ice is one purpose of refrigeration. Another purpose of refrigeration is Food Preservation. Many microorganisms, such as bacteria, yeast and mould cause spoilage of foods. At normal temperature, these microorganisms are active, causing food decomposition in varying degrees, the rate depending on the type and the condition of the food product as well as the extent and nature of microbial load. If the temperature of these foods is reduced sufficiently, the action of these microorganisms is retarded greatly making it possible to keep food longer. The preservation of food is therefore another highly important use of refrigeration.

Definition of Quick Freezing of Foods

Quick freezing process are "Ultra Rapid" freezing to a very low temperature (-30° to -40°C) designed to bring the inside of the product to a temperature of -18°C as quickly as possible. Rapid lowering of the temperature also makes it possible to pass quickly below the critical temperature of 0-5°C at which intracellular liquids freeze. When quick freezing process is used, these liquids solidify to form extremely small crystals of ice and the cellular structure is left intact, whereas in ordinary freezing process, where low temperature are reached more slowly, the texture of the product is altered, liquid forming large ice crystals which lacerate the tissues, which is responsible for poor texture of the frozen food after its thawing: Quick freezing is done by a process of lowering temperature food to much below their freezing point in a short period of time, say to -20°C in 3-30 min. The process involves direct immersion or indirect contact of foods with the refrigerant, and air blast over the foods being frozen.

Definition of IQF

IQF generally refers to Individual Quick Freezing" of solid food/ pieces/grains like green peas, cut beans, cauliflower pieces, meat fish etc. while "Quick Freezing" relates mostly to liquid, pulpy or semi liquid products like fruit juices, mango/papaya pulp/juice concentrates and purees etc. In IQF foods, there is no clumping together of pieces or grains. They remain individual separate pieces. Hence the term IQF.

Advantages of "Individual Quick Freezing' or IQF Technology Other Conventional Slow or Deep Freezing Technology

The main advantages of IQF are:

1. The ice crystals formed are much smaller, which therefore cause much less damage to the cell structure or texture of food.
2. The freezing period being much shorter, less time is allowed for the diffusion of salts and the separation of water in the form of ice.
3. The product is quickly cooled below the temperature at which bacterial mould and yeast growth occur thus preventing decomposition of foods during freezing.
4. The fourth and very practical reason in favour of IQF over "slow freezing" is the inherent speed and greater output and hence higher capacity for commercial freezing plants with the resultant cost reduction.

Steps Involved in Quick Freezing Preservation of Fruits

Selection of Varieties

Not all varieties of fruits are suitable for freezing. Peaches, apricots, berries, pineapple, apples and grapefruit respond well to freezing, but bananas, avocados, watermelons and grapes do not respond so satisfactorily. Even different varieties within the same fruit respond differently. Fruits should be harvested at optimum maturity (eating ripe maturity).

Desired Quality Characteristics

The variety of fruit suitable for freezing must have

1. Low browning tendency
2. Firm flesh
3. Rich and characteristic flavour
4. Attractive colour
5. Good yield

6. Resistant to insects and diseases and
7. Uniformity in maturity and ripening.

Handling, Washing and Cleaning

Since ripe and damaged fruits are subject to infection, a proper control measure is required. This includes chlorination of wash water, protection of fruit from bruising during handling and sorting.

Initial Sorting and Grading

Sorting is necessary to separate green, overripe and defective fruits. Sophisticated photo-electric devices have been developed for colour grading of oranges, lemons, apples, tomatoes, cherries and other coloured fruits, while floatation baths are used to grade fruits by density difference.

Pre-Freezing treatments

Some fruits require peeling which is usually done by lye-dipping followed by gentle rubbing and washing off, the peel. Some products can have the peel loosened by gentle warming or heating or freezing the surface. Most fruits are pitted by means by halving and pit removal (*e.g.* peaches and apricots)

Slicing, cutting, cubing or dicing

Many peeled fruits are cut into slices, cubes or dices in preparation for freezing. All these cut surfaces allow some loss of nutrients and browning. Most varieties of peaches, apples and pears darken quickly after peeling.

Blanching

Blanching is done to inactivate enzymes. Heat blanching is generally not used on fruits because of turgor loss from heat damage and the associated sogginess and juice drainage after thawing the individual quick blanch process is probably the most efficient technique for blanching and cooling.

Ascorbic acid may also be used to retard browning of fruits. Covering the slices with a sugar syrup containing a small quantity of ascorbic acid yield a product of superior flavour and texture.

Treatment of apples slices for 10 min. in a solution containing 400 ppm sulphur dioxide in the form of sodium sulphite, 500 ppm of ascorbic acid and 2500 ppm (0.25 per cent) of low methoxyl pectin has been reported to yield superior quality of frozen slices.

Some Other Protective Treatments for Fruit Freezing

Certain berries like blueberries and raspberries can be frozen directly without any protective pretreatment. Because of their high acidity and high ascorbic acid content, the stability of these fruits during frozen storage is excellent. Fruits are protected by the addition of sugar or immersion in sugar syrup that act as a protective agent. Different concentrations have been recommended *i.e.* 30°Brix for cherries, pears and plums, 40-50°C Brix for banana, and apples, and 40-60°Brix for strawberries, peaches, and apricots.

Industrial Freezing Methods (Freezers)

In order to remove heat, the product must be brought into contact with a cold medium. This can be cold air, in a blast freezer, cryogenic liquids or gases in a cryogenic freezer, or cold surfaces in a plate freezer. The heat transfer mechanism in cold air is through convection. This can be assisted by blowing the air over the product. Cryogenic liquids have a higher thermal density and remove heat more rapidly still due to an improved heat transfer capability. Heat transfer to a cold surface tends to be by conduction. Therefore the choice of appropriate freezer is in part, governed by the product size and in part, governed by the required freezing rate. Small fruits can be frozen by IQF method while fruits packed in large cans or drums will not freeze quickly due to heat transfer limitations and so are frozen in a cold room. Retail packs of approximately rectangular brick shape are often frozen in plate (conduction) freezers.

Table 20.2: Commercial Freezing Methods

Air Freezing	*Indirect Contact Freezing*	*Immersion Freezing*
Still air 'Sharp' freezer	Single plate	Heat exchange fluid
Blast freezer	Double plate	Compressed gas
Fluid bed freezer	Pressure plate	Refrigerant spray
	Slush freezer	

Selection of Suitable Freezing Plants/Freezing Techniques for Different Categories of Foods

Attempted below is a product wise break up of the products freezing techniques indicating the most suitable and effective product freezing range.

Air Blast Freezing (Tunnel and Tray)

Vegetable products, meat products with or without packing.

Belt Freezing (Tunnel and Belt)

Cut vegetables, cut meat pieces, small packed marine products.

Immersion Freezing (Brines and Glycols)

Poultry products and sheep products with packing only.

Plate Freezing (Vertical and Horizontal)

Flat, thin slicked vegetables, fruit pulps and juice concentrates.

Fluidized Freezing (Belt, Tray)

Very small spherical, particulate and irregular pieces of vegetables (peas), small fruits unpacked only.

Liquid Freezant –12 (Uniflow, Counter Flow)

Any product of vegetable, fruit and meat products.

Liquid Nitrogen Freezing (Spray and Immersion)

Only for very high cost products of marine and costly fruits of medium to small size.

Liquid CO_2 (Tubing freezer and Spiral Freezer)

Medium to small beef and other meat products and fruit products, packed as well as non-packed.

Freezer Storage

Proper storage of frozen fruits and fruit products is of paramount importance. In the beginning it was felt that a temperature of 15 to 20°F (–9° to –7°C) was satisfactory for long-term storage of frozen fruits. However, it was soon learnt by experience that the holding temperature as low as –20°F (0–29°C) was good for long storage. All frozen foods should be quickly cooled atleast to a temperature of 0°F (–18°C) in the centre of the product.

Purchasing Quick-frozen Foods

It takes 4–5 hours for a small packet of frozen product to thaw completely at room temperature. When frozen food is brought, it is unlikely that the purchaser will take that long to reach home. A normal shopping excursion usually takes 2–3 hours, during which time, packs of frozen food will only have started to soften on the outer surface. If this food is packed in the centre of the shopping basket, it will be insulated by other packages and will hardly thaw. Alternatively, one may invest in 'an insulated freezer bag' for this purpose.

In a food freezer operating at –18°C (0°F), purchased frozen food can be stored upto 3 months according to pack instructions. This allows for the fact that the food has already been in frozen storage during distribution.

Refrigerated Transportation

Transportation plays an important role in the frozen food distribution system. The success of frozen food industry depends on a safe and dependable refrigerated transportation system.

Future Scope

India has a great potential in exporting frozen pulps of mango, guava, papaya, litchi and other fruit products. Retention of fresh qualities of fruits is essential for the international markets. Quick frozen fruits have got advantages like (*i*) cent per cent edible portion (*ii*) available in ready to serve form (*iii*) lesser packing cost and (*iv*) reduced bulk during transportation. If the freezing techniques of producing fruit and fruit products with preservation of original fruit quality is commercialized with the easily available liquid nitrogen, it will widen the horizon of frozen fruit and fruit products industry of our country.

REFERENCES

Paruthi, J.S. (1999). *Quick Freezing Preservation of Foods* Vol. II. Allied Publishers Ltd. New Delhi.

Srivastava, R.P. and Sanjeev Kumar (1993). *Fruit and Vegetable Preservation*. International Book Distribution Co., Lucknow.

Somogyi, L.P., Ramaswamy, H.S. and Hai, Y.H. (1996). *Processing Fruits: Science and Technology.* Vol. I: *Biology Principles and its Applications*. Tectronic Pub. Co. Inc. Lancaster. Basel.

Potter, N.N. and Hotchkiss, J.H. (1996). *Food Science*. CBS Pub. and Distributors, New Delhi.

21

Dehydration of Fruits and Vegetables

S.S. Dhawan

Department of Horticulture, CCS Haryana Agricultural University, Hisar (Haryana)

Several fruits and vegetables are dried in the sun in different parts of the world. This is probably the oldest method of preservation of food and still it is used in many parts of the world because it is most economical method of drying. But it has several disadvantages as it is dependent on the elements, slow and not suitable for many high quality products, it also requires considerable space and the food being exposed to contamination.

Drying of fruits and vegetables under controlled conditions has assumed importance in recent years. The first patent for dehydration process was applied by Grayler in 1780. Artificial drying (Dehydration) of fruits is now an important method of preservation and production of a wide variety of products. Storage technologies are now available to keep fruit produce 'fresh' over extended periods of time without appreciable change in the original physical form of fruit. Drying, on the other hand, changes the physical and biochemical form of the fruit, leading to shrinkage and change of colour, texture, taste and so on. If water activity is reduced to appropriate levels depending upon fruit variety and sugar content, the dried product can have a shelf life exceeding one year if properly packed.

Fruits may be dried as whole (*e.g.* grape, various berries, apricot, plum, etc.), in sliced form (*i.e.*, banana, mango, papaya, kiwi etc.), in puree form (*e.g.*, mango, apricot etc.), as leather or as a powder by spray or drum drying. Depending upon the physical form of the fruit (*i.e.* whole paste, slices), different types of driers must be used for drying. Figure 21.1 illustrates the wide assortment of dryers that must be found in practice for drying of fruits.

Fruits are characterized by the following features which must be taken into account when selecting the type of dryer, as well as operating conditions.

1. Very high initial moisture content.
2. High temperature sensitivity (colour, flavour), texture, nutritional value subject to thermal deterioration.
3. High susceptibility to microbial attack

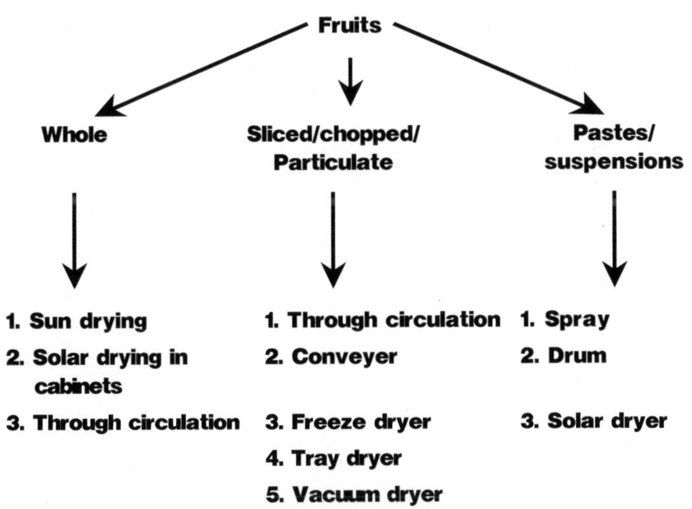

Figure 21.1: Various Types of Dryer for Drying of Fruits

4. High sugar content (problem of stickiness, fermentation etc.)
5. Presence of a 'skin' on some fruit that have poor permeability for water or moisture (*e.g.* grapes, blueberries).

Predrying Treatments

Preparation of raw materials for dehydration sundrying or other evaporation techniques is similar-some common steps applied to the important fruits are as follows:

1. Selection and sorting for size, maturity and soundness.
2. Washing
3. Peeling by hand, lye (sod hydroxide) solution or abrasions.
4. Cutting into halves, slices, cubes etc.
5. Alkali dipping used for grapes and prunes.
6. Blanching.
7. Sulphuring.

Sulphuring

Sulphurdioxide fumes act as a disinfectant and present the oxidation and darkening of fruits on exposure and thus improves their colour. This phenomenon is generally seen in sliced fruits which darken due to oxidation of the colouring matter. Sulphur dioxide also act as a preservative which check the growth of moulds etc and prevent cut fruit pieces from fermenting while drying in the sun. The whole fruits, slices or prices are exposed to the fumes of burning sulphur inside a close chamber known as "sulpur box" for 60-120 minutes. Sulphur box is an air tight container which can hold trays and sulphur powder is burnt @ 3-5 g/kg fruit inside the chamber sulphur dioxide released combines with the water and forms sulphurous acid which act as a preservative. Sulphur dioxide also bleaches the colour of the product and make the colour uniform and attractive.

Types of Dryers

Over 200 types dryers are available commercially to process the diverse physical forms of foods and required product characteristics

Menon and Majumdar (1987) have given detailed schemes for classification of dryers based on:

1. Mode of processing (batch, continuous, semi-batch)
2. Operating pressure (vacuum, atmospheric, high pressure)
3. Mode of heat transfer (Conduction, convection, radiation, directive and its combination)
4. Physical state of product being dried (*e.g.* stationary, moving, conveyed, fluidized, dispersed, vibrated, atomized, stirred etc.)

For drying of fruits and vegetables conventional types of dryers are commonly used because of their simplicity of construction and operation as well as low cost. Most such products are seasonal in nature, which means that dryers are operated only over a fraction of the year. Thus, the unit must have low capital costs for economic reasons.

Specific Drying Systems

Solar Drying

The drying of fruits and vegetables using solar energy direct and indirect. In the direct solar drying commonly called open sun drying or just 'sundrying' the product is placed directly on trays in open are where only the incident solar radiations is used as the energy source for drying. Such dryers have the following main disadvantages:

1. Large area requirement
2. Inability to control
3. Degradation of product by biochemical and microbiological reactions.
4. Insect infestation.
5. Uneven drying
6. Long drying time.

The indirect solar dryer use solar energy to heat drying air in special heat exchanges (Solar collectors) and then this air flows through over the product by natural or forced convection. The air conditions at the inlet of the collector (air humidity and temperature) vary with time depending upon the hour of the day and other factors such as the weather conditions the main advantage of these dryers is that the energy source is renewable, free and nonpolluting. Fruits commonly dried with solar energy are fig, plums, peaches, apricots and mainly grapes to obtain raisins.

Conventional Hot Air Drying

In this type of drying. The heat needed for drying is provided by convection with hot air in direct contact with the product. The most common of these dryers used for fruits are kiln, cabinet, tunnel and continuous belt dryers.

The kil on dryer is basically an oven heated by gas burners in which the product is placed in slotted trays. The equipment is operated in a batch mode and is still widely used to dry slices of apples and several other fruits and vegetables. It takes about 6-8 hours to dry a batch of slices of apples.

Cabinet dryer is also batch equipment, but there is a forced flow of air along or through the trays in which the material is placed. Heated air is entered in the cabinet. The operation can be controlled and drying is much more uniform. The unit is useful for small scale drying of cut fruits.

The tunnel dryer is the most used equipment to dry fruit. It is similar to the cabinet dryer, but the trays are made to move along a tunnel in which hot air flows in parallel or counter flow to the product movement. It is most suitable for industrial applications. This method is often used to dry apricots, peaches, pears, apples, figs, dates and so on.

Drum Drying

Fruit and vegetable juices and purees may be dried with drum dryer. These products are allowed to flow onto the surface of two heated stainless steel drums rotating in opposite directions with little clearance between them. The product dries on the drum and the dried product scrapped off by blades. This method has been replaced by other newer improved methods such as spray drying.

Spray Drying

This is preferred method for liquids, paste, and slurries. These products are sprayed into the top of a chamber simultaneously with hot air, which is also blown at the top. The cool moist air exists near the bottom and are collected by gravity flow or by the aid of scrappers.

Explosion Puffing

The incorporation of explosion puffing into hot air dehydration process leads to a final product with faster dehydration, rapid rehydration, and a highly porous structure. Explosion puffing has been successful with apples, blueberries, and banana slices. The process starts with conventional hot air drying of fruit pieces until a certain water content is attained. Then the pieces are placed in a closed chamber. The 'gun' where the pressure is increased by means of super heated steam. After some time, the pressure is suddenly released by opening the lid of the 'gun'. Due to the high temperature and the sodden decompression, water within the particles flashes explosively. Then the resulting solids are again placed in the air drying equipment to be dried to the desired final water content.

Microwave Drying

Conventional drying of foods is a slow process. Microwave heating increases the temperature of the interior, water parts of the solid. Microwave heating has three main advantages:

1. A penetrating quality that leads to uniform drying
2. Selective absorption by liquid water, which leads to uniform moisture profile within the particle
3. Case of control due to rapid response of such heating.

Apple, mango, pineapple and grapes more investigated recently for microwave drying under vacuum. Use of microwave applied in a vacuum for dehydration heats the product to cause water to vapourize, without causing changes in composition. Microwave energy penetrates deeply into food products and can reduce typical process time by 90 per cent. Microwave vacuum dehydration was first used for concentration of citrus juice in France.

Freeze Drying

Freeze drying (also called lyophilization) is characterized by heating (*i.e.* conduction, radiation, radio frequency, microwaves etc.) a moist frozen product in a vacuum chamber maintained at an absolute pressure below the vapour pressure of ice within the product. When the material is heated the ice sublimes.

Table 21.1: Schedule for Drying of Fruits

Fruits	*Preparation*	*Pretreatment Dipping in Lye etc.*	*Time of Sulphuring*	*Drying Temperature (ºC)*
Apple	Peel, core, trim and cut into 3 to 5 mm thick flices	NIL	30 min.	50-55
Apricot	Wash, halve, destone	NIL	30 min.	45-50
Peach	Halve, peel and pit	NIL	30 min.	45-50
Grape	Wash, cut into small burches	10-20 Sec. 0.3 per cent lye and wash	1-2 hrs	55-60
Prune	Wash, use whole fruit	3-5 seconds and wase	–	50-55
Litchi	Wash remove skin	Dip in 1 per cent KMS sol. for 16 hrs	–	50-55
Banana	Wash, peel, cut into 12 mm thick slices	NIL	2 hrs	55-60
Mango	Wash, peel, cut into 12 mm thick slices	NIL	2 hrs	45-50
Pineapple	Wash, punch, core, cut into 6 mm thick slices	NIL	2 hrs	55-60
Papaya	Wash, punch, core, cut into 6 mm thick slices	NIL	2 hrs	55-60
Fig	Wash	NIL	1 hrs	55-60
Aonal	Wash, grate/blach for 10 min. destova	Dip in 70ºB syrup for 10 hrs	1 hrs	55-60

Table 21.2: Drying Schedule for Vegetables

Vegetable	*Preparation*	*Pre-treatment*	*Maximum Drying Temperature (ºC) in Home Drier*
Potatoes	Peel and cut into 1 cm thick slices	Blanch in boiling water or steam for 3-4 minutes and sulphite in 0.125 per cent potassium metabisulphite (KMS) solution for 10 minutes using 0.5 kg. of solution per kg. of slices.	65
Carrots	Peel and cut into 1 cm. thick slices	Blanching in boiling water or steam for 3-4 minutes and sulphite in 0.25 per cent KMS solution for 10 minutes using 0.5 kg of solution per kg. of slices.	60
Cauliflower	Remove stalks, covering leaves and stems, break apart flowers and cut into 10-12 mm. thick pieces.	Blanch as above and sulphite in 0.125 per cent KMS solution using 0.5 kg. of solution per kg. of material	50
Onions	Remove tops and tails, peel and cut into 4 to 8 mm. thick shreds	NIL	55
Garlic	Peel the sloves, and cut into 6 mm. thick shreds	NIL	60
Okra(sliced)	Remove both ends with stainless steel knife and cut into 6 mm. thick slices	Blanch in boiling water for 4-5 minutes and sulphite in 0.25 per cent KMS solution for 10 minutes using 0.5 kg. of solution per kg. of slices.	60

Contd...

Table 21.2–Contd...

Vegetable	*Preparation*	*Pre-treatment*	*Maximum Drying Temperature (ºC) in Home Drier*
Okra (Small-sized 50-75 mm. Length, whole	Remove both ends with stainless steel knife and slit longitudinally along the ridges	Blanch in boiling water for 6-7 minutes	60
Bitter gourd	Remove both ends with steel knife and cut into 6 mm. thick slices	Blanch in boiling water for 8 minutes	60
Pumpkin	Cut into 50-75 mm. Wide, longitudinal strips, peel, remove the seeds and soft portions in contact with the seeds, prick uniformly using a stainless steel fork and cut into 6 mm. Cubes.	Blanch in steam for 10 minutes	65
Green peas	Remove pod and prick with a small nail or pin	Blanch for 3-4 minutes in boiling water containing 0.5 per cent KMS 0.1 per cent magnesium oxide and 0.1 per cent sodium bicarbonate	60
Spinach	Sort, wash thoroughly and cut into 10-12 mm. portions using a stainless steel knife.	Blanch for 3-4 minutes in boiling water containing 0.5 per cent KMS 0.1 per cent magnesium oxide and 0.1 per cent sodium bicarbonate	60
Fenugreek leaves	Sort, wash thoroughly and remove stalks and stems	Blanch for 3-4 minutes in boiling water containing 0.5 per cent KMS 0.1 per cent magnesium oxide and 0.1 per cent sodium bicarbonate	60

1. High flavour and aroma retention
2. High retention of nutritional value
3. Minimal shrinkage
4. Minimal change in shape, colour appearance
5. No damage in structure and texture
6. Porous final structure
7. Easily rehydratable

The industrial application of this method has been limited by the high capital and operating tools. The commercial application has been limited to exotic fruit and orange juices.

Osmotic Drying

Hot air drying reduces notably the quality of processed food stuff. Osmotic dehydration is an alternative technology to reduce the water content, as well as to improve the quality of final product. This process is being used in industry to dehydrate fruits, vegetables etc. Osmotic dehydration happens in the immersion of cut foods in concentrated solutions of sugars or salts. A flux of water out of the food and another of the solute into food stuff developed due to the difference in the osmotic pressure. The product thus loses some water to the external solution. Osmotic dehydration reduces the weight of slices upto 50 per cent prior to vacuum drying. Osmotic dewatering rate can be enhanced by increasing the concentration of the osmotic solution of temperature.

REFERENCES

Somogyi, L.P., Ramaswamy, H.S. and Hui, Y.H. (196). *Processing Fruits: Science and Technology Vol.1. Biology. Principles and Applications*. Technomic Publishing Co. Inc. Lancaster. Basel.

Potter, N.N. and Hotchkiss, H.H. (1996). *Food Science. 5th Edition.* CBS Publishing and Distributors, New Delhi.

Kadam, S.S. (1997). *Processing of Fruits and Vegetables*. Chaitanya Pub. Pune.

MeBean, D. McG. (1978). Drying and Processing tree fruits. *CSIRO Divison of Food Research Circular* No. 10, CSIRO, Australia.

Srivastava, R.P. and Sanjeev Kumar (1993). *Fruits and Vegetable Preservation: Principles and Practices*. International Book Distributing Co., Lucknow.

Dhawan, S.S. (1998). *Practical Manual on Home-scale Processing of Fruits and Vegetables*. Deptt. of Horticulture, CCS HAU, Hisar.

22

Nutritional and Medicinal Value of Under-exploited Fruits

Suneel Sharma

Department of Horticulture, CCS Haryana Agricultural University, Hisar (Haryana)

Worldwide a staggering number of people lack adequate amounts of foods that are rich in the vitamins and minerals needed for health and a productive life. Today, 790 million people in the developing world are chronically under nourished. The overwhelming majority of under-nourished people live in Asia and the Pacific. In India itself, there are 204 million under-nourished people. Malnutrition in the form of deficiencies of iron, calcium, iodine and vitamin A continues to cause related health problems such as blindness, osteomalacia, anaemia and infant mortality etc. that leads to severe illness or death of millions of people worldwide. Therefore, ensuring nutrition and feed security to the millions of the rural poor is a great challenge. Under-utilized fruit crops supplement and complement traditional food sources in achieving this objective.

Nutrition Security by Underutilized Fruits

A number of underutilized fruits are rich in vitamins, minerals, protein, carbohydrates and fats and are cheap source of these nutrients. They seem to provide an appropriate solution to the serious problems like malnutrition and related health disorders such as blindness, nutritional marasmus, osteoporosis, osteomalacia and infant mortality in the developing and underdeveloped world.

Sources of Vitamin A

Vitamin A is essential for normal vision, development of bones and growth. The requirement of vitamin A is the highest during infancy which gradually decreases with age. It's requirement for an adult is 300 to 4000 IU. Vitamin A deficiency is a major public health problem in more than 90 countries. About 3 million children world-wide are afflicted with severe vitamin A deficiency. It's deficiency is one of the major nutritional problems threatening children with blindness. Vitamin A is abundant in some of the underutilized fruits such as *Astrocaryum aculeatum* (51667 IU), *Eugenia stipitata* (12917 IU), *Gnetum gnemon*–melinjo leaves (10889 IU), *Bactris gasipals* or peach palm (7300 IU), persimmon (2710 IU), *Pouteria compechiana* or canistel (2000 IU) and *Lacuma abovata* (1500 IU). *Pereskia aculeate* or Barbados goose berry (3215 IU) and *Physalis peruviana* or cape gooseberry (1000-5000 IU) are also rich sources of vitamin A. (Pareck *et al.*, 1998).

Sources of Vitamin C

Vitamin C (ascorbic acid) is needed to develop resistance against diseases and other ailments in human body. Average requirement of vitamin C in the body varies from 20-27 mg. Its deficiency is widespread in the third world countries. The common symptoms are pin in joints swelling in limbs, bleeding of gums, tooth decay and rheumatism. The underutilized fruits are rich in vitamin C. Some of the extremely rich sources of vitamin C are *Malpighia glabra* or Barbados cherry (1500-5600 mg/100 gp pulp). *Myrciaria dubia* or camu-camu (2994-4000 mg/100 gp pulp), *Myrciaria eauliflora* or jaboticaba (700-2900 mg/100 gp pulp), *Hippophae rhamnoides* or seabuckthorn (600-2500 mg/100 gp pulp), *Eugenia tomentosa* or cabeludinha (931 mg/100 gp pulp), *Emblica officinalis* or Indian gooseberry (625 mg/100 gp pulp), *Ziziphus jujuba* or chinese juiube

(188-544 mg/100 gp pulp), *Actinidia cinensis* or kiwifruit (300 mg mg/100 gp pulp), (Pareck *et al.*, 1998).

Sources of Protein

Protein is essential for the growth and maintenance of human body but its deficiency is wide spread in the poverty ridden areas of the world leading to disorders like growth failure, oedema, loss of appetite, diarrhoea and nutritional marasmus. Although most of the common fruits are low in protein, some of the underutilized fruits and nuts are rich sources of protein. These includes: (1) Tropical and subtropical fruits and nuts such as *Macadamia integrifolia* (macadamia nut), *Buchanania lanzan* (cuddapah almond), *Canarium ovatum* (Pilinut), *Canarium indicum* (Java almond), *Butryospermum parkii* (Shea butter tree), *Bertholletia excelsa* (Brazil nut), *Treculia africana* (African breadfrui), *Trapa natans* (water chestnut), *Terminalia catappa* (Indian almond) *Sclerocarya caffra* (Marula nut), *Nelumbo nucifera* (Indian lotus) and *Gnetum gnemon* (Melinjo), (2) temperate fruits and nuts such as *Pistacia vera* (Pistachio nut), *Pinus edulis* (Pini nut), *Juglans nigra* (Black wal nut), *Juglans cinerea* (Butter nut), *Fagus grandifolia* (American beach), *Corylus avellana* (European filbert), *Corylus Americana* (American hazelnut or filbert), *Castanea mollissima* (Chinese hairy chestnut) and *Castenea dentate* (American or sweet chastnut). Among these, some are very rich in proteins, such as *Juglans cinerea* or butter nut (24.6 per cent). *Bertholletia excelsa* or brazil nut (14.3-17.4 per cent), *Buchanania lanzan* or Cuddapah almond or chironji (21.6 per cent), *Ceratonia siliqua* or carob (21 per cent) and *Ricinodendron rautanenii* or Mogongo nut (38 per cent). (Pareck *et al.*, 1998).

Sources of Minerals

Calcium and iron are important minerals required for healthy living. Calcium gives strength and firmness to the bones, forms healthy teeth and is also necessary for clotting of blood and movement of muscles. The daily requirement of calcium for an adult is 0.6-1.0 g. The rich sources of calcium are tamarind (0.74 per cent), Karonda (0.16 per cent), wood apple (0.13 per cent), Bael (0.09 per cent), Dates (0.07 per cent) and Aonla (0.05 per cent).

Iron is essential for development of blood and body cells. Average requirement of iron for an adult is between 15-30 mg. Iron deficiency is the world's most common nutrient deficiency. Two billion people suffer from iron deficiency "anaemia". Its deficiency in diet leads to anaemia, retarded growth in children, poor appetite and increases susceptibility to diseases, Karonda (39.1 per cent), Date (10.6 per cent) and Green Mango (4.5 per cent) are rich sources of iron. (Pareek *et al.*, 1998).

Medicinal Value of Underutilized Fruits

Value of the underutilized fruits in traditional medicines is well known. These have been a major source of raw materials for drugs since antiquity and have provided bulk of products used in the traditional system of medicine. In India, the fruits of *Emblica officinalis* (Aonla). *Terminalia chebula* (harar) and *T. bellirica* (bahera) are the most common, entering into 219 patented drugs. Similarly, *Berberis aristata* (resaut) and *B. asiatica* are used in 65 drugs and *Aegle marmelos* (bael) in 60. The demands for products that enter into the manufacture of popular medicines for common ailments and home remedies is truly large. "*Chavanprash*" based on *Emblica officinalis* (aonla) is an other excellent example. It consumes 20,000-30,000 tonnes of dry fruits of aonla annually. Bael (*Aegle marmelos*) pulp contains normalosin which acts as laxative and diuretic and in strong doses as eardic depressent of "Josh-anda" (an ayurvedic medicine used in chest trouble). Kernels are sedative, and as soporific are used to stop vomiting; also employed as an antidote to aconite poisoning and of abdominal pain in pregnancy. Seeds of *Gacinia kola* (bitter kola) are useful in treatment of coughs and hepatitis. The roots of *Nephelium lappaceum* (Rambutan) are used in decoctions for treating fever, the fruits for digestive problems and the leaves for headache. (Gupta, 1991, Kumar and Gupta, 1996).

The palms which are usually not considered significant medicinally, are used for their medicinal and chemical values. In South America alone, stems and roots are employed in the preparation of tea, poulticies, curative baths and therapeutic drops.

Besides the chemicals use of seabuck-thorn oil, some ten new drugs in the form of liquids, powders, plasters, films, pills, liniments, suppositories, aerosols, etc., have been developed in Russia for treating rectum mucositis, vaginal mucositis, cervical erosion, radiation damage, burns, sealds, duodenal ulcers, gastric ulcers, chillburns, skin ulcers, etc. (Rongsen, 1992).

It has been reported that the seeds of moreton bay chestnut (*Castanospermum australe*) yield castanosperine which appears to halt reproduction of the AIDS virus. (Anon, 1990). An alkaloid deoxyjirimycin extracted from the root of black mulberry (*Morus* sp.) has been reported to hinder the docking of the AIDS virus on human cells. Fruits of prickly pear (*Opuntia* sp.) and their products help in treating hypoglycemica, diabetes, high blood cholesterol and obesity. Under utilized fruit crops offers a great scope in providing nutrition and food security. (Hegwood, 1990).

Role of Antioxidant Photochemicals in Fruits and their Implications in Diet and Human Health

The discovery of photochemical in fruits and vegetables and their strong antioxidant potential in scavenging free radicals has generated tremendous attention among scientists.

Phytochemicals

Many phytochemicals found in fruits and vegetables act as powerful antioxidants, protecting cells and organs from damage caused by free radicals, neutralizing their damaging effects. They are the biologically active substance in plants that give them their colour, flavour, odour and protection against plant and human diseases. Consequently, hundreds of such plant substances are being investigated now for their role in preventing cancer and other degenerative diseases. Some of the promising phytochemicals which act as antioxidants are bioflavonoids (vitamin P), phenolics, lycopene, carotenoids, antioxidant vitamins (C and E) and glucosinolates (Prior and Cao, 2000; Roy, 2001).

Indirect Virus Direct Antioxidants

Direct antioxidants, such as vitamins C and E, neutralize dangerous free radicals to which cells are exposed or that are generated by cells themselves before free radicals can harm cell. In this process, a direct antioxidant molecule binds to a free radical molecule rendering it harmless and, thus, protecting cells from damage. Once the direct antioxidant reacts with a radical, the antioxidant is destroyed and can not be used again.

The indirect antioxidants work as a catalyst. It does not neutralize free radicals directly, but rather boosts the body's own elaborate antioxidant systems that exert ongoing and prolonged antioxidant activity. This is a broad spectrum of activity, cycling over and over again, that removes many free radicals. It is like an "army" of antioxidants, ready to neutralize free radicals over a period of time, and continuous to be effective even after the indirect antioxidant has left the body.

Citrus Fruits

Oranges, lemons, limes and grapefruits are the principal sources of vitamin C, folate and fiber. They are also high in a class of phytochemical called limonoids. This antioxidant has been found to be very effective against cancer.

Berry Fruits

Berries especially coloured like craneberry, strawberry, black berry, aronia, bilberry and whortle are rich sources of flavonoids and phenolics. According to a breakthrough research at Rutgers University, New Jersey, blue berries is the number one antioxidant fruit. It has been found to be beneficial in fighting off urinary tract infections by blocking or prohibiting the growth of bacteria. Concentrated extracts of anthocyanins also benefits visual activity as well as provide protection against mascular degeneration, glaucoma and cataracts.

Group and Wine

Flavonoids and other phenolics present in grape and grape products have been shown to possess anti-carcinogenic, anti-inflammatory, anti-thrombic and anti-oxidant effects. The ability of phenolic substances in prevention of atherosclerosis and risk reduction of cardiovascular disease has been aptly demonstrated by the benefits of red wine consumption often referred as French paradox. The paradox refers to epidemiological findings that in certain parts of France where wine consumption is high, coronary heart disease mortality is low despite relatively high cholesterol levels in population.

Apples

Apples according to scientists are a nutritional treasure chest. Apple skin helps ward off cancer through a unique mix of molecules called flavonoids and polyphenols. Studies show that apples checks proliferation of most liver cells and colon-cancer cells. The fruit supplies a range of nutrients such as iron, copper, vitamin C, malic acid and calcium. Eating an apple is a more efficient way of cleaning teeth than a toothbrush. It also plays a role in preventing heart disease. The fruit also prevents disease like rheumatism and arthritis.

Bael

Bael fruit is a source of riboflavin and ascorbic acid. The pulp of the fruit although a little acrid bitter is aromatic and acts as sweet cooling tonic. The pulp contains a large amount of mucilage and gum as a result of which it has the important curative properties as appetizer and is often prescribed for stomach problems.

Jamun

Jamun fruits known for their acid and astringent taste is useful for curing diarrhoea and diabetes. It is stomachic and diuretic apart from having coding and digestive properties. Jamun seeds are high in phenolics and are prescribed in diabetes, diarrhoea, ring worms and blood pressure.

Kokum

Kokum has fine flavour and makes excellent beverages. It is known to reduce obesity and regulate blood cholesterol level to keep heart healthy. It has fine flavour and makes an excellent beverage.

Aonla

It is rich in polyphenol and ascorbic acid. It is valued as an antiscobutic, diuretic, laxative and antibiotic. One or other part of the plant could be used in treating chronic dysentery, bronchitis, diabetes, jaundice, dyspepsia and cough. The fruits are processed into "Chavanprash" and "trifla" in Ayurvedic system of medicine.

Pomegranate

Every part of the pomegranate is used in the treatment of dysentery, diarrhoea, stomachache, dyspepsia, bronchitis and cardiac. These therapeutic properties are due to the presence of betulic acid and urrolic acid and different alkaloids–pseudop-elletierine, pelltierine, isopelleterine and methyl pelletierine. The unripe fruits and flowers are useful in inducing vomiting and the rind of the fruits is given in diarrhoea and dysentery. It is also useful in sore throat and eyes, brain disease and chest troubles.

Mahua

Both flowers and fruits of mahua are eaten in various ways. Mahua flowers are regarded as cooling tonic and demulcent. They are used in cough and colds. Tremendous interest in the antioxidant components in fruits and vegetables and their possible health-promoting effects has stimulated worldwide research to identify potential sources. In this context, there is a dire need to document the antioxidant potential of our indigenous wealth of fruits, vegetables and medicinal herbs in terms of their free radical scavenging to establish them as functional and health foods.

REFERENCES

Anon. (1990). Americans in Agiculture–Portraits of Diversity. *USDA Year Book of Agriculture*, 1990, p. 104-105.

Gupta, R. (1991). Trees in Indian Medicine. *Indian Farming*, 41 (8): 10-14.

Hegwood, D.A. (1990) Human health discoveries with *Opuntia* sp. (Pricklypeer). *Hort. Sci.*, 25(12): 1515-1516.

Kumar, Ashok and Gupta, P.N. (1996). Genetic Resources of Indigenous Tropical Fruits having medicinal value. In: *Genetic Resources of Tropical Fruits: Collection, Evaluation and Conservation Strategies*, (Eds.) P.N. Gupta, Mathura Rai, K.P.S. Chander, NBPGR, New Delhi. Pp 132-144.

Pareck, O.P., Sharma, Suneel and Arora, R.K. (1998). Underutilized edible fruits and nuts: An inventory of Genetic Resources in their regions of diversity. IPGRI office for South Asia, New Delhi, India. p. 11-13, 20-21.

Prior, R.L. and Cao, G. (2000). Antioxidant phytochemicals in fruits and vegetables: Direct and Health implications. *Hort. Sci.*, 35(4): 588-92.

Rongsen, Lu (1992). *Seabuckthorn: A Multipurpose Plant Species for Fragile Mountains*. ICIMOD Occasional Paper No. 20, Kathmandu, Nepal.

Roy, S.K. (2001). Food, Nutrition and Environmental Security through strategic postharvest management of fruits and vegetables *Indian Horticulture*, 45 (4): 4-5.

23

Utilization of Under-exploited Fruits for Value Added Products

Suneel Sharma
Department of Horticulture, CCS Haryana Agricultural University, Hisar (Haryana)

India is a home of wide array of minor fruits which are found growing in tropical, subtropical and temperate climatic conditions. Many of the under exploited fruits have nutritional, therapeutic and medicinal values and holds lot of potential to become as new fruit crop. However, there are some problems which makes them unpopular among growers inspite of several merits. Great opportunities also exist for their consumption in internal market as well as for the export of processed commodities to a number of countries by advertising their nutritional, medicinal and therapeutic values under "New Health" foods or high value foods or foods for Health Programme.

Among various fruits grown in the arid of semi-arid regions, pomegranate, ber, custard apple, aonla and tamarind, are some of the commercially important fruits. In the recent past, there has been substantial increase in the area under these fruits. Many times, this results in a seasonal glut, which leads to the substantial reduction in the market prices. Many of the fruits cannot be stored for longer time owing to perishable nature. Therefore, it is necessary to process these fruits into the acceptable value added products. Several reports on the utilization of fruits for the preparation of value added products are available in literature.

The techniques for the preparation of refreshing/health drinks or fruit sauce/pickles from these fruits in combination with conventional fruits have been standardized. The work on processing of these fruits have been described here.

Aonla

Aonla (*Emblica officinalis*) is one of the minor fruit crop of commercial significance. It has acquired wide popularity all over the world due to its medicinal properties. Fresh Aonla fruit are highly acidic and unsuitable for direct consumption. Hence, the fruits are processed into value added products such as preserves, pickles or powder. Preserves have been prepared by using optimum maturity fruit keeps longer and has better organoleptic qualities. Powder is prepared by mincing the fresh fruits and powdering slices after natural drying in the sun (Kadam, 2001). Recently, the technology has been standardized to prepare products such as herbal jam, herbal syrup, chutney and sauce. (Singh, 2001).

Bael

Bael fruit (*Aegle marmelos*) occupies an important place among minor fruits in India. It is quite popular for having important curative properties because of its laxative and therapeutic value. Since, the fruit has excellent aroma which is stable during processing. Hence there is ample scope for processing of various delicious products which are highly nutritive and therapeutic importance. A process for manufacturing RTS, Squash and Jam has been reported. Some work has been done at I.A.R.I. to make various products for this carbohydrate rich fruit like nectar, squash, toffee, pangiri, slabs, powder, leather etc. The technology has been standardized for preparing preserve (murabba) for green bael fruits, dehydrated green bael fruits slices can be prepared. (Roy and Singh, 1979).

Ber

Ber (*Ziziphus mauritiana*) is one of the ancient and common fruit of India which is an ideal fruit for growing in the arid and semi arid region. Ber fruit can be processed into different products such as Juice, RTS, Candy, dried ber, powder of wine (Kadam, 2001). Juicy varieties are better suited for extraction on pulp and juice, develop a process for extraction of ber juice. The juice can be used for preparation of ready to serve (RTS) beverage using 10 per cent juice. Dehydration of ber can be done by treating fruits with sulphur dioxide at 3.5-10 gm/kg for 3 hours followed by drying in the sun for 7-10 days or in a cabinet drier at 60°C for 20-35 hours. Untill the moisture content of the product was deduced to 15 per cent. Ber fruits can be utilized for preparation of candy by sugar syruping process. Like other fruit juices, ber juices can be used for preparation of wine. The Department of Horticulture, Haryana Agricultural University, Hisar the technology for processing of ber for preparation of wine preserve, dehydrated fruit and bottling of ber in 40 per cent sugar syrup (Dhawan and Sharma, 1996).

Cashew Apple

Cashewnut (*Anacordium Occidentale*) is grown throughout most of the tropics. In India it is mainly cultivated in Kerala, Tamil Nadu, and Maharashtra. The fruits are usually discarded in the fields after collecting the nuts. The fruit is very juicy and the expressed juice has a T.S.S. value of 12-14 per cent of about 0.35 per cent acidity. It is known for its ascorbic acid content. It contains an astringent and acrid principle which produces an unpleasant biting sensation in the tongue and throat when eaten. In addition, the fruits are highly susceptible to microbial deterioration. These drawbacks together with the difficulties experienced in collecting the fruits, have held back its commercial utilization. Only a very small quantity of fruit is used for the production of alcoholic liquor (brandy) called 'Feni' in Goa. However, extensive work carried out of CFTRI has shown the possibility of producing a number of products. Good quality cashew apple juice concentrate can also be prepared from the juice of cashew apple from which the tannin has been removed. Owing to highly perishable nature, it is usually transported after sulphating. Aroma recovery from the juice results in a high level of sulphur dioxide in the aroma concentrate. A method has been developed for selectively removing sulpher dioxide by ion exchange resin treatment.

Custard Apple

Custard apple (*Annonsa squamos*) is almost entirely consumed as a dessert fruit. The pulp has a pleasant texture of flavour of used in preparing delicious drink by mixing with milk. As the main season of harvesting for the fruit is very short (October to November), there is usually a glut, leading to wastage and distress sale. Extraction of the pulp is a major constraint in processing of custard apple. The prolem encountered during processing of custard apple are the development of bitterness, unpleasant–repulsive off flavour in the pulp of heating beyond 65°C and presence of gritty cells (Singh, 2001).

Although some attempts have been made to process the fruit into juice, squash and jam, on heating the pulp develops a bitterness, which has hindered their commercial exploitation. Also, the presence of gritty stone cells in the pulp is in undesirable characteristics. Work done at CFTRI has shown the possibility of centrifuging the pulp to remove the stone cells. Enzyme clarification of pulp was found to yield bitterness free juice which can be used to formulate nector, squash and wine (Anon, 1971-72).

Jack Fruit

Jack fruit (*Artocarpus heterophyllus*) grow in the wild and is also cultivated in some states. In India, over 26000 hac are cultivated with jackfruit, principally in Assam, Bihar, Tripura, Kerala, U.P., Karnataka, Tamil Nadu, and Andhra Pradesh. It is used in both its unripe and ripe stage. Ripe jackfruit are delicious, having a strong and pleasing aroma. It is a good source of vitamins and minerals. Raw jackfruit is utilized in curry preparation in the growing regions. The seasonal nature of the fruit with short storage life, even under low temperature conditions necessitate processing of ripe fruit. A number of products have been developed from the raw as well as ripe fruits and its seed.

Recently studies have been undertaken on preparation and packaging of curried raw jackfruit.

Jamun

Jamun (*Syzygium cumini*) is an important minor fruit of Indian origin. Jamun has never been exploited commercially for preparation of different products. However, it is reported that the fruits are used for making products such as RTS, Jam, jelly, beverages, wine, vinegar and pickles. The method of extraction of Jamun juice has been standardized by Raman Janaye, (1985). The pure jamun juice can also be preserved by pasteurizing (Khurdiya and Roy, 1985).

Karonda

Karonda (*Corissa corandas*) is grown extensively in many parts of India and at PARI, New Delhi. The fruits are used in pickles making being acidic in nature. Receipe for preparing a good quality preserves, candy, pickles of chutney was standardized from mature fruits of karonda. Ripe fruits may be processed into an excellent quality of syrup and jelly (Singh, 2001).

Ker

Ker (*Capparis decidua*) is a minor fruit crop which grow wild without much care in North West India. The unripe fruits of ker are generally used as fresh in vegetable or can be dried to yield 20 per cent dehydrated product. The receipe for preparing delicious pickle in oil has been standardized (Dhawan and Sharma, 1996).

Lasora

Lasora (*Cordia myrea* Roxb.) belong to the family Boraginaceae. In rural area it is considered important herbal fruit. The fruits are used medicinally as autheimatic, diuretic, demulcent and expectorant. Green mature fruits are utilized for making a good quality of the pickle. Ripe fruits may be used for the liquor. Lasora fruits contain a high percentage of the gum, but unfortunately no study has yet been made on this aspect (Singh, 2001).

Mahua

Mahua (*Bassia lotifolia* Roxb.) belongs to the family Sapotaceae. It is nature to both flowers and fruits are eaten by people in many ways. A great variability in the composition of the flowers, fruit of seed has been recorded. Four genotypes *viz.* NM–2,4,7 and 9 have been selected. Flowers may be utilized for making different food preparations, alcohols, bakery products, vinegar and syrup. Spent flower (after fermentation) may be used as a cattle feed. Fruit pulp and dry husk may be utilized for making the absolute alcohol (Singh, 2001).

Mulbery

Mulbery (*Morus indica* L) belongs to family moraceae. Fruits may be used for making the jam with its soft fruit and an excellent quality of the beverages such as squash and syrup.

Monkey Jack

Monkey jack (*Artocarpus lakoocha* Roxb.) fruit is a member of the family Moraceae. Mature fruits may be processed into pickle and jelly. Ripe fruits may be used for the preparation of ready-to-serve beverages and squash.

Phalsa

Phalsa (*Grewia subinaequalis*) plant is indigenous to India and is mentioned in the earliest vedic literature for its refreshing and medicinal properties. ARTS, beverages, squash and syrup from Phalsa fruit was formuled and standardized (Singh, 1997).

Pomegranate

Pomegranate (*Punica granatum*) is cultivated in almost all parts of India, but particularly in Maharashtra, Gujarat, Karnataka, U.P., A.P. and Tamil Nadu. Among the several varieties of pomegranate Badana (seed less) and Kandhari (hard seed) are considered best. Recently, a variety called "Ganesh" is being propagated on a large scale in the southern parts of Gujarat, Maharashtra and Northern parts of Karnataka. The grains from the fruit are fleshly, juicy slightly acidic sweet with pleasing pomegranate flavour. Pomegranate yields two crops during April-June and October-December.

The wild variety of pomegranate is extensively used for the manufacture of the product called 'Anardhana' which is obtained by sun drying the seeds along with pulp Anardana is a valuable commodity used in North-India as a condiment for acidifying chutneys and curries. Except for Anardana no other product from pomegranate is commercially produced. It is also reported to be used for tanning in iron. Recent work at C.F.T.R.I. has shown the possibility of extracting juice using a screw type juice extractor, which after removal of the tannins, makes a delicious beverage. Honey is very effective in precipitating the tannins and for clarifying the juice. A product locally prepared from the juice by adding the sugar and heating it to a thick viscous consistency called Anar Rub (Singh, 2001).

References

Anonymous (1971-72). Processing of Custard apple. *Annual Report.* Central Food Tech. Res. Inst., Mysore. pp. 19.

Dhawan, S.S. and Sharma, Suneel (1996). Potentiality of indigenous fruits of arid zone for processing. In: *Proceedings of Silver Jubilee National Symposium on Arid Horticulture* (Dec. 5-6, 1996, Hisar) organized by Horticultural Society of Haryana, CCS Haryana Agricultural University, Hisar. pp 306-309.

Khurdiya, D.S. and Roy, S.K. (1985). Processing of Jamun fruit into a ready-to-serve beverage. *J. Food Science Technol.,* 22: 27.

Ramanjaneya, K.H. (1985). Studies on some aspects of Jamun and its processing. *Ph.D. Thesis.* IARI, New Delhi.

Roy, S.K. and Singh, R.N. (1979). Studies on utilization of Bael fruit for processing. II Extraction of Bael fruit pulp. *Indian Food Packer,* 33 (1): 5.

Singh, I.S. (1997). Fruit beverage. *Res. Bull., NDUAT, Faizabad,* p. 3-14.

Singh, I.S. (2001). Minor fruits and their uses. In proceedings National Seminar on New horizons in postharvest management of tropical and sub-tropical fruits–Room Singh (Ed.) Horticultural Society of India. *Indian J. Hort.,* 58(1-2): 178-182.

24

Biochemistry of Fruit Ripening

Santosh Dhillon, Dharam Singh and Rakesh Kumar
Department of Biochemistry, CCS, Haryana Agricultural University, Hisar – 125 004, India

Fruits are an important source of vitamins and minerals in our diet. Besides their nutritional value, fruits also possess aesthetic qualities of flavour, colour and texture which are important from commercial or consumer preference point of view. All these characteristics of fruits are the result of several metabolic changes that occur during the ripening process. Colour changes are due to alterations in chlorophyll and carotenoid components of plastids or accumulation of anthocyanins in vacuoles while textural changes are brought about by alterations in cell wall metabolism leading to fruit softening that ultimately affect the shelf life of fruits. Alterations in metabolism of organic acids and generation of volatile compounds produce characteristic flavour and aroma. A thorough understanding of biochemical basis of ripening is, therefore, essential if improved quality, storage and processing of fruits are to be achieved. Such improvements by the application of genetic engineering have already started and scope for future advances is enormous (Smith *et al.*, 1988; Grierson, 1998). The various biochemical changes (Table 24.1) that take place during ripening of fruits have been reviewed by several authors (Brady, 1987; Tucker and Grierson, 1987; Tucker, 1993).

Table 24.1: Biochemical Change During Ripening in Fruits

Sl.No.	*Biochemical change*	*Process involved*
A.	Respiration and ethylene production	Increase in respiration and ethylene production, metabolism of carbohydrates and organic acids
B.	Flavour changes	Accumulation of sugars and organic acids, changes in lipids and amino acids, production of complex mixture of volatiles, phenolics
C.	Colour changes	Loss of chlorophyll, accumulation of carotenoids, synthesis of anthocyanins
D.	Texture changes	Alterations in composition of cell walls, degradation of pectins or cellulose, increased activities of cell wall hydrolases, degradation of starch
E.	Changes in gene expression	Disappearance of some mRNAs and appearance of ripening specific mRNAs, control by ethylene

Respiration and Ethylene Production

Fruits are divided into two broad categories, climacteric and non-climacteric based on whether or not a fruit exhibits a peak in respiration and ethylene production during ripening. The climacteric fruits display a characteristic peak of respiratory activity during ripening and are highly perishable. This category includes fruits such as banana, mango, apple, tomato, pear and avocado. In contrast, non-climacteric fruits such as lemon, orange, strawberry, grape and pineapple show a gradual decline in respiration during ripening. Climacteric and non-climacteric fruits may be further differentiated by their response to applied ethylene and the pattern of ethylene production during ripening. Ethylene biosynthetic pathway is now well known and thought to be highly regulated. Two key enzymes of the pathway are aminocyclopropane carboxylic acid synthase (ACC synthase) and aminocyclopropane carboxylic acid oxidase (ACC oxidase).Genes encoding both the enzymes have been cloned from a number of fruit crops. Antisense RNA technology has been used to specifically inactivate the expression of these genes in tomato. Transgenic tomato plants with inhibited ACC synthase or ACC oxidase gene expression showed severe reduction in ethylene production and had a prolonged shelf life. Similar results have been obtained in apple, melon and avocado (Grierson, 1998).

The major respiratory substrates in most fruits are sugars and organic acids which are found largely sequestered in vacuoles and contribute to overall flavour of the fruit. They are released in a controlled manner from the vacuole or alternately a separate pool is maintained for respiration. The respiratory pathways utilized by fruits for oxidation of sugars and organic acids are common to all plant tissues. These are glycolysis, oxidative pentose phosphate pathway, TCA cycle and malate oxidation. The activities of some key regulatory enzymes of these pathways increase during ripening in fruits indicating increased utilization of respiratory substrates (Tucker and Grierson, 1987).

Flavour Changes

The characteristic flavour of a fruit depends on the complex interaction of sugars, organic acids, phenolics and a wide range of volatiles. During ripening changes in the proportion of various components bring about flavour changes in fruits.

Carbohydrates

Sugars either free or combined with other constituents are important in attaining pleasant fruit flavour, color or texture. The carbohydrates accumulate during fruit development as starch, glucose and fructose, the relative proportions of each varying enormously between different fruits (Whiting, 1970). Thus some mature green fruits contain large amount of starch and these are exemplified by banana, apple and mango. Others contain relatively small amount of starch at the mature green stage, for example, tomato while fruits like pineapple, melon and grape contain very little or no starch and accumulate all the carbohydrate reserves as sucrose, glucose and fructose. As the fruit ripens, these forms undergo metabolic transformations both qualitatively and quantitatively. In general, the total level of sucrose, glucose and fructose increase during ripening either from further accumulation from the plant or in the case of starch containing fruit from breakdown of the stored carbohydrates. Mature green banana fruit contains 21 per cent starch and 2 per cent sugar. During ripening the level of starch declines until it reaches less than 1 per cent of the fruit fresh weight, at the same time the sugars accumulate to 20 per cent of the fresh weight. During early stage of the ripening process, sucrose is the predominant sugar while fructose and glucose predominate during later stages (Prabha and Bhagyalakshmi, 1998).

All the three starch degrading enzymes (starch phosphorylase, α- amylase and β-amylase) are found in ripening fruits and their activities increase with the progress of ripening process (Mattoo *et al.*, 1975; Prabha and Bhagyalakshmi, 1998). Sucrose synthetase is present in green banana fruit which increases during ripening and may be it is involved in breakdown of sucrose to glucose and fructose, however, invertase is thought to carry out this function in fruits (Mao and Kinsella, 1981).

Organic Acids

The non-volatile organic acids are among the major cellular constituents undergoing changes during ripening of fruits. In general, levels of organic acids decline during ripening due to their utilization as respiratory substrates (Ulrich, 1970). The main acids detected in mangoes are citric, malic and ascorbic acids. A considerable decrease in the acidity of the fruit occurs during ripening with a pH shift from 2.0 to 5.5. Levels of citric, malic and ascorbic acid decrease by ten, forty and 2.5 fold respectively in mangoes. Malic acid disappears first followed by citric acid suggesting possible catabolism of citrate via malate. In tomato, citric and malic acids account for 60 per cent of the total acid and the ratio of malic to citric decreases during maturation and ripening (Mattoo *et al.*, 1975). Citric, malic and oxalic are the main acids present in banana, levels of which increase during ripening.

Amino Acids and Proteins

There is a major turnover of amino acids in ripening fruits (Marriott, 1980). In mango, the amino acids alanine, tryptophan, isoleucine, valine and glycine increase progressively whereas lysine, proline and threonine are catabolised during ripening. At the climacteric stage, there is an increase of glutamic acid, glutamine, leucine and arginine, but at the onset of post-climacteric stage, the level of these amino acids decrease. Several fruits show a constant decline in amino acids during ripening with a drop in alanine, serine and glutamic acid. Of the nine free amino acids found in citrus, asparagine, glutamine and proline were predominant and the content of proline increases greatly during maturation. A high proline level is also found in oranges, lemons and mandarins. In banana, lysine and asparagine are the most abundant free amino acids in green stage, but their contents fall during ripening when the free form of histidine rises to become the predominant amino acid at full ripening and at the overipe stage. Isoleucine, leucine and valine, precursors of aroma compounds and serine also accumulate. Methionine, the precursor of ethylene has not been identified in the free form in banana. A small increase in proline content is observed during ripening of mangoes, avocados and tomatoes. A high turnover of amino acids and proteins occurs resulting in qualitative changes in soluble proteins.

Lipids

Lipids play an important role in maintaining texture, flavour and colour in fruits. The lipid content of most fruits is generally low (except avocado) and does not increase during ripening. However, there is considerable increase in the levels of total lipids and fatty acids in ripening mangoes. The main fatty acids present are palmitic, stearic, oleic, linoleic and linolenic. Unsaturated fatty acids increase more than saturated ones during mango ripening. These changes were found to correlate with aroma and flavour, a lower palmitate/ palmitoleate ratio indicates higher aroma (Bandyopadhyay and Gholap, 1973). In tomato, relatively large amount of neutral lipids (triglycerides, diglycerides, sterolesters, free fatty acids and hydrocarbons) are found in green stage. These decrease during colour development and increase to some extent at the full ripe stage. The percentage of linoleic acid and oleic acid decrease during color development. The phospholipids content appears to increase during early stage of maturation, but gradually decreases at the colour break (pink) stage and during storage. The percentage of galactolipid and chlorophyll markedly decrease at the beginning of colour development and finally disappear after the fruit color turns red (Mattoo *et al*, 1975).

Avocado fruit contains especially high content of lipids which can reach over 20 per cent of the fresh weight in some cultivars. The fat content increases during development of the fruit, changes during ripening are much less marked than those during growth and development. Increase in monoglycerides and free fatty acids are observed which coincide with flavour development in the fruit.

Volatile Compounds

A number of volatile compounds have been isolated from ripening fruits, which give characteristic aroma and flavour to the fruit. The flavour volatile profile of any fruit is usually very complex. For example, analysis of volatiles present in apple and orange indicates 230 and 330 different compounds, respectively (Van Straten, 1977). The nature of volatile is also very diverse and includes alcohols, aldehydes, esters and many other chemical groups (Nursten, 1970). The biosynthetic pathway of such a wide range of volatiles is also very diverse. The alcohols and aldehydes are derived from metabolism of their corresponding amino acids and oxosugars. Other flavour volatiles may be synthesized via mevalonate/isoprene pathway. Another possible pathway is via the action of lipoxygenase on membrane lipids following maceration of the tissue which has been implicated in the formation of distinctive flavour notes in tomato.

The quantity of aroma compounds emitted depends on the stage of maturity as well as environmental factors such as temperature and humidity. Keeping bananas for a longer time at 10°C reduces their overall production of volatiles while ethylene has a favourable effect on aroma. Except for some fruits such as banana in which major volatile compound is isoamylacetate, precise roles for any particular compound in the development of flavour and aroma of most fruits have not been identified.

Phenolics

Although sugars and organic acids constitute the major components in the fruit flavour, astringency to a certain extent is determined by phenolic compounds in fruits (Morton and Mc Cleod, 1990). They are derived from phenylpropanoids and include a large variety of compounds. Banana and plantain fruits contain high level of phenolic compounds (tannins) especially in the peel that are responsible for astringency in banana. As fruit ripens, their astringency becomes lower which is associated with a change in structure of tannins rather than reduction in their levels. Generally polymerization of phenolics take place. Phenolics are also responsible for oxidative browning which occurs upon slicing of the fruit pulp and is due to the reaction catalysed by polyphenoloxidases.

Declining levels of tannins and gallotannins are observed during ripening in several cultivars of mango (Selvaraj and Kumar 1989). Gallotannins are present in high quantities in unripe fruits and are toxic. The concentration of total phenols in grapes generally rises for a short time during ripening and then shows a steady decline.

Colour Changes

The external colour of the fruit is an important factor in consumer preference. Fruit ripening is associated with changes in colour which is due to changes in composition of pigments such as chlorophyll, carotenoids and anthocyanins.

Chlorophyll

For majority of the fruits, the first sign of ripening is the disappearance of green colour. The chlorophyll content of ripening fruits deceases slowly. Generally a small amount of green pigment remains in the internal portion of the fruit as has been observed in pears, apples and tomatoes (Mattoo *et al.*, 1975). The precise mechanism of chlorophyll degradation is unclear but may involve both enzymatic and chemical reactions. A maximum of chlorophyllase activity has been observed in apples and bananas at the time of climacteric and may be responsible for degradation of chlorophyll in fruits. However, no chlophyllase activity has been detected in ripening tomatoes while chlorophyll is rapidly decreasing.

Carotenoids

The carotenoid pigments in plants are chiefly β-carotene and its derivatives and are localized within the chloroplast which then becomes the chromoplast. The amount of carotenoids synthesized during ripening can be very large and often leads to formation of pigment crystals (Tucker and Grierson, 1987). This is especially true in tomato, where crystals of the predominant carotenoid pigment lycopene can be seen in the chromoplast. The principle carotenoids in mango fruit are β-carotene, xanthophyll esters and xanthophylls. The composition of carotenoids in mango changes during ripening with γ-carotene being most abundant in unripe fruits and phytofluene in ripe fruits. β-carotene is the predominant pigment present at all stages of ripening (Gomez-Lim, 1997).

Carotenoids are terpenoid compounds and are derived from Acetyl CoA via the mevalonic acid pathway. Phytoene is the primary carotenoid produced from this pathway which is further metabolized to give lycopene and β-carotene. Several tomato ripening mutants are known to be deficient, to varying degrees, in carotenoid biosynthesis during ripening (Rich, 1987) yet these mutants do not affect β-carotene synthesis in green tissue. Also a temperature of 30-35°C is known to inhibit lycopene synthesis while no effect is observed on β-carotene accumulation in tomato fruit (Goodwin, 1980) suggesting a ripening specific pathway for lycopene synthesis.

Anthocyanins

Anthocyanins are a very diverse range of pigments localized within the vacuole of the plant cell (Timberlake, 1981). They can give rise to colours from red to blue and often occur in a wide range of pigment types in individual fruit. For example, orange fruits are known to contain about 30 different anthocyanin type pigments. In apple, cherry, plum and blackberry the colour of the fruit is mainly due to anthocyanin pigment cyanidine-3-galactoside while peonodine-3-galactoside is the only anthocyanin identified in ripe mango. Anthocyanins are derived from flavonoid compounds and are synthesized from the aromatic amino acid phenylalanine.

Texture Changes

Ripening of many fruits is characterized by softening of the flesh and is a major quality attribute that often dictates shelf life of fruits. Fruit softening could arise from any of the three mechanisms *viz.* loss of turgor, degradation of starch or breakdown of the fruit cell walls. In general, change in texture during ripening of most fruits is thought to be largely the result of cell wall degradation.

Cell Wall Changes

Fruit cell walls are structurally very similar to primary cell walls from other organs except that they contain a larger proportion of middle lamella to primary wall and are pectin rich. Dissolution of pectin rich middle lamella of cell walls has been observed under electon microscope in ripening fruits of tomato, avocado and pear (Fischer and Bennett, 1991). In biochemical terms, major changes are observed in pectin polymers of the cell wall. During ripening there is a loss of neutral sugars, mainly galactose in most fruits but some loss of arabinose also occurs (Tucker and Grierson, 1987). There are also major changes in acidic pectin or rhamnogalacturonan fraction of the cell wall. As the fruit ripens, content of soluble pectates and pectinates increases while the total substances decrease. This trend is found in many fruits such as melon, bananas, citrus, strawberries and mangoes.With the change in pectin content, fruit firmness decreases. Two processes bring about this change during ripening, 1) depolymerization or shortening of chain length and 2) de-esterification or removal of methyl groups from polymer.

In tomatoes, depolymerization of pectic substances is indicated by a decrease in the viscosity of the sols as softening advances. The degree of esterification of the polyuronide fraction also changes during ripening; tomato polyuronide is about 75 per cent esterified in green fruit and this declines to about 55 per cent during ripening. Changes in cell wall components other than pectins are poorly understood. However, in tomato there is some evidence for depolymerization of hemicelluloses during ripening (Cheng and Huber, 1997).

Cell Wall Hydrolases

Several types of cell wall hydrolyzing enzymes are known to be present in fruits, the major being pectinesterase (PE), polygalacturonase (PG), β 1- 4 glucanase or cellulase and β-galactosidase (Huber,1983; Tucker and Grierson, 1987). Pectinesterase acts to remove methyl groups from C-6 position of a galacturonic acid. PG hydrolyses the α 1-4 link between adjacent demethylated galacturonic acid residues. Cellulase hydrolyses the β1-4 link between adjacent glucose residues while β-galactosidase in fruits acts on native galactan polymers. Activities of these enzymes have been observed to increase during ripening and contributes to softening in several fruits such as apple, banana, mango, pear and tomato. The increase in PG activity was found to be correlated with softening in tomato and avocado (Brady, 1987).However, in mango softening of the fruit was more closely related to β- galactosidase activity than to PG and PE activities (Ketsa *et al.*, 1998). Pectinmethylesterase (PME) activity though detected in various ripening fruits, except in tomato, little correlation is found between softening of fruits and PME activity (Fischer and Bennet, 1991). Increase in cellulase activity has been reported during ripening of fruits such as avocado, mango and banana.Many of the above enzymes exist as several isoforms within the fruit tissue and novel isoforms often appear during ripening (Tucker and Grierson, 1987). Other enzymes thought to be involved in fruit ripening are α -galactosidase, α and β mannosidases, mannanase, xylanase and β-xylosidase, however their roles are not clearly understood.

Control of Ripening

Ripening, like other plant senescent processes is under strict genetic control. Analysis of mRNA and protein synthesis during ripening of fruits such as avocado, melon and tomato has shown the synthesis of distinct ripening related proteins indicating that ripening, at least is partially controlled at the level of gene expression. There is evidence that plant hormone ethylene is involved in the control of these processes. This has been shown by transformation experiments in tomato where use of antisense cDNA for enzymes involved in ethylene biosynthesis such as ACC synthase, ACC oxidase and ACC deaminase resulted in delayed ripening of transgenic tomato fruits (Hamilton *et al.*, 1990).Studies with ripening mutants in tomato have demonstated that individual ripening events can be inhibited without any effect on other events (Rich, 1987). Thus ripening can be considered as a coordinated rather than as a linked set of biochemical pathways.

Conclusion

In conclusion, fruit ripening is a genetically programmed, highly coordinated set of biochemical pathways that provides optimum blend of colour, taste, flavour and texture to the edible fruit. These pathways are most likely to be under hormonal control for their initiation and coordination, however, much remains to be understood regarding complexity of the controls. With the availability of molecular biology techniques, it is now possible to turn off the expression of specific genes in transgenic plants permitting their identification and assessment of their function in ripening.Genetic transformation studies have already shown their potential in enhancing shelf life of fruits and thus have considerable commercial advantage. Such manipulation is currently restricted to tomato fruit.

It is hoped that in near future, more aspects of ripening will be understood and many other fruits may be manipulated to our advantage.

REFERENCES

Bandyopadhyay, C. and Gholap, A.S. (1973). Relationship of aroma and flavour characteristics of mango (*Magnifera indica* L.) to fatty acid composition. *J. Sci. Food Agric.*, 24: 1497-1503.

Brady, C.J. (1987). Fruit ripening. *Ann. Rev. Plant Physiol.*, 38: 155-178.

Cheng, G.W. and Huber, D.J. (1997). Carbohydrate solubilization of tomato locule tissue liquefaction during ripening. *Physiol. Plant.*, 101: 51-58.

Fischer, R.L. and Bennett, A.B. (1991). Role of cell wall hydrolases in fruit ripening. *Ann. Rev. Plant Physiol. Plant Mol. Biol.*, 42: 675-703.

Gomez-Lim, M.A. (1997). Postharvest physiology In: *The Mango: Botany, Production and Uses*, (Ed.) Richard E. Litz. CAB International, p. 425-445.

Goodwin, T.W. (1980). *The Biochemistry of the Carotenoids. Vol 1: Plants*, 2nd Edition. Chapman and Hall.

Grierson, D. (1998). Manipulation of fruit ripening by genetic modification In: *Transgenic Plant Research*, (Ed.) Lindsey, K. Harwood Academic Publishers, p. 109-124.

Huber, D.J. (1983). The role of cell wall hydrolases in fruit softening. *Hort. Rev.*, 5: 169-219.

Ketsa, S., Chidtragoal, S., Klein, J.D. and Lurie, S. (1998). Effect of heat treatment on changes in softening, pectic substances and activities of polygalacturonase, pectinesterase and β-galactosidase of ripening mango. *J. Plant Physiol.*, 153: 457-461.

Mao, W.W. and Kinsella, J.E. (1981). Amylase activity in banana fruit: properties and changes in activity during ripening. *J. Food Sci.*, 46: 1400-1409.

Marriott, J. (1980). Physiology and biochemistry of storage and ripening for optimum quality. In: *Bananas and Plantains*. CRC Crit Rev. Food Sci. Nutr., 13: 41-48.

Mattoo, A.K., Murata, T., Pantastico, E.B., Chachin, K., Ogata, K. and Phan, C.T. (1975). Chemical changes during ripening and senescence In: *Postharvest Physiology, Handling and Utilization of Tropical and Subtropical Fruits and Vegetables*, (Ed.) Pantastico, E.B. AVI Publishing, Westport, p. 130-156.

Morton, I.D. and McCleod, A.J. (1990). *Food Flavours, Part C: The Flavour of Fruits*. Elsevier.

Nursten, H.E. (1970). Organic acids. In: *The Biochemistry of Fruit and their Products*, (Ed.) Hulme, A., Vol. 2, Acadmic Press, p. 89-113.

Prabha, T.N. and Bhagyalakshmi, L. (1998). Carbohydrate metabolism in ripening banana fruit. *Phytochem.*, 48: 915-919.

Rick, C.M. (1987. Genetic resourse in Lycopersicon. In: *Tomato Biotechnology*, (Ed.s) Nevins, D.J. and Jones, R.A. Alan Liss, New York, p. 17-26.

Selvaraj, Y. and Kumar, R. (1989). Studies on fruit softening enzymes and polyphenoloxidase activity in ripening mango (*Mannifera indica* L.). *J. Food Sci. Technol.*, 26: 218-222.

Smith, C.J.S., Watson, C.F. and Ray, J. (1988). Antisense RNA inhibition of polygalacturonase gene expression in transgenic tomatoes. *Nature*, 334: 724-726.

Tucker, G.A. (1993). Introduction. In: *Biochemistry of Fruit Ripening*, (Eds.) Seymour, G.B., Taylor, J.E. and Tucker, G.A. Chapman and Hall, London, p. 1-51.

Tucker, G.A. and Grieersonm, D. (1987). Fruit ripening. In: *The Biochemistry of Plants: A Comprehensive Treatise*, (Ed.) Davies, D.D. Vol 12, Academic Press, p. 265–318.

Ulrich, R. (1970). Organic acid. In: *The Biochemistry of Fruits and their Products*, (Ed.) Hulme, A. Vol 1, Academic Press, p. 89-118.

Van straten, S. (1977). *Volatile Compounds in Food*. TNO. Zeist

Whiting, G.C. (1970). Sugars. In: *The Biochemistry of Plants: A Comprehensive Treatise*, (Ed.) Davies, D.D. Vol 12, Academic Press, p. 1-32.

25

Postharvest Management of Vegetable Crops

J.L. Mangal and S. Lal

Department of Vegetable Crops, CCS Haryana Agricultural University, Hisar (Haryana), India

Vegetables occupy an important place in diversification of agriculture and have played a vital role in food and nutritional security of ever growing population of our country. Vegetables are very valuable for adding quality component to the food as they are rich source of minerals, vitamins and other nutrients. Besides that they act as antioxidant which protect, by destroying free radicles produced in our body from heart ailments.

India with variable soil and climatic conditions has excellent potential for growing both tropical, sub-tropical and temperate vegetables. More than 70 types of important vegetables are grown in our country besides a number of minor vegetables growing in different parts of the country (Singh, 2001). Since independence, India has emerged as the second largest producer of vegetables in the world with a total production of 84 million tones from 6.2 million hectare having 2.6 per cent annual growth. However, per capita availability of vegetable is far below the requirement of 285 g/capita/ day for a balanced diet. The vegetable requirement of our country is estimated at 100 million tones by the turn of the century and 225 million tones by 2020 (Paroda, 2000).

Vegetables are also being exported to other countries. The Agricultural and Processed Food Products Export Development Authority (APEDA) has identified traditional vegetables like onion, potato, okra, bitter gaurd chilli and non-traditional vegetables like asparagus, cavalry, stinect, pepper, sweet corn, baby corn, green peas, French bean, cucumber and ghertcinand cherry, tomato having good export potential. Vegetables are also being exported in processed form. According to APEDA the country exported 498861.69 MT fresh vegetables valued for Rs.34115.61 lakhs and processed vegetables 88675.54 MT valued for Rs.23107.54 lakhs during 1996-97.

Only less than 2 per cent of the produce is used for processing compared to 70 per cent in Brazil, 65 per cent in USA, 83 per cent in Malayasia and 50 per cent in Israel. There are great losses in different vegetables after harvesting because of their highly perishable nature. The postharvest losses in different vegetables varies to great extent (08-40 per cent) depending upon the various factors such as improper harvesting, sorting, grading, packaging, and lack of transportation and storage facilities to enhance the shelf-life of vegetables (Table 25.1). Assuming an average loss of about 25 per cent, there is a loss of Rs.8000 crores per annum to our country (Annon 1998).

There is great demand of vegetables to meet the requirement of ever increasing population as well as requirement of processing industries and to earn more foreign exchange. To meet these challenges the productivity and production of vegetables as well as reducing postharvest losses in vegetable crops are prime need of present days. A lot of emphasis is being given to increase the productivity and production of vegetables but not much emphasis is being given to reduce postharvest losses in vegetable crops. Therefore, more emphasis need to be given to postharvest management of vegetable crops. Postharvest losses occur in postharvest chain of harvesting, sorting, grading, packaging, transportation, storage, and marketing. Hence, more attention should be given towards the postharvest management of vegetables.

Table 25.1: Postharvest Losses of Vegetables (mean range of all the regions and seasons)

Vegetables	*Postharvest Losses (%)*	*Remarks*
Onion	6.01-40.0	Storage and transportation
Potato	30.0-40.0	Harvesting and storage
Garlic	0.9-2.7	Storage
Tomato	6.7-33.5	Transportation and Marketing
Brinjal	3.7-13.4	Marketing
Cabbage	3.0-15.0	Marketing
Cauliflower	10.0-13.2	Transportation and Marketing
Beans	7.45	—
Smooth gourd	8.0-15.0	—
Bottle gourd	7.0-10.0	Transportation and Marketing
Okra	5.0-10.0	Marketing
Chilli	4.0-35.0	Farm and storage
Beet root	10.0-15.0	Marketing
Radish	3.0-5.0	Marketing
Carrot	5.9	–
Pointed gourd	19.81	Marketing storage
Garlic	0.9-2.7	Marketing storage

Causes of Postharvest Losses

Improper Postharvest Operations

The postharvest losses may be due to the practices currently used for that postharvest handling, packaging, transportation and storage are improper and outdated that contribute directly to the postharvest losses.

Further, loss of water, high respiration rate and metabolism and microbial spoilage limit the shelf-life these highly perishable commodities.

Water Loss

Most of vegetables contain 80-90 per cent water and as little as 5 per cent loss in water causes to appear wilted or shrivelled and eventually unmarketable. Moisture loss adversely affects the appearance, texture, flavour and weight of commodities. Most noticeable effect of moisture loss is the softening of tissues caused by loss of turgidity.

Respiration and Metabolism

The continuous high rate of respiration leads to depletion of stored food and consequently the loss of quality. The rate of respiration is a good index of the postharvest life of vegetables. The low respiring commodities have a long shelf-life *e.g.* potato and onion.

Control of Postharvest Losses

To achieve our target of feeding population as well as machine the requirements of the processing industries and export trade, increasing production and productivity will not be enough. A lot of emphasis has to be given to postharvest management of these highly perishable crops. We shall have to adopt the latest technology for reducing postharvesting losses in vegetables. In order to minimize the loss rate, all efforts are in two directions:

1. To strengthen the postharvest facilities of the country
2. To develop technological needs

Strengthen the Postharvest Facilities

1. Packaging Station
2. Primary Processing
3. On-Farm Storage
4. Polletisation
5. Containerization and Transport
6. Packaging and Pre-cooling
7. Cool Store
8. Establishment of Cold/Cool Chain System
9. Irradiation
10. Control Atmosphere/Modified Atmosphere Storage
11. Cooperative System of Marketing
12. Innovations in Organized Marketing
13. Rural, Urban and Major Terminal Markets

14. Price Support Operations and Marketing Information Service
15. Export Promotion
16. Developing Processing Industries/Plants.

Development of Technological Needs

1. Suitable Cultivars with Longer Shelf-life.
2. Cultural Practices
3. Harvesting and Curing
4. Sorting and Grading
5. Protective Treatments
6. Packaging
7. Transportation
8. Storage
9. Marketing
10. Processing

26

Role of Biotechnology in Enhancing Shelf-Life of Fruits

Randhir Singh
Department of Biochemistry, Chaudhary Charan Singh Agricultural University, Hisar – 125 004, Haryana

Fruits contain a very high percentage of their fresh weight as water. Consequently, fruit exhibit relatively high metabolic activity when compared to other plant derived foods such as seeds. This metabolic activity continues postharvesting and thus makes most fruit highly perishable commodities. It is this perishability and inherent short shelf life that presents the greatest problem to the successful transportation and marketing of fresh fruits. For the major commercial crops the post harvest problems have been largely solved by harvesting at the immature or mature green stage, and/or by using refrigeration and controlled atmosphere storage. However, many other fruits cannot be handled successfully using these methods, and this precludes the maximum commercial exploitation of these fruits. For instance, mangoes harvested at full maturity do not store well, but if harvested immature, fail to ripen properly. Solutions to post harvest problems of fruits may come from a better understanding of the biochemistry and molecular biology of fruit ripening.

The advent of molecular biology techniques has provided a powerful tool to probe the biochemistry of fruit ripening. These techniques are being used to investigate the role of hydrolases in cell wall degradation and hence fruit softening. They can also be applied to investigations of other ripening events such as ethylene biosynthesis, pigment production and starch degradation etc. Accordingly, it is possible now to manipulate ripening, which could have sufficient commercial advantages. Such manipulation is currently restricted to tomato fruits where fruit softening has been influenced by the down regulation of the polygalacturonase gene. Ethylene synthesis can also be manipulated and this again has a significant effect on the fruit's shelf life. The ability to apply this technology to various fruits is expanding as the number of fruit bearing plants that can be transformed and regenerated is increasing steadily. The results of some of these studies are described here to indicate how biotechnology could help us to increase the shelf life of fruits.

Genetic Modification of Fruits

Almost all of the work on genetic modification of fruits has been carried out on tomato. This is because it has many scientific advantages for studying the physiological and biochemical changes that occur during ripening. Moreover, it is easy to transform tomato genetically. The tomato is also an important crop, now second only to citrus as a fruit crop. The history of cloning of tomato ripening genes has been reviewed well by Gray *et al.* (1992, 1994). Between 20 and 30 genes have been identified so far. Only those genes that have been shown to be involved critically in ripening changes or have featured in transgenic plant experiments will be discussed here.

Ethylene Biosynthesis Genes

The tomato, banana and apple, are climacteric fruit. This means that ripening is associated with a burst of respiratory activity at the onset of ripening. All climacteric fruits synthesize ethylene at the start of ripening and the gas functions as a hormone to initiate and co-ordinate expression of ripening genes and the development of quality attributes. Briefly, there are two enzymatic steps dedicated to ethylene production. In the first reaction, S-adenosyl methionine is converted to the cyclic amino acid l-amino cyclopropane-l-carboxylate (ACC), by ACC synthase (ACS). The ACC is then metabolized by ACC oxidase (ACO), which requires O_2, Fe^{2+} and is activated by CO_2, to produce ethylene. During ripening, expression of ACS and ACO genes and the activity of their encoded enzymes governs the rate of ethylene production. Each enzyme is encoded by multigene family in tomato, and there are at least 9 ACS and at least 3 ACO genes. The ACO1 and ACS, genes are specifically expressed during ripening and increases in gene transcription are associated with onset of ethylene production. The process of ethylene production is autocatalytic, *i.e.* once ethylene production begins, it stimulates expression of a range of genes involved in ripening. Included among these are ACO and ACS and their activation thus results in a surge of ethylene production that promotes rapid and near synchronous ripening' throughout the fruit. For those interested in fruit storage, it is important to prevent ethylene being formed, or to remove it quickly once it begins to build up in the atmosphere surrounding stored fruits, otherwise they will ripen rapidly and deteriorate. Conversely, it is possible effectively to stimulate ripening of some unripe fruits (such as banana and tomato) by treating them with ethylene gas, or supplying compounds such as ACC or ethephon, that lead to ethylene production *in planta.*

Genes Involved in Ethylene Perception and Signal Transduction

The ethylene receptor was first identified in *Arabidopsis.* A gene (ETR) encoding a dominant mutation rendering seedlings insensitive to the inhibition of growth by ethylene was obtained by map-based cloning. DNA sequencing showed the gene encoded a membrane-spanning protein with homology to receptor proteins from other organisms that contain a characteristic histidine kinase domain. Subsequently, a second receptor gene (ERS) was identified, which has homology to ETR, but lacks part of response domain. A second mutation in a gene called CTR, which encodes a protein with homology to Raf kinase, renders seedlings incapable of responding to ethylene. The rust tomato ethylene receptor was identified by cloning and sequencing the tomato "Never-ripe" (NR) gene from a ripening mutant that has reduced sensitivity to ethylene and thus reduced ripening. The polypeptides encoded by NR gene has 70 per cent homology to ETR. There are known to be at least five genes which hybridize to tomato ETR and it appears that at least two of these are expressed during ripening. There are indications that the ethylene signal transduction pathway during ripening is similar to that found in *Arabidopsis.*

Genes Encoding Enzymes Involved in Tomato Cell-wall Metabolism

The first ripening cDNA to be cloned and sequenced, originally called TOM 6, encoded polygalacturonase (PG). This is a cell wall modifying enzyme that catalyzes the hydrolysis of polygalacturonic acid chains in unmethylated regions of pectin. There are at least two different PGs (endo- and exo-acting), which are synthesized in a variety of locations including pollen, abscission zones and ripening fruit. There are three isofom1s of this enzyme termed as PG-1, PG-2A and PG-2B. The PG gene (there appears to be one gene) is transcriptionally activated at the onset of ripening and the mRNA and proteins are extremely abundant in ripening fruit.

Pectinesterase (PE) is a second cell wall modifying enzyme that removes the methyl groups from methylated polygalacturonic acid in pectin. Its action may be a pre-requisite for PG attack. There are several isoforms of PE and these are now known to be encoded by at least two distinct multigene families. The cDNA and genomic sequences encoding two PE types do not cross hybridize under standard conditions. One group of PE isoforms is expressed in unripe and ripe fruits, and the members of the second group are expressed in roots, stems, leaves and unripe fruit. In addition to PE and PG, other cell wall modifying enzymes encoded by multigene families also occur in tomatoes, including cellulase and galactanase, but the precise role of each during ripening remains to be established.

Carotenoid Biosynthesis Genes

The gene for tomato phytoene synthase (PSY) was the first gene for a carotenoid biosynthetic enzyme to be isolated from plants. The original cDNA was named as TOM 5 and the gene expressed in ripening fruit is called Psy1. There is at least one another related gene (Psy2) expressed in green organs and flowers which is responsible for chloroplast carotenoid production prior to ripening. The concentration of Psyl mRNA increases dramatically during ripening as the chloroplasts are transformed into chromoplasts.

There is a major increase in Psyl mRNA at the onset of respiratory climacteric in tomato and it seems that the level of PSY protein is the main factor in determining lycopene production in ripening fruits.

Other Ripening Related Genes

Several other cDNAs encoding mRNAs that increase in abundance during tomato ripening have been characterized. These include glutamate decarboxylase, histidine decarboxylase, an integral membrane channel protein, and a polypeptide with homology to a low molecular weight heat-shock protein. The precise role of these proteins is still not understood.

Ripening Genes from Other Fruits

Ripening genes have been cloned from several different climacteric fruits including ACO from melon and avocado, cellulase from avocado, PG and a Thaumatin like protein from avocado and PSY from melon. A novel protein called fibrillin has been cloned from pepper which appears to be associated with the carotenoids deposited in the chromoplasts of the ripening pepper fruits. A similar protein has been detected in tomato and it is possible that this represents a new class of proteins required for carotenoid deposition as fruit chloroplasts are transformed into chromoplasts. In addition, cDNA cloning for ripening genes from other climacteric fruits, such as banana, mango, papaya and star-fruit, is in progress. Less attention has been paid to the non-climacteric fruits, in which ethylene seems to take little or no part in ripening. Major studies on ripening related cDNA clones from strawberry, however, are in progress and promise to open up this area of research.

Ripening Gene Promoter Studies

Four fruit gene promoters have been used in transgenic tomato experiments, in addition to CaMV 35S promoter. These are the 2All fruit-specific promoter, the E8 ethylene-regulated ripening promoter, the ACO1 promoter, and the PG gene promoter.

Genetic Modification of Ripening in Transgenic Fruits

The processes of texture change, hormone synthesis and colour production have all been manipulated genetically and several commercial products based on these modifications have already reached the market place. These modifications have been generated mainly by gene silencing, a process which involves the inactivation of a specific gene in a targeted manner. The plant gene could be inactivated either by antisense technology or by sense-suppression or co-suppression technology. There are at least two types of sense suppression, both involving the transfer of additional copies of sense genes to transgenic plants. Inhibition of gene activity may occur by transcriptional inactivation, if there is DNA sequence homology between the promoters of the transgene and the endogenous gene, or post-transcriptional RNA degradation, if there is sequence homology between the two transcripts.

Manipulation of PG Gene Expression

The inhibition of PG gene expression was first achieved in transgenic plants using antisense gene driven by a constitutive promoter. PG antisense plants had reduced levels of PG mRNA and enzyme activity. However, inhibition of PG activity had no effect on other ripening attributes such as colour change and ethylene synthesis. Sense genes could also be used to achieve this goal. In an initial study, a 730 bp fragment, representing one half of the PG cDNA was transferred to plants under the control of CaMV 35S promoter. The transgenic tomato plants generated a truncated PG mRNA which, in fruits, inhibited the accumulation of normal full length PG mRNA and caused a major reduction in PG enzyme activity. The phenotype of these low PG fruits was identical to that of PG antisense fruits. There are quite major effects on texture change in low PG fruits when they ripen. They are much more resistant to cracking and splitting than controls, and they are better able to withstand damage from handling and transport. At the biochemical level, transgenic fruits in which PG was reduced to less than 1 per cent of normal were shown to have longer polygalacturonic acid chains compared to controls. This retention of long chain polygalacturonic acid can influence the apparent firmness of fruit and also affects cell adhesion and makes the fruit more robust. This means that they can be left to ripen on the plant for longer, because the risk of subsequent decrease in storage life is reduced. Most importantly, longer chain polygalacturonic acid has a major effect on the consistency (viscosity) of fruit extracts, which is beneficial for processing. The main conclusions to be drawn from these transgenic plant experiments are:

1. PG does playa significant role in cell wall metabolism and texture change during ripening.
2. It is not the sole or even the major determinant of softening.
3. Other cell wall metabolizing enzymes such as cellulases or galactanases are likely to be involved in softening.
4. Low PG fruits offer some benefit for both fresh market and processed tomatoes.

Inhibition of Pectin Esterase Expression

There are three closely related genes encoding the major form of pectin esterase expressed in tomato fruit. When their expression is inhibited by antisense gene silencing in transgenic plants, the production of PE is inhibited by approximately 90 per cent in ripe fruits. The degree of pectin esterification is enhanced but there is no significant effect on softening of fruit. The reason why fruit PE activity is only inhibited by about 90 per cent is that there is a second form of PE encoded by a distinct set of genes with a significantly different DNA sequence. These are not susceptible to gene silencing by the transgene used presumably because the DNA sequence homology is too low. The expression of this second set PE genes accounts for 10 per cent of the activity in fruit, but is responsible for all of the PE activity in leaves and roots.

Regulation of Ethylene Synthesis and Storage Life

Inactivation of genes involved in ethylene synthesis was first achieved in tomato. Two small multi gene families encoding ACC synthase (ACS) and ACC oxidase (ACO) control the biosynthesis of ethylene. These genes can effectively be down-regulated with antisense or sense genes, and fruit, vegetables and flowers can be produced with different level of ethylene. Almost complete inhibition of ACS and, therefore, ethylene production using antisense genes inhibited the change in colour and texture of tomatoes. Inhibition of ACS has also been achieved in tomato using sense genes. Inhibiting ACO by 95 per cent permitted normal development of ripening attributes of fruits attached to the plant, but prevented the extreme softening, cracking and spoilage normally associated with over ripening.

Another way of reducing ethylene production is to overexpress foreign genes for S-adenosyl methionine hydrolase (SAMase) or ACC deaminase, which deplete plant cells of the required precursors. Although a complete inhibition of ethylene was not achieved, delayed ripening characteristics in these transgenic tomatoes were observed. In an alternate strategy, a gene promoter based on elements from the ethylene regulated E8 tomato gene promoter was used to drive expression of the bacterial SAMase gene in transgenic cherry tomatoes. The reduction in ethylene production led to protracted ripening of the fruits, and prevented over-ripening.

The first successful demonstration that the ripening behaviour of a fruit other than tomato can be genetically modified was carried out with melon by Ayub *et al.* (1996), who inhibited ACO gene expression using antisense techniques. The melon ACC oxidase gene was transferred as an antisense construct to melon using an agrobacterium transformation vector. ACC oxidase activity in fruit was virtually undetectable, and ethylene production was reduced below 1 per cent of the control level. After 10 days storage at 25°C, the low ethylene fruits were still green and retained their shape, whereas the control fruits had a shrivelled yellow rind, showing signs of fungal infection, with soft flesh and squashed shape.

Manipulation of Carotenoid Levels

The first committed step in the biosynthesis of carotenoids is catalyzed by phytoene synthase, located in the chromoplasts. Bird *et al.* (1991) produced transgenic tomato by using an antisense gene, which inhibited the expression of cDNA clone (TOM 5) for phytoene synthase. The transgenic fruits with low psy 1 were yellow, rather than red, due to the lack of β-carotene and lycopene. In other experiments, psy 1 has been over-expressed in wild type tomatoes under control of35 S promoter. However, it did not result in much increase in carotenoids. During the ripening of bell pepper, a 32 kDa protein called fibrillin accumulates in the chromoplasts. It has been proposed that fibrillin surrounds the carotenoids. A similar protein and mRNA have been reported in tomato fruit, and it is possible that fibrillin-like proteins are intimately associated with carotenoid accumulation. If this is the case, it may be advantageous to over-express the tomato equivalent of fibrllin, together with psy 1, in order to increase fruit carotenoid content.

Commercial Application of Transgenic Fruits

The first genetically modified plant food to enter the market was the "Flavr Savr" tomato. This is low PG fresh market tomato, first sold by Calgene in 1994 in the USA. The low PG trait was produced by an antisense PG gene. The claimed benefits relate to the alterations in texture caused by low PG, which enables the tomatoes to be left to ripen longer on the plant before harvesting, without the attendant risk of subsequent rapid over-ripening and deterioration. For tomato processors, the texture change reduces waste and leads to an increase in viscosity of the processed products. There is great potential for improving storage life and preventing deterioration of fruits by reducing ethylene production genetically. A low ethylene variety produced by sense-suppression of ACS was marketed for a short time in the USA in 1995. Reduction in ethylene is likely to be beneficial for a wide range of climacteric fruits such as apple, pear, banana, peach, papaya, mango, tomato and melon. Several important issues remain to be resolved concerning the precise levels of ethylene to be aimed for in particular crops. If fruit remain attached to the plant for longer, they may accumulate more sugars or organic acids, which would improve the flavour. If the ethylene production is too low, however, it is possible that progression of specific facets of ripening, such as the production of flavour volatiles, might be impaired. This suggests that a significant amount of fine-tuning of ethylene levels will be required, in order to get right balance of ethylene, either produced in plant, or supplied after picking, in order to maximize beneficial ripening changes, while capitalising on the ability to reduce the deleterious aspects of ripening.

References

Grierson, D. (1998). Manipulation of fruit ripening by genetic modification. In: *Transgenic Plant Research,* (Ed.) K. Lindsey. Harwood Academic Publishers, p. 109-124.

Grierson, D. and Schuch, W. (1993). Control of ripening. *Physiological Transactions of the Royal Society of London,* 342: 241-250.

Grierson, D., Hamilton, A.J., Bouzayen, M., Kock, M., Lycett, G.W. and Barton, S. (1992). Regulation of gene expression, ethylene synthesis and ripening in transgenic tomatoes. In: *Inducible Plant Proteins,* (Ed.) J.L. Wray. Cambridge University Press, Cambridge, p. 155-174.

Howie, W., Lee, K., Joe, L., Bade, C., McGugin, C. and Bedbrook, J. (1996). Suppression of ACC synthase. Commercialization of ripening-impaired fresh market in tomatoes. In: *Mechanisms and Applications of Gene Silencing,* (Eds.) D. Grierson, G.W. Lycett and G.A. Tucker. Nottingham University Press, Nottingham, p. 119-125.

Klee, H.J. and Tieman, D. (1997). Potential applications of controlling ethylene synthesis and perception in transgenic plants. In: *Biology and Biotechnology of Plant Hormone Ethylene,* (Eds.) A.K. Kanellis, C. Chang, H. Kende, D. Grierson. Kluwer Academic Publishers, Dordrecht.

Kramer, M.G., Kellogg, J., Wagoner, W., Matumura, W., Good, X., Peters, S., Clough, G. and Bertwick, R.K. (1997). Reduced ethylene synthesis and ripening control in tomatoes expressing S-adenosyl methionine hydrolase. In: *Biology and Biotechnology of Plant Hormone Ethylene,* (Eds.) A.K. Kanellis, C. Chang, H. Kende, D. Grierson. Kluwer Academic Publishers. Dordrecht.

27

Freeze-drying Technique for Food Preservation

Saleem Siddiqui

Department of Horticulture, Chaudhary Charan Singh Agricultural University, Hisar – 125 004, Haryana

The freeze-drying process consists of removing moisture from foods in the frozen state by sublimation under high vacuum. The low temperature used in the process inhibits undesirable chemical and biochemical reactions and minimizes loss of volatile aromatic compounds. The dried product is light in weight and can be stored in airtight containers for long periods without refrigeration. Freeze drying causes very little shrinkage of the product, and consequently allowed nearly complete rehydration. Because of the absence of free liquid water during freeze drying, there is no migration of dissolved solutes carried by capillary liquid flow and no spattering or undesirable frothing resulting from liquid entrainment in the water vapor.

The important steps in making freeze-dried products include raw material selection, grading, size separation, washing, peeling or removal of undesirable parts, sorting, slicing, blanching, freezing, freeze drying, inspection, screening, packaging and marketing.

Freezing

In the manufacture of freeze-dried products, the raw material is pretreated and frozen before drying. There are two methods of freezing namely, pre-freezing and self-freezing. In each case, the moisture in the raw material should be converted to very fine ice crystals. Freezing results in a change in elasticity and moisture-keeping ability of the raw material. If the raw material is frozen at a temperature higher than 14°F (-10°C), ice crystals will be formed slowly, and the crystals will be coarse and large. The coarse ice crystals will damage the food texture. In addition, denaturation of the cell membrane or protein components may occur, and enzymatic reactions may continue to take place until the raw material is completely frozen. When quick freezing process is used, these liquids solidify to form extremely small crystals of ice and the cellular structure is left intact.

Self-Freezing

The majority of blanched vegetables can be frozen by direct evaporation of moisture from their surfaces. In this process, the latent heat of vaporization is drawn from the material, causing a decrease in temperature. The reduction in pressure must be sufficiently rapid to accomplish self-freezing. Deformation and foaming, which may occur during evaporation, may be minimized by adjusting the vacuum at a fixed level. The process cannot be adapted to products requiring very good shape and appearance. It can only be utilized in freeze-drying powdered products.

Pre-freezing

Another process for freezing the raw material before sublimation is pre-freezing. Such freezing can be accomplished by (*a*) cold-air blast, (*b*) immersion in cold brine, (*c*) contact with cold metal plates and (*d*) quick freezing by spraying with supercooled liquid nitrogen or Freon or dipping in the refrigerants. For better texture in the final products, the quick-freezing process is preferred. In general, fruits and vegetables start to form ice crystals at a relatively high temperature and can be prefrozen easily.

Drying

After pretreatment and freezing, the raw materials are sent to a freeze drier. The temperature is maintained below the triple point of the constituent aqueous solution so that water vapor can be sublimated from the frozen solution. There is a direct transfer from the solid state to a vapor without passing through the liquid phase. The heat required to sublimate a given quantity of ice at any temperature is equivalent to the heat of fusion of ice and the heat of vaporization of water plus the heat necessary to raise the temperature of the ice to its melting point. The quantity of heat required is the same whether the process is carried out slowly at ordinary pressures or rapidly under a high vacuum. Since the vapor pressure of ice increases with temperature, the higher the temperature is, the faster the drying process and the lower the cost will be. Obviously, the food tissue should not reach 0°C until nearly all the water has been removed, because ice melts at this temperature.

It is necessary to remove the vapor evolved when drying at very low pressure. This can be done either by condensation, pumping, or absorption with a dessicant. In order to condense the water vapor, the temperature of the condensing medium must be below that of the frozen product being dried.

The weight load on a unit area of drying tray is an important factor influencing the drying time in a freeze drier. Because the raw material is dried from the outer to the inner layer, the thicker the raw material is, the longer it takes to complete the process. For this reason, raw materials are usually sliced to 0.05-0.75 in (6-19 mm) in thickness.

Freeze drying is generally conducted at the lowest possible pressure (high vacuum). According to mass-transfer theory, a micro particle of water vapor may accelerate heat conduction in the chamber during the process of drying. However, during the drying process, attention must be paid to control the vacuum and prevent the frozen material from melting. Even partial melting will cause a change in the quality of the dried product. As drying progresses, the amount of vapor coming out decreases, and the cooling by latent heat decreases. The temperature of the product surface in contact with the heating surface should be controlled so that it will not exceed the limit. These factors influence of end point of freeze drying.

During drying, attention must be paid to the heat supply and the temperature, which will affect product quality during storage. Immediately after drying, products freeze dried under normal conditions and those freeze dried at temperatures higher than the limiting temperature all look alike, but they exhibit changes as time goes by. Products with a high fat content dried at higher temperature will exhibit more severe oxidation than those dried at lower temperature.

Freeze-dried vegetables are usually dried to 2 or 3 per cent moisture; thus they absorb moisture when exposed to air. After drying is complete, if the vacuum in the chamber is broken with humid air, the product will absorb moisture and become a less desirable product for long term storage. It is advisable to break the vacuum of the chamber with nitrogen and to package freeze dried products under nitrogen gas.

Freeze Drying of Foamed Material

Foaming of liquid foods before or during freezing prior to freeze-drying has been recommended. Foaming before freeze-drying is a means of controlling product density and also gives a higher internal mass transfer coefficient during freeze-drying. The higher mass transfer coefficient may make it possible and may increase the drying rate unless the insulating properties of the dried foam layer become an important heat transfer rate limit.

Additives to Improve Freeze-Drying

Liquid foods rich in fructose and, to a lesser extent, glucose are particularly difficult to freeze dry because of the tendency of the material to collapse into a viscous or glassy state. Thus, apple juice, grape juice, and honey are highly resistant to freeze-drying by ordinary methods. For freeze-drying adding substantial quantities of either lactose or corn syrup solids to fruit juices rich in fructose or invert sugar is recommended so as to prevent loss of structure. The addition of certain calcium compounds was also useful.

Properties of Freeze-Dried food

1. Freeze-dried foods are porous and low in moisture, usually below 3 per cent. The raw material and the freeze-dried product differ greatly in volume weight and specific gravity. In general, freeze dried products are lighter in volume/weight.
2. One of the major benefits of freeze-drying for many foodstuffs is the relatively good retention of volatile flavor

and aroma components with the process. Other methods of dehydration are as a rule inferior to freeze-drying in this respect.

3. Freeze drying causes little change in the chemical components of food. There is very little difference in the nutritive value of a fresh fruit and its freeze-dried product.
4. Freeze-dried foods are easily reconstituted, and the quality of the reconstituted products is excellent.

REFERENCES

Judson, King C. (1971). *Freeze Drying of Foods*. CRC Press, London.

Pruthi, J.S. (1999). *Quick Freezing Preservation of Foods: Principles, Practices and R&D Needs*. Allied Publishers. New Delhi.

Rothmayer, W.W. (1975). Basic knowledge of freeze drying. In. *Freeze Drying and Advanced Food Technology*, (Eds.) S.A. Goldblith, L. Rey and W.W. Rothmayer. Academic Press, London.

28

Food Packaging

M.K.Garg

Department of Agricultural Processing and Energy,
College of Agricultural Engineering and Technology,
CCS HAU, Hisar – 125 004

Introduction

Packaging plays a vital role in the conservation, preservation and transportation of food products. Packaging is considered as an index of civilization. As the country progresses, the use of packaged products increased considerably. Packaged products create confidence with the consumer with respect to quality, quantity, safety and protection. There is the growing sophistication and changing market trends, indicating the requirements of convenience, use of materials, which are eco-friendly, recyclable and reusable.

With the advancement in preservation technologies coupled with introduction of suitable packaging materials and systems, it has been possible to introduce new generation products into the market.

The packaging lines generally occupy about 50 per cent of the floor space, and the packaging and related activities engage about

60 per cent of the 5 million labour force encountered with the Indian Food Industry. The money spent on food packaging and packages alone amounts to nearly Rs.20,000 million. Another Rs.10,000 million is spent on packaging operations and related activities making a total of Rs.30,000 million. This constitutes nearly 1 per cent of the total value of the primary processed foods and 15 per cent of the value of secondary and tertiary processed foods processed in the country. It is estimated that the amount will double by the next five years. The packaging materials may be broadly grouped as paper and paper-based, metal, glass and plastic. Generally the cost of packaging of a packaged food product may range from 10 to 25 per cent, which includes cost of packaging material and packaging operations.

In the present era of consumerism, supermarkets and electronic media, packaging has assumed a multifunctional role by serving as a symbol of value addition, an assurance on quality and quantity, a conveyer of convenience and ultimately a marketing tool.

Food packaging is a coordinated system of food for ensuring safe delivery to the consumer in sound condition through techno-economic functions at minimum cost. A coordinated system of preparing goods for transport, distribution, storage, retailing and end-use. A means of ensuring safe delivery to the ultimate consumer in sound condition at minimum cost. An techno-economic function aimed at minimizing costs of delivery while maximizing sales and hence profits.

Functions of Packaging

1. Packaging serves as a material handling tool containing the desired unit amount of food within a single container and may facilitate the assembly of several such units into aggregates. For example, some fluids are packaged in bottles, which may be placed in boxes, and these boxes in turn can be assembled into easily handled pallets.
2. The package may also serve as processing aid. The metal can be used in heat sterilization of many food items serves not only a protective function but, by its dimensional stability, assures that when fully packed the food maintains a certain shape and location for which heat penetrations can be calculated.

3. A package is a convenience item for the consumer. A cold drink bottle serves as the drinking utensil as well as a process, storage, and distribution container.
4. The package is a marketing tool. The sales appeal and product identification aspects of packaging are particularly important to the sales and marketing branches of food companies.
5. Packaging can be a cost-saving device. Certain packages have obvious economic benefits such as prevention of spills, ease of transporting, prevention of contamination, reduction of labour cost etc.
6. Protection of the product.
7. Value addition.

Principles

1. It must contain the product efficiently throughout the journey.
2. It must provide protection against external climatic conditions and containments.
3. It must be compatible with the product.
4. It must be easily and efficiently filled and closed.
5. It must be efficiently handled by the appropriate mechanical or other means.
6. The package must be inexpensive.
7. The package must be readily disposable or be reusable.
8. To protect an item against mechanical damage resulting from shock and vibration encountered during transportation and storage.
9. To protect the contents against dehydration, oxidation, light, flavour and other loss.
10. To protect the contents against environmental factors.
11. To provide handling facilities for loading and storage and in many cases provide sufficient protection to act as its own warehouse for long periods of time.
12. To protect the contents against pilferage.

Forms of Packaging

1. Rigid: Cans, glass
2. Semi-rigid: Thermoformed cups
3. Flexible: Films

Methods of Food Packaging

Vacuum Packaging

Vacuum packaging provides anaerobic condition to the product and hence aerobic organisms are completely destroyed by vacuum packaging. Further, the vacuum packaged products are found to have better flavour retention. This is also found more suitable for beef. However, due to anaerobic condition, it is not safe for fish and fish products with pH over 4.5.

Modified/Controlled Atmosphere Packaging

The undesirable changes in the food influence by various factors can be controlled/minimized by this method. The internal atmosphere of the package is controlled by other agents or modified by some inert gases. In this packaging technique, there are two kinds of methods are followed.

Carbon Dioxide Packaging

To control moisture loss and to prevent drying out without promoting mould growth in baked products, this has been found to be more useful.

Nitrogen Packaging

Protection of fatty and vitamin containing foods against oxygen is more difficult to achieve in flexible packages than is protection against moisture vapour. This is because the quantity of oxygen, which can cause spoilage, is usually very small. However, flushing with an inert gas such as nitrogen can afford some protection against oxidation.

Aseptic Packaging

It preserves food by preventing microbial contamination from the raw state to the finished product. Aseptically packaged products require sterilization of the starting material, packaging material and sterile surroundings. Aseptic packaging of liquid foods such as milk and fruit juices is found to be more acceptable in the market.

Flexible Packaging

Flexible packaging provides an alternative solution to the distribution of many types of foods for which crush protection is not important. If functions primarily in retaining the goods, separating them from their environment whilst identifying and displaying them to advantage. Dried foodstuffs and precooked dehydrated foods are packed by this method.

Successful Package

For the development of successful package, following considerations should be kept in mind:

1. Facts about the product
2. Facts about the transport hazards
3. Facts about the market
4. Facts about packaging materials, forms, machinery and labour costs.

Facts About the Product

1. The nature of the product.
2. Size and shape of the product.
3. Its weight and density.
4. Its weakness–which part will break, move about, become loose or scratch or abrade easily.
5. Its strength–which parts will withstand loads or pressures and which might be suitable for locating the product in the pack.
6. The effect of moisture and temperature changes on the product, and whether it will absorb moisture or corrode.
7. Compatibility–Whether the product is likely to be affected by any of the possible packaging materials, which items can be packed together, with protection if necessary and which item must not be packed together under any circumstances.
8. Possibilities for dismantling complex products, how far stripping down may be carried out to reduce the package size to a minimum, and whether the required assembly,

installation and use instructions will be such that the customer can handle them.

Facts About the Transport Hazard

1. The type of transport–road, rail, sea or air
2. The degree of control over the transport–is it private or public transport
3. The form of transport–bulk, freight container, postal, passenger train etc.
4. The mechanical conditions and duration of storage
5. The nature and intensity of the mechanical and climatic hazards in transport, storage, retailing and use
6. Whether handling aids are available for loading and off-loading at all points between maker and user
7. The importance of minimum cube(volume) in relation to transport costs

Facts About the Market

1. The product–What is the competition?
   (a) Packages used
   (b) Quantity sold
   (c) Price bracket
2. The retail service–Self-service, Department store, Mail order, Doorstep
3. The customer–Age, Sex, Income group, Social level, Location
4. Primary package–Size, shape, weight, bag, envelope, pouch, sachet, rigid or fold box, metal container, glass container, and plastic container
5. Transport package: Size, shape, weight, number of units, wooden case or crate, sack, metal drums, plastics.

Facts About Packaging, Machinery and Labour Costs

1. Packaging material prices, machinery cost, machine efficiency, machine speed, line efficiency, single or multiple packaging line, labour cost, inflation.

Properties of Various Packaging Materials

Paper and Board

1. No barrier without coatings
2. Good stiffness
3. Absorbent
4. Low density
5. Tear easily
6. Not brittle
7. Creasable

Metal

1. Perfect barrier, if sealed
2. Rigid
3. High density
4. May react with foods
5. Good tensile strength
6. Need joints and closures to form packs

Plastics

1. Wide range of barrier properties
2. Low density
3. Flexible creasible
4. Transparent
5. Low stiffness
6. Tensile and tear strength variable

Glass

1. Perfect barrier, if sealed
2. Rigid
3. Brittle
4. Inert towards foods
5. High density

6. Needs separate closure
7. Transparent

Graphics and Design

1. Brand name, logo, maker's name, product name, content, colour code, legal requirement, instructions, price panel, colours and relations to product, legibility, visibility, impact and personality.

Evaluation of Functional Properties

Various types of flexible packaging materials and their laminates with wide range of physico-mechanical properties are being produced in the country. This situation has made the task of selection of appropriate packaging materials for packaging specific food products is difficult and demand extensive investigations. To assess different types of flexible packaging materials for their functional properties, the following tests have to be carried out:

1. Thickness
2. Tensile strength
3. Tear strength tests
4. Bursting strength tests
5. Seal strength
6. Compression strength
7. Crease stiffness
8. Folding endurance
9. Water vapour permeability
10. Gas transmission rate
11. Migration

Eco-friendly Packaging

The packaging scenario and the requirements world over are constantly undergoing change with respect to materials, forms, systems, machinery and also legal aspects which are linked with environmental equilibrium and maintenance. These requirements vary for different countries.

The Indian Exporters of processed food products are advised to use eco-friendly packaging for their products. They have to follow the following priorities:

1. Wherever possible, certain materials and packages like secondary packaging materials may be eliminated.
2. Reduction in weight and volume of packaging materials.
3. As far as possible, reusable and recyclable packaging materials may be used.
4. Possibility of energy recovery by incineration without creating any air pollution.
5. Biodegradable materials may be considered.
6. Packaging materials like bitumanised paper/paperboard and coated materials, which are difficult to recycle, have to be restricted.
7. Conventionally processed jute/hessian material should be substituted by hydro-carbon free materials.
8. All plastic films/materials must be clearly designated by internationally accepted marking for recycling.

Transport Packaging

For transport packaging, the following points have to be followed:

1. Wooden boxes or crates are to be avoided.
2. All types of plastic tapes should be avoided.
3. Bitumanised CFB boxes should be substituted.
4. Non-reusable pallets are to avoid.

29

Factors Affecting Potato Chips Quality

Bhupendar Singh Khatkar

Department of Food Technology, Guru Jambheshwar University, Hisar – 125 001, Haryana

Introduction

Food processors have learned through years of experience that high quality products never fail to sell. This is due to the fact that the consumer recognizes brands that maintain their quality at reasonable price. Repeat of sales are, thus, the outgrowth of uniform quality. Thorough understanding of factors affecting quality of a product is essential to achieve uniform quality of that particular product. Factors affecting potato chips quality can be classified into three major groups namely; raw material, unit operations and packaging material.

Raw Material

The raw material requirements of potato chip processors are dictated by their desire to maximize chips yield, avoid extra processing costs and maintain product quality standards. Control over quality begins with the selection of raw material and extends right through the manufacturing process to package design, type and quality. The raw material for potato chip manufacturing include potato, oil, salt and other additives.

Potato

Potato is the most important raw material for achieving good quality chips. A wide range of factors influence the suitability of potatoes for processing into chips. Most important factors are: Specific gravity or total solids, reducing sugar content, enzymatic activity (tyrosine content), chlorogenic acid and citric acid content, tuber size, shape, defects, and peel characteristics (Gould, 1986).

Specific Gravity

Specific gravity of potatoes greatly influences oil content of chips and the processing efficiency *i.e.* chips yield. Potatoes with specific gravity greater than 1.080 increase yield and decrease oil content of chips. Therefore, ideal potatoes for chips processing will have specific gravity greater than 1.080 and solids greater than 21.2 per cent (Gamble and Rice, 1989).

Reducing Sugar Content

Reducing sugar content is a primary factor of potato chip colour and taste. The most important influencing factors on the level of reducing sugars are: variety, maturity of tuber at harvest and storage conditions. There are several ways to correct the negative effects of high level of reducing sugars on chip colour such as by blanching to leach out the reducing sugars, sulphating to block the maillard reaction and by closely regulating storage temperature (not less than 12°C) and humidity. Reconditioning of stored potatoes 2–3 weeks prior to processing (by bringing them slowly to room temperature) can moderate reducing sugar levels and recover dry matter content. Potatoes with a reducing sugar content in excess of 0.25 per cent should be avoided for the manufacture of chips (Coffin *et al.*, 1987).

Enzymatic Activity

The tyrosine content of tuber causes discolouring of peeled and sliced potatoes. Discolouring can be avoided by holding peeled and sliced potates under water or in a weak solution of sulphur oxide (Gray and Hughes, 1978).

Chlorogenic and Citric Acid Content

High levels of these acids cause after cooking discolouration. Some controls can be achieved by applying chemicals like sodium and pyro-phosphate.

Potato Sugar and Size

Potato shape and size greatly affect peeling losses and hence in plant efficiency. Thus, reasonable sized (35 to 45 mm) tuber with round or spherical body is preferred. Tuber size also affects packaging of chips. Too small or too large sized tubers should be rejected to avoid packaging problem (Gould, 1986)/

Potato Defects

Raw potato defects may be of 4 types *i.e.* disease/pathological defects, insects infestation, mechanical defects and excess dirt. Absence of defects is important in terms of amount of trimming needs resulting in losses or low recovery and, if not trimmed it will result in defects in the finished product.

Oil/Fat

Frying oil is the another important raw material which largely affects the taste, appearance, texture and shelf-life of potato chips. There is a wide range of different oils or fats which could be used for frying. In particular groundnut oil, cotton seed oil and palmolein, mostly in blends, are used for potato chips frying. Iodine value and free fatty acids (FFA) determine the quality of oil. High iodine value and free fatty acid content will turn the oil rancid faster which will drastically reduce the shelf-life of chips (Buck, 1984). Free fatty acids of fresh oil should not be more than 0.05 per cent. The FFA value will increase with use and can be usually go upto 0.5 per cent without serious effects to the quality of the chips. However, as the FFA value goes up beyond 0.5 per cent, the oil will have detrimental effects to the shelf-life of chips.

The oil content of potato chips is particularly an important consideration for the chips process. Oil is an expensive raw material representing an extremely important cost factor for the manufacture. If it is absorbed into the chips in larger quantities, not only the cost of chips increases, but it also influences the quality and taste of chips. A high oil content may make the chips to appear greasy.

Chips with low oil content can be processed by using potatoes of high specific gravity, thin slice thickness, partial drying prior to frying, reasonably high temperature (170–180°C) of the oil with less frying time and good quality oil (Brown, 1960).

Salt and Other Additives

Salt and other additives like monosodium glutamate are added to chips primarily to enhance natural flavour of the basic ingredients like potato and oil. However, excess quantities and poor quality of these additives may prove detrimental to flavour and overall acceptability of chips.

Unit Operations

Grading, washing, peeling slicing, blanching, dewatering (surface drying), frying, salting and packaging are the major unit operations affecting quality of chips (Gadekar, 1988). Grading for uniform tuber shape and size is desirable to minimize peeling losses. Washing removes dirt adhering to the surface of potatoes. Peeling operation primarily affects appearance, taste and acceptability of chips. The slice thickness has to be extremely accurate, as the slice thickness determines the frying time and hence oil content of chips. If the slice thickness varies widely, some of the slices will be under cooked while others will be burnt. Slices with high reducing sugar levels are blanched after slicing operation to correct the negative effects of high reducing sugars on the product taste, colour and frying oil life. Blanching partially cooks the slices, an essential step in reducing frying time and its associated costs (heating, frying oil usage) as well as to improve the texture of the chips. The balanced slice are dewatered by a vibrating screen conveyor before frying to minimize heating and deterioration of frying oil (Gamble and Rice, 1987). Next critical unit operation is frying which influences oil content, crispness, taste, appearance and overall quality of chips. Heating frying oil causes changes in its physical and chemical properties. As oil combines with oxygen at higher temperature (170–180°C), oxidation occurs, thus potentially causing a rancid flavour. This reaction can be controlled by good "Oil house keeping" including: choosing especially prepared oils, avoiding excessive heat and aeration of the oil, rapid oil replacement or turnover, filtering, avoiding contamination (Augustin and Berry, 1983). After frying, the chips are salted and/or flavoured and packaged.

Packaging Material

Apart from general requirements in accordance with food legislation (health-related aspects), specific requirements of the product should be considered in selection of packaging material. The product under consideration is sensitive to water vapour (loss of crispness), light (rancidity of fat) and mechanical damage. In addition, loss of aroma may occur during prolonged storage or foreign odour may enter into package. Packs for chips must, therefore, be sufficiently light-proof and impermeable to water vapour and gases. Packaging materials with a water vapour permeability less than 1 g/m^2 day are recommended *e.g.* polypropylene, oriented polypropylene, and laminates like polyester/LDPE, BOPP/LDPE and metallized polyester/LDPE and LDPE/Al Foil, LDPE. Use of inert-gas (nitrogen gas) in packaging gives further protection of chips.

Conclusion

Ideal potatoes with specific gravity greater than 1.080, reducing sugar content less than 0.25 per cent, reasonably sized (35–45 mm) round or spherical shape without any major defects, through washing, proper peeling, accurate slice thickness, right time and temperature of blanching water, partial pre-fry drying, right temperature of frying oil, good oil-houses keeping, and sufficiently light, vapour and gas proof packaging material are the key factors for obtaining good quality chips.

REFERENCES

Augustin, M.A. and Berry, S.K. (1983). Efficacy of the antioxidant BH and BHT in palmotein during heating and frying. *J. Am. Oil Chemists Soc.*, 60(8): 1520-1523.

Brown, H.D. (1960). Problems of potato chip industry: Processing and technology. *Adv. In Food Sci.*, 10(18): 181-279.

Buck. D.F. (1984). Food antioxidants: Application and uses in snake foods. *Cereal Foods World*, 29(3): 301.

Coffin, R.H., Yada, R.Y., Parkin, K.L., Grodzinski, B. and Stanley, D.W. (1987). Effect of low temperature storage on sugar concentration and chip colour of certain processing potato cultivars and selections. *J. Food Sci.*, 52(3): 639.

Gadekar, S. (1988). Potato chip technology. *Food Service*, p. 73-76.

Gamble, M.H. and Rice, P. (1987). Effect of pre-fry drying on oil uptake and distribution in potato crisp manufacturing. *Ind. J. Food Sci., & Technol.,* 22(5): 535.

Gamble, M.H. and Rice, P. (1989). Effect of initial tuber solids content on final oil content of potato chips. *FST Abstracts,* pp. 2175.

Gould, W.A. (1986). Procurement of chipping potatoes: A practical guide to chipping potato procurement. *Chipper/Snacker,* 43(1): 86-88.

Gray, D. and Hughes, J.C. (1978). Tuber quality. In: *The Potato Crop,* (Ed.) P.M. Horris. Chapman and Hall, London, pp. 504.

Part IV
MEAT AND MEAT PRODUCTS

30

Modified Atmosphere Packaging Technology of Meat Food Products

J. Sahoo
Department of Animal Products Technology,
CCS Haryana Agricultural University, Hisar – 125 004, India

Introduction

The term "modified atmosphere" (MA) and "controlled atmosphere" (CA) mean that atmospheric composition surrounding a perishable product is different from that of normal air. Both MA and CA commonly involve manipulation of carbondioxide, oxygen and nitrogen levels. However, other gases such as carbonmonoxide, ethylene, propylene and acetylene are sometimes included. MA differs from CA, only in how precisely gas partial pressures are controlled, while CA is more precise than MA.

Use of modified atmosphere packaging (MAP) technology in meat industry had shown phenomenal increasing trend during the past two decades in various countries, especially in Australia, New Zealand, U.S.A., U.K. and many European countries. This commensurate with the increasing consumer demand of fresh and chilled meat without preservatives. MAP greatly extends the shelf-life of muscle foods under refrigerated storage and maintain colour, texture, sliceability and flavour of the product for a considerably longer time. Modified and controlled gas atmospheres have received attention for their potential benefit to maintain the quality of fresh and processed meats packaged in flexible films.

History

The first practical use of modified atmospheres containing elevated levels of carbondioxide, as a preservative in the handling of fresh meat, was in the shipment of whole chilled beef carcasses from Australia and New Zealand to Great Britain in the 1930s. By 1938, 26 per cent of the beef from Australia and 60 per cent from New Zealand was being commercially shipped under CO_2 atmospheres. It is only during the 1970s that developments in the packaging materials and techniques had made this new technology both practically and economically feasible. A big turning point in MAP technology came in 1981, when Marks and Spencer Co in U.K. introduced a wide range of fresh meat products packaged under MA. The impetus behind these products came mainly from an increased consumer demand for preservative-free fresh and chilled products. There are numerous advantages of using MAP technology, but some basic problems involved, must also be considered.

Vacuum packaging (VP) can delay the growth of aerobic spoilage micro-organisms and retard oxidation of lipids in fresh meats. However, disadvantages such as deformation of cuts by film tightening, temporary and unsightly discolouration and weight losses from purge have prevented widespread vacuum packaging of retail cuts.

Use of appropriate gas mixtures for MAP offers an alternative to VP. Gases most commonly used in MAP are oxygen, carbondioxide and nitrogen, usually in combinations. In gas mixtures containing high concentrations of oxygen, the meat myoglobin is readily oxygenated to give a desired bright red colour of fresh meat. However, such atmospheres cause enhanced growth of aerobic microorganisms and development of oxidative rancidity in meat, thus decreasing shelf-life. Moreover, high proportions of oxygen (*i.e.*, more than 50 per cent) constitute serious industrial and transportation fire hazards. Carbondioxide in high concentration has been shown to inhibit spoilage bacteria in red meats. Carbondioxide-induced discolouration of red meats is still controversial. Nitrogen is used commonly as a filler gas to balance MAs and is generally considered inert to meat quality. The nitrogen gas (100 per cent) in the packaging of frankfurters has been shown to inhibit the growth of yeasts and moulds and to reduce greening and off-odour in the product.

More recently, the *captech* process has proven to be most promising for chilled meats. This system relies upon the integration of hygienic processing, optimum storage temperature (-1.5°C), low residual O_2, high concentration of CO_2 and gas impermeable packaging films. The captech process has been applied to extend the storage life of venison, beef, pork and lambs for periods approaching 6 months. Also, commercial application is evident in the sea shipment of chilled New Zealand lamb to North American markets. Low temperature storage of pork in CO_2 has prolonged shelf-life to 2 to 3 months and the captech process has further extended the quality to 5-6 months.

Another method called *"master pack"*, is more commonly used for MAP of fish. In this type of packaging, the fish product is first overwrapped with PVC film and is then placed in a large partial barrier film, master pouch bag that has usually been flushed with CO_2. Prior to retail display, the consumer package is taken out of the master package and once removed, has a shelf-life of approximately 3 to 4 days.

Modified Atmosphere Technology

Several distinctions can be drawn in discussing atmosphere technology. In CA systems, the selected atmosphere concentrations of gases are maintained throughout the storage process, whereas in MA systems, only an initial charge is made, consistent with the expected requirements of the commodity and transportation duration. Dynamic changes due to gas permeation, leakage, absorption, and/or biochemical conversion via respiratory activity must be considered in both the systems. In many instances, CA is technically superior to MA, but often it is not sufficiently advantageous to balance the lower cost advantage of MA, resulting from the latter's less complex mechanical requirements.

Oxygen partial pressures can be reduced by purging with nitrogen, burning, using catalytic converters or using silicone membranes. Water, lime and molecular sieves are found to be effective scrubbers for removal of CO_2. Pressurized cylinders or dry ice can be used for addition of CO_2 into the holding chambers. Carbonmonoxide can also be added from pressurized cylinders.

In commodity-modified atmospheres, an actively respiring and metabolizing product reduces the oxygen and increases the carbondioxide in the ambient air within a chamber in which various barriers and restrictions to gas exchange exist. Various techniques used to accomplish desired commodity-modified atmospheric conditions include waxes and other surface coatings, plastic covers with diffusion windows, manipulation of shipping containers and vehicles, pallet shrouds, polyethylene liners in shipping containers, packaging in film wraps or bags and air-tight cold storage rooms.

Equipments and Films for MAP

The success of MAP technology depends upon the use of proper packaging films and equipment. A number of films are available, but considerations for the choice lie with the machinery and system being used for packaging rather than the type of food involved. Machines used in U.K. for gas flushing are adapted vacuum machines, manufactured by Multivac and Swissvac. Vacuum packaging machine manufactured by Roschermatic (Germany) is also commonly used.

Meat requires different conditions for packaging, while beef needs high oxygen content to maintain a bright brick-red colour, which is popular among the customers. Chicken in comparison needs little oxygen and reduced amount of carbondioxide, which can flavour the meat. Pork with its high fat content needs less oxygen and some nitrogen for glow. In general, the packaging films should have good barrier and sealability properties and must have optimum clarity and anti-fogging properties so that packs maintain their clear appeal in or out of the chill display.

Effect of Gases on MAP Foods

Oxygen, nitrogen and carbondioxide are the three main gases used commercially, although trace gases such as carbonmonoxide, nitrous oxide and sulphurdioxide are mentioned as possible gases for MAP of foods.

Effect of Oxygen (O_2)

One of the major functions of O_2 in MAP meats is to maintain the meat pigment, myoglobin in oxymyoglobin state responsible for the bright red colour, which most consumers associate with fresh meat. Tissue foods utilize oxygen by respiration, oxidation or microbial action. O_2 stimulates the growth of aerobic bacteria and inhibits the obligate anaerobes. Lowering oxygen level in MAP environment helps in extending the shelf-life of this product by reducing metabolic and chemical oxidation rates, but it can stimulate the outgrowth of anaerobic pathogens. Restricting oxygen will generally retard the growth of *Pseudomonas*, while enhancing the growth of *Lactobacilli* or *Brochothrix*.

Effect of Nitrogen (N_2)

Nitrogen is an inert tasteless gas, which displays little or no anti-microbial activity of its own. Because of its low solubility in water, the presence of N_2 in MAP food can prevent pack collapse that can occur, when high concentration of CO_2 is used. In addition, N_2 by displacing O_2 in the pack can delay oxidative rancidity and also inhibit the growth of aerobic microorganisms.

Effect of Carbondioxide (CO_2)

Carbondioxide is both water and lipid-soluble. It is mainly responsible for bacteriostatic effect on the microorganisms found in MA environment. The gas selectively inhibits the growth of gram negative bacteria, such as *Pseudomonas* and related psychrotrophs, which grow rapidly and produces off-odours and-flavours in raw meat and poultry. Lactic acid bacteria, such as *Streptococci* and *Lactobacilli* are less affected by elevated levels of CO_2. The level of CO_2 that gives maximum inhibition of the common spoilage organisms is 26 per cent CO_2 in air.

There have been many theories regarding the mechanisms by which CO_2 acts on bacterial cell, which include alteration of intracellular pH, disruption of the internal enzymatic equilibria (interfere with dehydrogenating enzymes of the cells), enters mass action equilibria for enzymatic decarboxylation, changes in cell membrane functions and consequently, nutrient uptake and absorption, affecting the physico-chemical properties of proteins and toxicity of carbonic acid in its undissociated form.

For maximum anti-microbial effect, the storage temperature of a MAP product should be kept as low as possible, because the solubility of CO_2 decreases dramatically with increasing temperature.

MAP Foods Commercially Available

In North America, the foods being packaged under MAP include fresh raw meats (sliced bacon, steak; beef hearts, pork kidneys, ox tails); cooked meats (hamburgers, beef jerky, sausage rolls, sliced meats, cretons or head cheese, wieners). Poultry (whole carcass, nuggets, peeled hard cooked eggs); cheese, prepared salads, pasta and in Canada, the commonly distributed MAP foods are fish and various types of sandwiches. The expected shelf-life of most of these products is 1 to 2 months.

Effect of MAP on the Quality of Fresh Meats

Red Meats

Carbondioxide alone or in combination with oxygen extend the shelf-life of meats. For fresh meat in which red colour maintenance is important, 60 to 80 per cent oxygen is generally used so that formation of metmyoglobin, which is responsible for undesirable brown colour of meat, is retarded. CO_2 is generally used in concentration ranging from 15 to 40 per cent. Optimal inhibition of meat spoilage bacteria is achieved at 40 to 60 per cent CO_2. But at a higher concentration, CO_2 causes discolouration of meat due to denaturation of meat proteins. Carbondioxide (20 to 30 per cent) or even 10 per cent, may be sufficient to retard bacterial growth. Vacuum packaging produced beef rib eye steaks with highest overall acceptability and protected colour and lipids against oxidation, while modified atmospheres were more effective in prolonging shelf-life, reducing purge value and increasing cooking yield of steaks. Fresh pork chops had acceptable odour upto 9th day stored in air, N_2 and N_2O, upto 13th day in vacuum and high vacuum packed samples and even upto 16th day for CO_2–enriched packs. N_2 or N_2O did not inhibit the rate of bacterial growth.

The loss of red colour in lean ground beef was greatest in PVC packaged samples and during frozen storage of above samples, vacuum packaging was the most protective for most characteristics studied. The type of gas and film pouch used in MAP of frozen (IQF) turkey strips influenced its organoleptic quality.

Many workers had studied the microbiological quality of meat stored under MAs. Depending upon storage temperature and gas mixtures used, the shelf-life of beef, pork and lamb varied from 2 to 8 weeks. In atmospheres composed of 20 per cent O_2 (similar to air) plus 20 or 40 per cent CO_2, total bacterial numbers on beef steaks did not reach log 7.00 until 5 weeks of storage at 2-4°C. Superior performance of high CO_2 atmosphere over vacuum packaging was due to marked inhibition of some pathogenic psychrotrophs, belonging to *Enterobacteriaceae e.g., Yersinia enterocolitica*. A shift from an initial gram negative aerobic spoilage flora to a predominantly gram positive facultative anaerobes, especially *Lactobacillus* species occurs during MA storage of beef and pork loin.

The various technical points and precautions to be followed during the storage life of vacuum packaged and CO_2 CAP of chilled meat for export purposes and for supermarkets have been reported. Meat pH was found to be an important factor influencing the shelf-life (total counts, coliform counts) and weight loss of meat during processing and storage period in the vacuum packs. Oxygen-free packaging (less than 0.1 per cent O_2) of retail meat cuts, using natural meat enzymes or by incorporating a catalytic system and storing at 0-2°C even upto 3 to 4 weeks had lower drip loss and better red colour of meat. Beef and lamb cuts stored in an atmosphere containing 80 per cent O_2 and 20 per cent CO_2 at 5°C had a storage life of three times more than those kept in conventional over wrap trays. Fresh meat cuts when treated with phosphate, ascorbic acid or ascorbate and citric acid or citrate and packaged in MA containing CO_2 and O_2 in a critical ratio, retained fresh meat colour for 20 days and delayed microbial spoilage.

Trigon packaging system (U.K.) Ltd. and Marks and Spencer jointly developed a new bulk package for fresh meat cuts, combining vacuum skin packaging (VSP) with modified atmosphere packaging (MAP) to retain bloom of beef steak, pork chops and lamb chops and extend their shelf-life. In this process, individual meat cuts are first vacuum skin packaged in permeable 3-ply co-extruded film and then bagged in preformed nylon pouches (with a polyethylene sealant layer) injected with a gas blend of 80 per cent O_2 and 20 per cent CO_2 and sealed. In the similar line, United States developed a patent consisting of a peelable O_2 barrier layer inside where remains a O_2 permeable skin layer enclosing the product. Minced beef when vacuum packaged in the presence of solid CO_2 maintained its bloom and had reduced total microbial numbers during storage and display. Irradiation combined with MAP had little scope for beef primal cuts, whereas, it improved the sensory and microbial quality of pork, mutton, sliced luncheon meat and poultry meat.

Poultry Meat

Earlier research in 1970s by various workers had established that the shelf-life of MAP chicken carcasses/pieces had 2 to 3 times more than air-packed counterparts. Halved ready-to-cook chicken carcasses stored under CO_2 had shelf-life of 26 days and frozen turkey strips stored in MAP upto 84 days had highest sensory scores in respect of aroma, appearance and structure. Increased levels of CO_2 were synergistic with the lethal effect of irradiation of fresh minced chicken and increased its shelf-life. For a shelf-life of 28 days of ground chicken meat, it was necessary to apply CO_2 concentration of 60 to 80 per cent while under the same concentration of CO_2, the shelf-life of chicken quarters extended to 35 days at 2°C.

Fish

The three major modes by which fish can be stored in MAP environments, enriched with CO_2 are large van, master pack and individual consumer package. The shelf-life of MAP fish is substantially increased as compared to air control at the same temperature. Generally, CO_2–MAP provides for consistently improved stability of fresh fish. Vacuum skin packaging (VSP) of weakfish significantly reduced lipid rancidity development and retarded lipid hydrolysis as compared to traditional overwrapping.

Effect of MAP on the Quality of Process Meats

Information regarding factors affecting the quality of ready-to-eat MAP meats, cured meat products and packaging requirements is available. Use of MAP for a range of meat products *viz.*, mortadella, sausages, salami and pipperoni slices due to their advantages of shelf-life extension, separation of meat slices and attractive product appearance is reported.

Nitrogen backflush produced the lowest purge value than that of vacuum packaged controls and CO_2 treatment yielded better taste panel scores of MAP ground beef patties during frozen storage. MAP

(80 per cent N_2/20 per cent CO_2) of ground beef patties gave better colour and decreased weight loss by exudation over VP counterparts during storage at 36°F. Roast beef was still acceptable after 28-35 days and hamburger only after 14 days storage at 4°C packaged in gas impermeable film with 50 per cent CO_2 and 50 per cent air. But exclusion of O_2 extended storage life of hamburger from 14 to 35 days. An atmosphere containing 75 per cent CO_2, 15 per cent N_2 and 10 per cent O_2 was the most effective in simultaneously inhibiting *Clostridium perfringens, Salmonella typhimurium, Staphylococcus aureus* and *Pseudomonas fragi* in roast beef. The pH and lipid composition of pre-cooked pork and turkey under CO_2 and N_2 storage changed little, but moisture content had decreased after 3 months at –20°C. VP turkey was darker and redder than other samples.

Dip treatment in a solution containing 70 per cent ethanol, 3 per cent L-ascorbic acid and 0.08 per cent DL-tocopherol and MAP in 20 per cent CO_2 + 80 per cent O_2 of beef loin steaks retained acceptable surface colour throughout the storage period of 13 days at 4°C and had least variation of pH and TBA values. VP dried beef was acceptable in terms of sensory and microbial quality during storage at 32-33°C for upto 6 weeks. CO_2 flushing and packaging under slight overpressure had been found to eliminate discolouration, caused by photo-chemical pigment degradation of VP sliced ham. The bacterial counts and colour of the processed ham packaged in 35 per cent CO_2 + 65 per cent N_2 remained more or less same during the entire storage period of 30 days at 3-5°C.

Safety Concerns of MAP Muscle Foods

Considerable interest in CAP/MAP muscle foods has resulted in extended shelf-life refrigerated (ESLR) foods and has become a new area of concern for food microbiologists. Despite the increased shelf-life of the above foods, Food and Drug Administration expresses concern that if proper controls are not in effect, some *sous vide* foods may pose a potential public health hazard. The concerns involved microbiological safety of these products. On principle, low oxygen and high CO_2 levels in MA environment induce lengthening of lag phase/or log phase (generation time) of aerobic spoilage organism. On the other hand, it gives chance to the growth of anaerobic pathogens. In the past, psychrotrophic non-proteolytic strains of *Clostridium botulinum* had been the major safety concern associated with MAP foods. However, new concern, have been raised with some of the newly recognised psychrotrophic pathogens such as *Listeria monocytogenes*, which may grow well in refrigerated MAP products.

The most common types of *Clostridium botulinum* causing human illness are A, B and E. They produce neurotoxin. *C. botulinum* could produce toxin in nitrogen packed hamburger sandwiches at 12°C, but not at 8°C and also toxin was not produced in turkey or sausage sandwiches. So, keeping the storage temperature as low as possible is quite important. In MAP fish products under certain conditions of temperature abuse or even at 4°C under anaerobic MA *botulinum* toxin can be detected well before organoleptic deterioration. So, National Academy of Sciences in U.S. has recommended not to store VP/CAP fish products at refrigerated temperature.

Another culprit is *Yersinia enterocolitica* which causes gastrointestinal disorders and arthritis in man, can grow in high pH beef packaged under 100 per cent CO_2 at 5 and 10°C temperature, but unable to grow at 0, 2 or -2°C. However, this organism was not able to grow in minced beef at 4°C in a MA of 20 per cent CO_2 and 80 per cent O_2.

Recently, much attention is paid to *Listeria monocytogenes* by many research workers. It affects mainly pregnant women, foetuses, newborns, the elderly and immunodeficiency subjects. It is a very hardy organism, psychrotrophic pathogen, can grow in cook-chill MAP products and appears to be more heat tolerant. The disease *listeriosis* is associated with high mortality rate. These organisms grow in CO_2- packed beef at 10°C and VP beef at 0, 2, 5 and 10°C but they do not survive well in raw minced chicken in MA containing 75 per cent CO_2 and 25 per cent N_2. However, if 5 per cent O_2 is added, the bacteria can grow even at 4°C.

Furthermore, the pathogens capable of growing between 5-12°C in MAP products due to temperature abuse are *Salmonella* spp, *Clostridium perfringens, Bacillus cereus, Staphylococcus aureus* and *Vibrio parahaemolyticus*. Besides, refrigerated conditions may allow *Campylobacter jejuni* and *Brucella* spp to survive for longer periods.

Another potent pathogen, *Escherichia coli* 0157: H7 strain has caught to the attention of food microbiologists. Ground beef, pork chop, chicken, turkey, lamb and even unpasteurized milk are the most important sources of infection by the above pathogen. It causes bloody diarrhoea and renal failure in humans. None of the gas mixtures tested ($CO_2/O_2/N_2$:0/5/95, 0/10/90, 5/10/85, 5/20/75, 10/5/85 and 10/20/70) significantly affected growth of the bacterium at 5, 10 and 20°C. It was also observed that MA did not significantly affect the tolerance of *E. coli* 0157: H7 to 3 per cent NaCl. The *E. coli* grew at 10 and 20°C and survived at 5°C.

In spite of the above potential hazards, there is nothing to fear in consuming MAP products, unless the product is temperature abused, because sufficiently high numbers of pathogen(s) must be ingested to cause illness. Therefore, strict adherence to temperature control must be mandatory not only for MAP food processors, distributors and retailers, but also for the consumers.

Cost Benefit Relationship

Though attention to good temperature control of MAP/CAP products is essential, the achievable benefits can be well worth it. One of the most important benefits is that the MAP commodity being transported can result in better handling throughout the distribution chain, augmenting the returns available from use of MAP. MAP can be used to extend shelf-life and to maintain high quality in the marketing channels for considerably longer time. Various methods for employing MA *e.g.*, rail cars, sea vans, pallet-in-a-bag, bag-in-a-box, portion bags and *in situ* MA generation and related cost of the product had been reported. Loss/value relationship of the commodity justifies the use of MA. Valuable products such as meat and fish easily justify MA costs.

Conclusion

The success of MAP technology depends upon initial products quality, hygienic slaughtering of food animals, correct packaging material and equipment, correct gas mixtures to be used and strict control of temperature. It should be remembered that CAP or MAP is not a substitute for good sanitation. When spoilage organisms are low in MAP products, pathogens grow due to lack of competitors in case of temperature abuse. In such situations, additional hurdles/barriers, such as use of lactic acid starter culture, water activity, competitive flora and redox potential may be of great use in MAP foods in future. These hurdles may interact directly or synergistically to maintain microbial stability of a particular product. Further research is required in microbiological safety concerns (potential health hazards) of MAP foods. Although there is a fair chance to abuse MAP foods in this regard, there have been no reported illness of food borne diseases linked to the consumption of these products. Cost involved in MAP food is superceded by its high quality maintained for long time. So, by conducting necessary research in future, a wide variety of new MAP products will be evolved by regulatory agencies for its wide marketing to meet the increasing consumer demand for fresh meat products without any harmful preservatives.

31

Use of Antioxidant Vitamins in Meat and Meat Products

J. Sahoo

Department of Animal Products Technology,
CCS Haryana Agricultural University, Hisar – 125 004, India

Introduction

Lipid oxidation in general and specifically oxidation of phospholipids has been found to be responsible for the characteristic rancid off flavour developing in pre-cooked meat. Furthermore, it has been found that development of warmed over flavour (WOF) is related to the content of myoglobin in the muscle, chicken dark muscle being more unstable towards oxidation due to more myoglobin content than white muscles. Haematin compounds catalyze the decomposition of hydroperoxides, in effect propagating further lipid oxidation. Heat denatured myoglobin appear to be more effective catalyst than inorganic iron for the cleavage of hydroperoxides in muscle tissues. The lipid oxidation processes lead to discolouration, drip losses, off-odour and off-flavour development and the production of potentially toxic compounds. The oxidation effects on proteins, peptides and amino acids can impair texture, flavour development and the production of potentially toxic compounds.

The low oxidative stability of meat and of pre-cooked and restructured meat products is a problem for all those involved in the meat production chain, including the primary producers, processors, distributors and retailers. Understanding and controlling the processes which lead to lipid oxidation is a major challenge for meat scientists. The above oxidative effects in meat can be minimized or prevented by chemical/synthetic antioxidants, but for the carcinogenic and other adverse effects of BHA, BHT, TBHQ and other synthetic antioxidants. The health conscious meat consumers now disapprove the use of chemical preservatives in meat foods because of their residual effect. On the other hand, vitamins C and E are the two potent naturally occurring antioxidants.

Keeping this in view the use of ascorbic acid and tocopherol acetate to improve the quality, especially the oxidative stability of meat and meat products is discussed below:

Vitamin C (Ascorbic Acid)

Ascorbic acid was first isolated by the American biochemists King and Waugh in 1932. It is a lactone of a sugar acid. It is required in the diets of man, monkey, guineapigs and insects. It is not present in microorganisms nor does it seem to be required. Ascorbic acid is a strong reducing agent, readily losing H^+ to become dehydroascorbic acid, which also has vitamin C activity. However, vitamin C activity is lost, when the lactone ring of dehydroascorbic acid is hydrolysed to yield diketogluconic acid. In food, it is largely destroyed by cooking.

Use in Meat Processing

It is used as a biological antioxidant to stabilize meat colour by way of pre-slaughter injection in beef, exogenous addition in ground beef, dip treatment or surface application of beef steaks. In cured meat products such as cooked ham and smoked beef sausages, it reduces the requirement of nitrite, reduces nitrosamine formation and also reduces the time of processing. In dry fermented sausage, it greatly enhances colour stability.

Function in Meat System

Ascorbic acid acts as an antioxidant by inhibiting radical formation at double bonds of mono or polyunsaturated fatty acids, quenching free radicals, scavenging oxygen and serving as a reductant. Ascorbic acid is a good antioxidant for peroxidations initiated in the aqueous phase, but does not trap peroxy radicals in the lipid phase. In aqueous systems containing metals, ascorbic acid can act as a peroxidant by reducing the metals, which become more active catalysts of oxidation in their lower valence state. In the absence of added metals, ascorbic acid is an effective antioxidant at high concentration. In non-aqueous media, ascorbic acid and esters are not good antioxidants. Ascorbyl palmitate is a good antioxidant in lipid phase. The multiple effects of ascorbic acid and ascorbyl palmitate include:

1. Hydrogen donation to regenerate the stable antioxidant radical,
2. Metal inactivation to reduce the rate of initiation by metals;
3. Hydroperoxide reduction to produce stable alcohols by non-radical processes and
4. Oxygen scavenging.

Pigment Oxidation

Sodium ascorbate, when added to ground beef at 500 ppm level, showed lower surface and extract metmyoglobin content of 26.7 and 24.3 per cent, respectively during 7 days illuminated display at 4°C, whereas, control samples had corresponding values of 60 and 70 per cent, respectively. Surface spreading with 0.1 ml of 10 per cent vitamin C solution to 20 g beef steaks had lower MMb (40 per cent) as compared to the control sample (60 per cent) after 4 days of refrigerated display. In an aqueous carotene linoleate solution, vitamin C upto 176 ppm acted as a pro-oxidant, while at higher than 176 ppm level it acted as antioxidant. The pro-oxidant activity of vitamin C increased with ferric (Fe^{+++}) iron concentration. Vitamin C incorporation maintained good colour in ground beef upto 5 days refrigerated display, when added at concentrations of 500 ppm and 1000 ppm as determined by visual colour scores and spectrophotometric colour analysis. The colour stability of sodium ascorbate treated fresh beef muscles, particularly *M. psoas major* and

M. gluteus medius held at 5°C was significantly better than the control samples. The magnesium salt of ascorbic acid showed a similar protective effect to sodium ascorbate against fading of cooked cured meat colour. This magnesium salt of ascorbic acid-3-phosphate was hydrolysed to AA by acid phosphatase contained in the water soluble fraction of pork, beef and mutton. A comparative study between sodium ascorbate and sodium erythorbate for the technological properties and processes of cooked cured pork loin was made. Residual ascorbate or erythorbate concentration, nitrite and nitrate concentration, colour properties and sensory quality were assessed at intervals during storage. No significant differences were observed between samples made with ascorbate and those made with erythorbate.

Sodium erythorbate, a sterioisomer of sodium ascorbate is cheaper to manufature than sodium ascorbate. It crystalizes in the monohydrate form and its antioxidant activity develops, when it is dissolved in water. It is a highly effective antioxidant owing to its ene-diol structure in various meat products. Ground buffalo meat containing 500 ppm SA had significantly higher colour score, Lovibond tintometer red colour units and lower metmyoglobin contents as compared to other levels of sodium ascorbate. The metmyoglobin was positively correlated with TBARS number. Addition of reducing agents, *i.e.*, ascorbic acid, isoascorbic acid and their salts improve colour stability and extend storage shelf life of the meat products.

Lipid Oxidation

Ascorbic acid treatment showed inhibition of lipid oxidation in beef steaks stored in a cold room at 4°C for 13 days. 50 ppm vitamin C addition to ground beef caused greater TBA value than controls. Exogenous addition of ascorbate delayed lipid oxidation in ground beef. Vitamin C treatment showed high lipid stability in ground beef, recording TBA value of 1.52 as compared to control samples (4.0) during illuminated display at 4°C for 7 days.

Combination of ascorbate and phosphate minimised flavour deterioration in PVC packaged refrigerated cooked beef-carrageenan patties. Reduced fat patties containing ascorbate or phosphate had much less lipid oxidation than their counterparts without additives. SA showed a protective effect against photoxidation (measured by

TBA test), but it did not have an influence on the patties samples in darkness. Ground buffalo meat treated with 500 ppm SA had significantly higher odour score and lower TBARS number as compared to control and other levels of SA. The TBARS number was inversely correlated with odour scores. High levels of AA (1000, 10000 ppm) inhibited, whereas lower level (3-500 ppm) increased warmed over flavour of muscle tissue. On the other hand, AA catalysed development of lipid oxidation at lower levels upto 250 ppm but inhibited the reaction at high level (500 ppm). At high concentration, ascorbates reduced TBARS number in beef patties, whereas at low concentration (250 ppm) increased the TBARS number from 1.99-6.55 mg malonaldehyde/kg. At low concentration, 0.02 per cent ascorbyl palmitate along with 1.5 per cent sodium chloride increased TBA value in turkey steaks from 0.49 to 1.24 mg malonaldehyde/kg. Concentration of asocrbic acid and tripolyphosphate reduced TBA from 1.24 to 0.60 mg malonaldehyde/kg.

Physico-chemical Quality

The improving action of L-ascorbic acid on gel formation of fish may be due to the oxidation of sulphydryl groups in fish protein by dehydro-L-ascorbic acid. In addition to colour stabilization and lowering nitrate levels, ascorbic acid can reduce cooking losses in both fat-free and fat containing meat products. Ground buffalo meat containing 500 ppm sodium ascorbate had significantly higher pH, lower cooking loss as compared to other levels of sodium ascorbate. Addition of sodium ascorbate to brine was shown to suppress N-nitrosodimethylamine (NDMA) from added DMA in experimental curing of pork middles. NDMA is formed due to canning or freezing of the cured meats. However, lowest amounts of NDMA were found in the bacon cured in the presence of ascorbate. Ascorbate addition proved beneficial in cured meat products (cooked ham, smoked beef sausages), as it helped to reduce nitrite contents to less than 50 per cent and the residual nitrite could be maintained at a lower level. Use of sodium ascorbate was proved to be superior to sodium erythorbate in the above products. The heat stability of ascorbic acid in food during cooking was studied. It was observed that ascorbic acid was oxidised faster in utensils made up of standard aluminium than in enamelled cast iron or standard aluminium coated with pure aluminium. Therefore, cooking utensils made up of standard aluminium and coated with aluminium of 99.99 per cent purity was recommended.

Sensory Quality

In the salami type sausages, made from 74 per cent lean pork and 26 per cent fat ripened for 45-90 days under different temperatures and ventilation conditions, the aroma and taste of sausages (i) containing 25 mg $NaNO_3$ and 100 mg ascorbic acid/kg were found superior to those of sausages, (ii) made with 250 mg sodium nitrate and 150 mg sodium nitrite per kg. No significant difference was observed between the cooked cured pork loin samples made with ascorbate and those made with erythorbate in respect of residual ascorbate or erythorbate concentration, nitrate and nitrite concentration, colour properties and sensor quality during storage.

Microbiological Quality

No significant effect was observed on the microbiological (aerobic mesophiles count, psychrotrophic plate counts) quality of ground buffalo meat treated with sodium ascorbate at different levels. It was also reported that microbial numbers were not different for ascorbic acid treatments in ground beef. Ascorbic acid did not appear to inhibit bacterial spoilage of beef.

Vitamin E (α-tocopherol Acetate)

Vitamin E was isolated from wheat germ and was given the name tocopherol (a Greek word, tokos meaning child birth). Vitamin E is not synthesized by the animal/human and so a constant dietary supply is required. Its activity is derived from a series of compounds namely, the tocopherols and tocotrienols from plant origins. These have been called nature's antioxidants. Tocopherols include α, β, γ and δ homologs and the corresponding tocotrienols. The most active and abundant is α-tocopherol. Tocopherols are present in appreciable quantities in all vegetable oils. On the other hand, animal fat contains almost no tocopherol.

Use in Meat Processing

Vitamin E as an antioxidant has now received much attention for its ability to maintain meat colour, extend shelf life, improve taste, reduce drip loss and offer health benefits. Reduced heart attack risks in uses of vitamin E supplements and reducing susceptibility of food lipoprotein to oxidation have encouraged many food scientists to have concerted research efforts in the area of vitamin E supplementation either through diet of meat animals or by exogenous addition or by directly adding to the muscle foods for increasing lipid stability during storage and improving flavour and taste of meat products. The *in vitro* addition of vitamin E appeared to be over 400 times more efficient than *in vivo* supplementation to enhance the quality and stability of rendered fowl fat. Vitamin E is very widely used in restructured meat products. In the processed meats such as cooked rabbit meat, minced beef, cured turkey products, pork chops, ground buffalo meat and buffalo meat nuggets, vitamin E had been used successfully to enhance product quality and to extend shelf life during storage conditions.

Mode of Action in Meat System

Vitamin E acts as biological antioxidant in the process of lipid oxidation. Rate and extent of lipid oxidation are dependent on the level of polyunsaturated fatty acids (PUFA) present in the muscle system. It is now generally recognised that the phospholipid present in the sub-cellular membrane, *i.e.*, microsomes, mitochondria, rather than triglycerides are responsible for initial development of oxidised flavour in raw and cooked meat products during storage because phospholipids fraction is highly unsaturated and contains fatty acids with more than 3 double bonds. In addition, sub-cellular membrane contains an enzyme system capable of initiating the peroxidative reactions and are thought to be the origin of free radicals to initial lipid oxidation.

α-tocopherol is associated with biomembranes and protects them by neutralizing oxidative induced free radicals. It acts as an antioxidant by donating a hydrogen atom to a free radical. A logical hypothesis is that α-tocopherol quenches free radicals originating from lipid oxidation and in turn, protects oxymyoglobin oxidation. α-tocopherol inhibits free radical oxidation by reacting with peroxyl radicals to stop chain propagation and with alkoxyl radicals to inhibit the decomposition of the hydroperoxides and decreases formation of aldehydes. α-tocopherol behaves as a chain breaking antioxidant by competing with the substrate for the chain carrying peroxy radicals, normally present in the highest concentration in the meat system. It was suggested that vitamin E would protect lipids from oxidation through formation of a semi-quinoid structure.

Lipid Oxidation

Vitamin E treatment resulted in moderate increase of TBA values from 0.45 to 2.16 as compared to control from 1.25 to 4.6 in ground beef during refrigerated storage (4°C) for 7 days. Several investigators reported that lipid oxidation was delayed with increased concentration of α-tocopherol in meat. α-tocopherol provides a direct antioxidation effect in the lipid phase and indirectly protects water soluble components such as oxymyoglobin from pro-oxidant free radicals. The different levels of vitamin E used in the processed meat products by different researchers were 200 ppm in restructured pork, 100 and 200 ppm in ground pork, 50 ppm in ground beef, 800 ppm solution for dipping beef steaks, 10 ppm in ground buffalo meat and buffalo meat nuggets. High concentration of vitamin E has been shown to exert a pro-oxidant effect. Hence, too much addition of vitamin E to ground meat may accelerate lipid oxidation as well as pigment oxidation. Added α-tocopherol showed oxidation both in raw and cooked ground pork during refrigerated or frozen storage. Similar observation was reported in fresh ham slices and loin chops. Mitochondria and microsome lipids isolated from pork loins from swine receiving the long-term treatment (200 mg tocopherol/kg diet for 6 weeks) were more stable to oxidation than from control ones. Retention rates of vitamin E in the meat at 30 and 60 days of storage (2°C) were 61.5 per cent and 28.9 per cent, respectively. The vitamin E contents in pork during storage at –2°C decreased sharply. Vitamin E supplementation improved immune response, increased serum and muscular α-tocopherol levels, causing delayed metmyoglobin formation in a dose-dependent manner. Supplemental vitamin E was more effective in meat from beef steers than from dairy cattle. In the muscle from vitamin E supplemented pigs, the development of TBARS remained low during storage at 4°C. The tocopherol levels of muscles tissue did not change during storage. Vitamin E supplementation (250 mg/kg) or spraying directly onto turkey meat after slaughter resulted in lower TBARS values than controls during storage at 2-4°C. Exogenously added α-tocopherol reduced lipid oxidation and decreased myosin denaturation during heating. Use of 100 ppm a-tocopherol for pre-blending of ground buffalo meat significantly reduced TBARS number and increased odour score during refrigerated storage. Cured reformed cooked turkey ham and cured restructured cooked turkey patties manufactured from turkeys breast meat obtained from turkeys supplemented with dietary α-tocopherol acetate showed better oxidative and colour stability, when compared to non-supplemented control during refrigerated storage at 4°C. α-tocopherol levels in plasma and muscle were significantly higher in the supplemented group of rabbits, which also showed an increased oxidative stability in both raw and cooked meats.

Pigment Oxidation

Supplementary swine diet with α-tocopherol acetate and 15 per cent ground flaxseed improved pigment stability in the fresh ham slices and loin chops throughout a 6 days retailed shelf life display. Increased colour stability and decreased drip loss were observed on keeping pork chops previously frozen for 3 months at 4°C under fluorescent light for 10 days. Vitamin E supplementation to the beef cattle caused delay in metmyoglobin formation of beef in a dose-dependent manner during storage. Longissimus muscles for Holstein steers supplemented with vitamin E at 500 or 2000 mg per day showed less surface metmyoglobin (MMb) accumulation than control during 12 days storage at 4°C.Vitamin E also diminished the adverse effect of temperature abuse of muscle at 25°C for 24 h. Vitamin E supplementation or spraying directly on turkey meat resulted in a lower myoglobin oxidation ($P<0.05$) and a higher Hunter 'a' value at day 2 of storage at 2-4°C. Ground buffalo meat samples treated with 10 ppm α-tocopherol acetate significantly lengthened the desired visual colour, had higher Lovibond tintometer red colour units, chroma and lower MMb contents, indicating pigment oxidative stability during refrigerated storage at 4±1°C. There is a positive correlation between MMb content and TBARS number.

Physico-chemical Quality

Muscle tenderness or muscle pH was not affected due to α-tocopherol acetate in ham slices and loin chops during retail display conditions. TA at 10 ppm level in ground buffalo meat showed lower cooking loss during refrigerated storage. Dietary α-tocopherol acetate supplementation (200 mg/kg diet) to the hybrid rabbits improved the physical traits of the meat, significantly reducing shear force and increasing WHC, n-3 fatty acids in raw as well as cooked meats, while decreasing thrombogenic index.

Addition of antioxidant tocopherols (natural) or BHT (synthetic) prior to irradiation of chicken meat showed synergistic effect in lowering FFA content and TBARS number during chilled storage. Adjusting pH values of ground raw poultry meat to neutral or alkaline showed increase in TBA values during refrigerated storage.

Microbiological Quality

No difference ($P<0.05$) in bacterial load was observed in longissimus muscle from Holstein steers supplemented with vitamin E at 500 or 2000 mg/day during storage. Similar observation was made in ground buffalo meat treated with 10 ppm α-tocopherol acetate during refrigerated storage. In general, use of high vitamin E beef versus control beef in patty manufacture had no major effect on populations of aerobic bacteria, coliforms, sorbitol negative bacteria or *L. monocytogenes* in ground beef patties displayed at 4 or 12°C.

Sensory Quality

In case of pigs and poultry, high n-3 fatty acids concentration in meat are associated with fishy flavours, whose development can be prevented with high dietary levels of the antioxidant vitamin E. There is a relationship of TBA values to the development of off flavours.

Interactive Effect of Vitamins C and E

The mixture of TA and AA exhibiting a strong synergistic effect is well recognised. Evidence based on pulse radiolysis technique and electron spin resonance studies supports a redox mechanism, involving reduction of the tocopheroxyl radical intermediate by ascorbic acid to regenerate α-tocopherol. By this synergistic mechanism, TA and AA can mutually reinforce one another by regenerating the oxidised form of the other. Another mechanism for synergism involves the metal inactivating effect of ascorbic acid. These are readily decomposed in the presence of metals and their activity is significantly improved by adding a metal inactivator. The lipophillic derivative ascorbyl palmitate is also known for its synergistic activity with natural tocopherols in vegetable oils, where it is more active than BHT and BHA. The mixture of TA, AA or ascorbyl palmitate and phospholipids are also known for their good synergistic activities.

Pigment as well as the lipid oxidation was inhibited by combined treatment of TA and ascorbate in oxymyoglobin liposome model and in ground beef. Ascorbate alone enhanced lipid oxygen radical propagation but its prooxidant effect was reversed, when TA was incorporated. Vitamin E acted as primary antioxidant to scavenge free radical in lipid phase and ascorbate effectively regenerated α-tocopherol acetate, reacting with vitamin E radical and thus prolonged the antioxidant life of α-tocopherol acetate.

Dipping the beef loin steaks into ethanol containing solution of α-tocopherol acetate and ascorbate delayed lipid oxidation significantly, compared to either α-tocopherol acetate or ascorbate alone during 13 days in cold room at 4°C. 10 ppm vitamin E + 500 ppm vitamin C showed the strongest synergism to inhibit pigment and lipid oxidations of beef.

The use of vitamin E and/or C (ascorbic acid or ascorbyl palmitate) for oxidative stability of meat products under typical processing and storage conditions was reviewed. Experimental studies were used to obtain a more detailed understanding of the interaction mechanism of these vitamins and to further optimise the use of these antioxidants in processed meats. The benefits of combining natural tocopherols with vitamin C in turkey meat products are demonstrated in ground buffalo meat, the combined treatment of 500 ppm SA, 10 ppm TA and 0.5 per cent sodium tripolyphosphate improved pigment and lipid stabilities and functional properties during refrigerated storage. Similar observations were made in pre-blended vacuum-packaged ground buffalo meat. The above workers in another experiment with buffalo meat nuggets (BMN) treated with the above additives noticed a significantly ($P<0.05$) higher pH, W-B shear force, moisture, protein, better sensory attributes and lower free fatty acid contents in comparison with control samples. Use of natural antioxidants and vacuum-packaging extended the shelf life of BMN from 10-30 days under refrigerated storage.

Conclusion

Lipid oxidation, particularly oxidation of phospholipids limits the keeping quality of the meat and meat products during storage, especially of pre-cooked meats due to the development of unacceptable warmed over flavour (WOF) by autooxidation processes. The meat containing more myoglobin is more prone to such oxidative changes, because the oxidised meat pigment is the more potent catalyst than inorganic iron for the initiation reaction of the autooxidation process. Moreover, different free radicals generated during such oxidative changes, are injurious to human health. There is an urgent need to prevent the pigment and lipid oxidations by use of natural antioxidants, because of a public disapproval of chemical/synthetic antioxidants, *e.g.*, BHA, BHT, TBHQ etc. in meat due to their residual effect. Vitamins C and E are two such naturally occurring potent biological antioxidants. Vitamin C is more effective in aqueous phase and E in lipid phase to inhibit oxidative changes in meat. Ascorbic acid reduces tocopheroxyl radical to form α-tocopherol, thereby increasing the antioxidant potency of vitamin E. Therefore, a synergistic antioxidant effect is observed, when vitamins C and E are used in combination in meat products. However, the optimal level of these vitamins to minimize the pigment and lipid oxidations is different in different meat systems, which needs further research.

Part V
ADDITIVES

32

Enzymes in the Modification and Processing of Plant Foods

Dharam Singh, Santosh Dhillon and Rakesh Kumar

Department of Biochemistry,
CCS Haryana Agricultural University, Hisar – 125 004, India

In food industry, plant products are an essential source of proteins, lipids and starch. The plant raw materials that are converted to foods and food materials contain hundreds of different enzymes which are important agents in the growth, maturation and ripening of food crops. Some of these enzymes continue to be active even after harvest until all the substrate is exhausted or the enzyme is denatured due to change in pH or any other reason. Foods as we know are largely composed of starch, proteins and lipids and smaller molecules such as amino acids, sugars, vitamins and minerals. Of these, macromolecules can be degraded and modified by the endogenous enzymes (Poultanen, 1997). This is the prime reason for the importance of enzymes in food industry. In fact, there is no food without enzymes. They can be used to change colour, texture, flavour, aroma and nutritional qualities of foods. Enzymes are used at large scale to modify and process various food constituents to improve digestibility, palatability, keeping quality, ease of preparation and yield of edible fraction from the raw material. This is particularly more true of fruits and vegetables.

The raw material at the start of processing contains a wide range of enzymes. Some of these will be active during and after the process of modification which could be either beneficial or harmful to the processing of food. A good example of this is the action of pectinases which are often useful in the extraction and clarification of fruit juices. However, excessive pectinase activity in canned fruits can result in over softening and deterioration of the product. Other such undesirable enzyme activities are seen in proteolysis of dough proteins, rancidity in fat containing foods and autolysis in fruits and vegetables kept in cold- stores. Endogenous enzymes are also well known for their adverse effects in preharvest sprouting. Along with endogenous enzymes, many food modifications and processors use exogenous enzymes to improve the end products. Exogenous enzymes are a good alternative to chemical additives in baking where the application of microbial enzyme is expected to increase.

Enzymatic Hydrolysis/Modification of Proteins

Proteins, as a result of the cleavage of peptide bonds are broken down into peptides of different sizes and free amino acids. This degradation termed hydrolysis can be carried out by enzymes, acids or alkali. Acid and alkaline hydrolysis tends to be difficult processes to control and yield products with reduced nutritional qualities. Chemical hydrolysis can destroy L-form amino acids, produce D-amino acids and can form toxic substances like lysino-alanine. Enzymatic hydrolysis is developed under mild conditions of pH (6–8) and temperature (40–60°C) avoiding the extremes usually required for chemical and physical treatments. The overall amino acid composition of enzymatic protein hydrolysates is similar to that of the starting material. Besides, protein hydrolysates show advantages such as improved solubility, heat stability and relatively high resistance to precipitation by many agents such as pH or metal ions. The excellent properties of plant protein concentrates and isolates as substrate for proteolytic enzymes are well known (Clemente *et al.*, 1999). A high content of proteins and low levels of polyphenols, sugars and protease inhibitors facilitate the control of hydrolytic process and increasing yield. The main drawback of using legume protein as source of hydrolysate for meeting nutritional requirements is their low levels of some essential amino acids (S-containing).

The functional properties of proteins in food matrix are highly influenced by their molecular structure, protein interactions with other components *e.g.* water, other proteins, carbohydrates, lipids and ions, etc. Apart from sensory properties, typical functions of proteins in food systems include adhesion and cohesion, emulsification, gelation, foaming, texturization, water, lipid and flavour binding and retention. Enzymatic modifications are used to facilitate desirable changes in properties such as solubility, flavour enhancement and in case of medical, dietary, infant and animal foods an increased digestibility. The most common effect of any modification of a protein is to reduce the size of the protein molecule by proteolytic cleavage and thereby alter the primary, secondary and tertiary structure (Kinsella, 1976; Clemente, 2000).

Table 32.1: Protein Modifications Using Enzymes

Protein	*Change*	*Benefits*
Soya	Solubility	Improved digestibilities and nutrition
Wheat	Flavour, hydration, rheology and solubility	Improved baking performance and digestibility
Pea	Flavour, solubility	Food ingredients
Maize	Solubility	Animal feed

The enzyme modification/hydrolysis studies are extremely delicate and require a strict control. To date improvements in one or more of the functions of emulsifying, whipping, foam stability and gelling have been modest. The process of enzymic hydrolysis is not commercially viable due to lack of method to halt enzymatic action at a precise stage of hydrolysis. As a result some of the desirable functions of proteins can be lost or greatly reduced during the processing of foods. These losses need to be avoided because many modern foodstuffs require protein with high degree of functionality. Recently, Motoki and Seguro (1998) have reported new foods with unique protein modification of functional properties through the action of microbial transglutaminase (MTGase). The following examples illustrate some practical uses of MTGase in food protein modification:

1. Maintenance of smooth texture of 'tofu' for a long time.
2. Improved strength of 'yuba', edible soybean film, a traditional delicacy of Japan.

Role of Endogenous Enzymes in Food Quality

Colour

It is probably the first attribute that is linked with quality and acceptability of foods by the consumers. The quality of many fresh vegetables and fruits is judged on the basis of their greenness. On ripening the green colour of many fruits decreases and is replaced by red, orange, yellow or black colours. In such fruits and vegetables, chlorophyll content decreases with maturity and all these changes are a result of enzyme action. Two main enzymes responsible for chemical alterations of pigments in fruits and vegetables are lipoxygenase and polyphenol oxidase (Richardson *et al.*, 1996).

Lipoxygenase

This enzyme has six important effects on foods, some desirable and other undesirable. Two desirable functions are:

1. Bleaching of wheat and soybean flours
2. Formation of -s-s-bonds in gluten during dough formation

Undesirable effects of lipoxygenase in foods include:

1. Destruction of chlorophyll and carotenes.
2. Development of oxidative off-flavours and off-aromas.
3. Oxidative damage to vitamins and some proteins
4. Oxidation of essential fatty acids; linoleic, linolenic and arachidonic acids.

Polyphenol oxidase

Action of polyphenol oxidases on various phenolic substrates result in the formation of melanins, the polymerization product. In fact, this product of polyphenol oxidase action is responsible for the undesirable brown discolouration of many fruits (bananas, apples, peaches, potatoes, mushrooms, etc.) and the desirable brown and black colours of tea, coffee, raisins and human skin pigmentation. Changes in texture, taste and nutritional quality also result from browning reaction. It is estimated that up to 50 per cent of some tropical fruits are lost due to enzymatic browning. This reaction is also responsible for the deterioration of colour in juices and fresh vegetables such as lettuce. Elimination of oxygen and phenols can prevent browning. Secondly, addition of reducing agents, such as ascorbic acid, sodium bisulfite and thiol compounds can also prevent browning due to reduction of the initial products, o-benzoquinone back to the substrate that is by preventing melanin synthesis (Richardson *et al.*, 1996).

Table 32.2: Some Complex Carbohydrate Degrading Enzymes Important in Food Texture

Enzymes	*Remarks*
Pectin methylesterase	Increase textural strength due to cross bridges between calcium and carboxyl groups of pectic acid
Polygalacturonase	Decrease textural strength of some raw foods, such as tomatoes by hydrolyzing pectic acid
Cellulases	Causes softening of green beans, still a matter of controversy
Pentosanases	Hydrolyze xylans, arabans and arabinoxylans to small components and causes softening of cell wall
Amylases: three types α-amylases; β-amylases and Isoamylases (α-1,6)	Their action influence texture by decreasing viscosity of starch due to hydrolysis, products are linear dextrins.
Pullulanase (α-1, 6)	Acts on pullulan to give maltotriose and on starch gives linear dextrins
Cyclomaltodextrinase (α-1,4)	Acts on cyclodextrins and linear dextrins to give maltose and maltotriose
α-glucosidase (α-1, 4)	Produces α-glucose

Texture and Shelf Life

Texture is a very important quality attribute in foods. In fruits and vegetables, texture is primarily due to the complex carbohdyrates: pectic substances, hemicelluloses, celluloses, starch and lignin. There are one or more enzymes that act on each of the complex carbohydrates that are important in food texture. Proteases are important in the softening of animal tissues and high protein plant foods. Most of the published data on the use of enzymes in baking relates to the improvement of bread volume, texture and shelf life. α-amylases have a long history of use as flour improvers. One suggested explanation of positive effects of α-amylase in reducing staling is that the enzyme produces low molecular weight branched chain starch polymers as hydrolysis products which interferes with amylopectin recrystallization (Akers and Hoseney, 1994; Poutanen, 1997)

Proteases

Texture of food products is changed by hydrolysis of proteins both by endogenous and exogenous proteases. Gelatin will not gel when raw pineapple is added because the pineapple contains, bromelain, a protease. Protease activity on gluten protein of wheat bread doughs is important in rising and mixing characteristics. Generally proteases are used to modify gluten properties and therefore, specific action is needed. Gluten properties are governed by complex associations between two types of proteins, gliadin and glutenin.

Flavour and Aroma Changes

Chemical constituents contributing to the flavour and aroma of foods are numerous but it is very difficult to determine their critical combinations. Also, it is equally cumbersome to identify the enzymes associated in the biosynthesis of flavours and in the development of undesirable flavours. Enzymes cause off-flavours and off-aromas in foods, particularly during storage. Improperly blanched foods, such as green beans and green peas, corn, broccoli and cauliflower develop off-flavours and off-aromas during frozen storage. The roles of enzymes in the development of bread flavour were recently reviewed by Martinez-Anaya (1996). Both endogenous and added enzymes as well as those of yeast and starter cultures provide precursors of aroma compounds. Degradation products of carbohydrates and proteins may react with one another and proteins may be converted into bitter peptides. In some cases, specific exogenous enzymes are added primarily to improve aroma (Noomhorn and Bandola, 1994). Flavour aspects of tea are also partly under enzymic control. Lipoxygenase activity produces the leaf aldehyde-'hexenal' which is important in tea aroma. Many of the 200 or so known tea volatiles are terpenes which probably emanate from enzymic breakdown of carotenoids during fermentation.

Lim *et al.* (1989) and Vilasco *et al.* (1989) demonstrated that lipoxygenase is responsible for off-flavours and off-aroma development in green peas, green beans and corn and that cysteine lyase is the primary enzyme responsible for off-flavour development in broccoli and cauliflower. Naringenin known for bitter taste of grape fruit and its juice can be destroyed by treating the juice with naringinase. Another bitter principal, the triterpenoid limonin, is also found in citrus, and particularly in Navel oranges which can be removed enzymatically (Lea, 1995).

Bitterness in Proteins

Native food proteins do not contribute significantly to flavour, their large molecular size suggests that they cannot interact with taste receptors. The products of enzyme hydrolysis, peptides and amino acids have distinct tastes such as sweet, sour, salty, umami and bitter. Many of these flavours contribute in a positive way to food flavour while presence of bitterness has been a major limiting factor in the use of protein hydrolysate in food and health care products. Bitter peptides contain typically 3-15 amino acids and are characterized by the presence of hydrophobic amino acids, such as leucine, isoleucine, proline, valine, phenylalanine, tyrosine and tryptophan. Almost all peptide containing these amino acids are bitter and the intensity of bitterness has been found to be proportional to the number of amino acids and the size of peptide.

Peptidases and Debittering

The food industry has employed a range of methods to overcome bitterness in hydrolysed proteins such as masking with other flavourings (salt, sugar, monosodium glutamate, binding with polyphosphate and gelatin and extraction with charcoal) and limiting hydrolysis. For some food products, these methods can achieve the desired effect, however, they all have their limitations. The use of peptidases to reduce or eliminate bitterness has increased significantly in recent years (Pawlett and Bruce, 1996). The effectiveness of peptidases (aminopeptidases and carboxypeptidases) in debittering will largely depend on their specificity. Obviously those peptidases that can cleave hydrophobic amino acid residues and proline are most valuable. Casein hydrolysates produce a high level of bitter peptides, therefore, large amounts of peptidase or very long reaction times are required for debittering.

Nutritional Quality

Relatively little data is available with respect to the effect of enzymes on nutritional quality of foods. Lipoxygenase oxidation of polyunsaturated fatty acids (linoleic, linolenic and arachidonic acids) certainly decreases the amounts of these essential fatty acids in foods. The free radicals generated by lipoxygenase catalyzed oxidation of PUFAs decrease carotenoids (vit A precursors), tocopherols (vit E) ascorbate (vit C) and folate content of foods. The free radicals are also damaging to cysteine, tyrosine, tryptophan and histidine residues of proteins. Ascorbic acid is destroyed by ascorbate oxidase found in some vegetables, such as squash. Thiaminase destroys thiamine, an essential cofactor involved in carbohydrate, fat and amino acid metabolism. Polyphenol oxidase caused browning decreases the available lysine content of proteins (Richardson and Hyslop, 1985).

Enzymes Used as Modification and Processing Aids

Enzymes are ideal for producing key changes in the functional properties of food removal of toxic constituents and producing new ingredients. This is because they are highly specific, act at low temperatures (25-45°C) and do not produce side reactions. There are some major successes in the use of food related enzymes. Production of high fructose syrups (HFCs) is one example. This invovles a relativley heat stable α-amylase, glucoamylase and glucose isomerase. Glucose or glucose-fructose mixture is used as sweetner or glucose is recycled to produce more fructose.

A second example is the use of aminoacylases to separate racemic mixtures of DL-amino acids. A third example is the use of specific lipases to tailor lipids with desirable chemical and physical properties with respect to melting point, unsaturation or specific location of fatty acids in a triacylglycerol. For example a 1,3-specific lipase catalyzes the transesterification of fatty acids at the 1- and 3-positions of glycerol, a 2-specific lipase catalyzes transesterification only at the 2-position of glycerol, and a nonspecific lipase catalyzes transesterification at the 1-, 2- and three positions of glycerol. An example using 1,3- specific lipase is as follows:

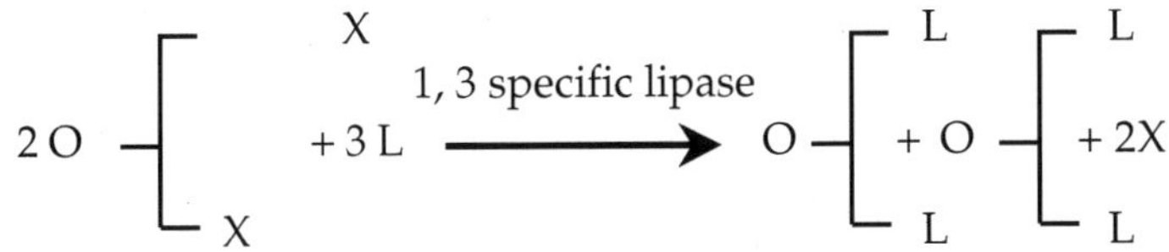

Fourth example is enzymatic degumming *i.e.* enzymatic hydrolysis of phospholipids. Phospholipids and phosphatides cause problems in many processes. In seed oil processing, they cause gumminess and different processes have been developed to remove them from the oil. Some phosphatides can be removed by water and those remaining need to be removed by chemical degumming process. Many methods have been suggested for the removal of iron from edible oils including acid treatments and use of chelating agents such as EDTA. Currently used methods for degumming oils include distillative deacidification and alkaline deacidification. Neither of these processes is ideal as they can cause formation of stable emulsions. The structure of a common phospholipid (lecithnin) is shown with four different sites of four different phospholipases:

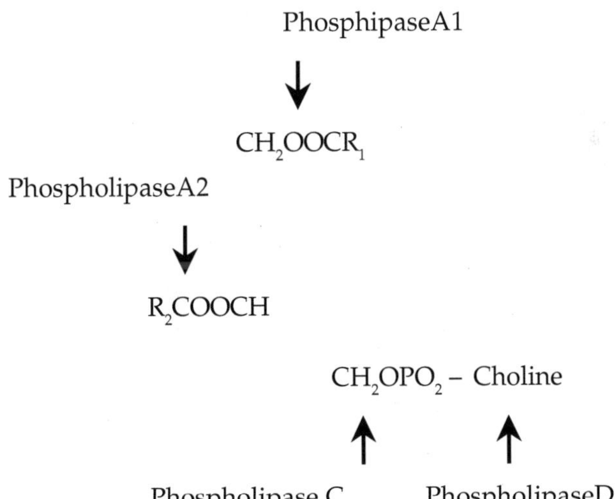

Phospholipase A1 removes fatty acid R1, A2 cleaves fatty acid R_2, phospholipase C removes phosphate/nitrogen moiety to produce 1, 2-diacylglycerol and phospholipase D removes the nitrogen moiety to produce a phosphatide. Theoretically speaking, phospholipases

A1, A2 and C all could be used to degum oils but at this time only A2 is commercially available. Ideally, the best phospholipase to use would be C as the phosphate/nitrogen moiety is very water soluble and hence easily removed (West, 1996).

Enzymes and Fruit Juices

Among the food industries, the fruit juice industry has to deal with the most delicate problems. Today the main aim of this industry is to process fruits at the lowest cost while maintaining rather improving the organoleptic quality and stability of the finished product. Juices are concentrated to decrease the volume and to improve their stability by preventing spoilage. This objective has been made possible through the use of enzymes at different stages of the process as it is not possible to produce clear concentrated juice without enzymes. In recent years, new activities have been discovered in relation to fruit biochemistry. These include new pectinases *i.e.* rhamnogalacturonases, pectinacetylesterases and a new glycosidase (*i.e.* apiosidase) for apple (Grassin and Fauquembergue, 1996).

Exogenous enzymes are the tools of choice to modify fruits and vegetables and to improve the quality of juices and their stability. Enzymes are used during pulp maceration and liquefaction or depectinization. The addition of exogenous enzymes *e.g.* pectinases give smooth texture and at the same time preserve colour and vitamins. Pectinases, hemicellulases and cellulases added to the fruit can be used to achieve almost complete liquefaction. Yields reach 95 per cent as a result of solubilization of cell wall polysaccharides. The dry matter content of juices also increase.

Table 32.3: Technical Role of Enzymes

Enzyme	*Technical role*
Pectin methylesterase (PE)	Apple juice clarification
Polygalacturonase (PG)	Fruit and vegetable maceration
PE+ PG	Juice extraction, depectination and clarification
Pectin lyase (PL)	Fruit maceration and juice extraction
PE+PG or PC+hemicellulases+ cellulases	Cell wall disruption, fruit liquefaction

Enzymes as Antioxidants and Detoxifying Agents

The spontaneous oxidation of food stuffs by atmospheric or dissolved oxygen is often a problem, causing losses in nutritional value and organoleptic quality and formation of toxins, for instance oxidation of PUFAs is a common problem. Enzymes such as glucose oxidase, superoxide dismutase, catalase, glutathione peroxidase can be used as antioxidants to remove oxygen or other reactive species from foods. Glucose oxidase is the commonest enzyme used as an antioxidant to prevent changes in colour, flavour of foods during processing and storage. Raw food materials often contain toxic or antinutrient compounds that are sometimes removed by proper heat treatment, extraction or by enzymatic reactions (Table 32.4). There are more than 12000 plants that may have potential as food sources. Many are not used because of undesirable properties, some of which could be overcome by the proper use of enzymes. Toxins or antinutritional factors can be formed by endogenous enzymes during processing of foods. Alternatively endogenous toxic substances, such as goiterogenic, thiocyanates and isothiocynates can be degraded enzymatically (Cheetham, 1996).

Table 32.4: Removal of Unwanted Endogenous Constituents by Enzymatic Reactions

Enzymes	*Unwanted Constituents*
α-Galactosidase	Raffinose, stachyose
β-Galactosidase	Lactose
Glucose oxidase	Glucose, O_2
Phytase	Phytic acid
Thioglycosidases	Thioglycosides
Catalase	H_2O_2
Superoxide dismutase	Superoxide anion
Sulfhydryl oxidase	Oxidized flavours
Urease	Carbamates
Pepsin, chymotrypsin, carboxypeptidase A	Bitter peptides
Naringinase	Bitter compound, naringin in citrus fruits
Proteases	Phenylalanine
α-amylases	Amylase inhibitors
Proteases	Protease inhibitors

Enzymes in Brewing

In the brewing of beer, enzymes are used at two stages, mashing and conditioning. In mashing, proteases, pullulanase, α-amylase and β-glucanase are added to the barley malt which has been finely grounded to increase hydration and enzyme penetration together with malt extract, cereal adjuncts and hot water to form the mash. The brewing industry has also used amyloglucosidases recently to make "light" beer. Amyloglucosidases hydrolyze α-1, 6 glycosidic bonds of amylopectin fraction, permitting the complete fermentation of starch. Use of β-glucanases may solve the high viscosity/slow filtration rate problems caused by mannans from cell walls. The most exciting advance is the use of acetoacetate decarboxylase and dehydrogenase which are used to shorten the fermentation time by avoiding diacetyl formation (Cheetham, 1996).

Table 32.5: Enzymes in Brewing

Enzymes	*Application*
α- and β-amylases	Convert non-malt starch to maltose and dextrins
Proteases (endo and exo)	Hydrolyze proteins to amino acids that are used by yeast Papain Degrade protein component of haze, complex of polyphenols, carbohydrate and protein, thus chill proofing the beer and extend shelf life
Amyloglucosidase	Hydrolyze 1,6-linkage of amylopectin permitting complete fermentation of starch (light beer)
β-glucanases	Degrade glucans to reduce viscosity and enhance filtration
Tannase	Degrade polyphenol component of haze, used for chill proofing
Acetoacetate decarboxylase	Decrease fermentation time by avoiding diacetyl formation

Enzymes in Baking

Traditional methods of baking are based on the presence of endogenous enzymes which catalyze natural changes during growth, ripening and storage. They also carry out the saccharification of starch prior to fermentation and the degradation of gluten which is very important determinant of the rheological properties of the dough. In modern bakeries, several enzymes are used but addition of amylases and proteases has been common for years (Hamer, 1991). Several enzyme preparations are available for reducing the rate of staling in bread. These include (a) a debranching enzyme (α-1,6-splitting; pullulanase type) along with α-amylase; (b) a genetically engineered α-amylase that rearranges the branched chains of amylopectin to give a linear polymer; (c) a mixture of cellulase, β-glucanase and pentosanases and (d) a mixture of cellulase, hemicellulases and pentosanases used in conjuction with fungal α -amylase. Preparations containing pentosanases are reported to increase the moisture content of rye bread products by decreasing high pentosan content of rye flour. Rye flour contains pentosans which prevent gluten formation during dough mixing (Scott, 1989; Vandam and Hilde, 1992; Lince *et al.*, 1997).

Table 32.6: Enzymes in Baking

Enzymes	*Uses*
Amylase	To maximize fermentation process, prevent staling
Proteases	By improving handling and rheological properties, improves bread quality
α-glutamyl transferase, glutathione oxidase and cysteinyl glycine-dipeptidase	Improve dough elasticity, leaf volume, crumb structure and retention of bread softness on storage
Pentosanases	In rye bread products decrease dough development time and power requirements increased moisture.
Sulfhydryl oxidase	Strengthen weak doughs by forming -s-s-links
Lipoxygenase	Causes whiteness by bleaching cereal pigments, such as carotene
Phytase	Remove phytate, an antinutrient

Several enzymes are now especially engineered and produced on a commercial scale which has markedly increased production yields and reduced cost (Vandem and Hilde, 1992; Linco *et al.*, 1997). In addition to improved dough handling properties, other advantages of adding enzymes as processing aids in baking include increased bread volume, improved crumb structure and increased shelf life. Virtually all the microbial enzymes used have a positive effect on the quality of baked foods. However, Bruce Wasserman stated in 1990, "The challenge lies in understanding the molecular basis of food functionality flavour, texture, colour, nutritional value and appearance. This effort is pre-requisite to the successful application of modern enzymology and biotechnology tools to food systems".

Medical Considerations-allergenicity

There are several well documented medical conditions that are sometimes considered to be allergies to proteins or protein fragments. These include unpleasant gastric reactions, urticaria and eczema in response to ingestion. We know very little about what makes individuals respond adversely to food constituents, especially proteins. Wheat gluten, egg-white protein and milk proteins all cause significant allergic reactions in susceptible individuals. There is a substantial industrial activity to produce protein hydrolysates that are hypoallergenic. It is reported that hydrolysates containing 40-60 per cent amino acids and peptides upto 12 amino acid residues are hypoallergenic. However, the reports are conflicting because different individuals sensitized differently to the same protein and their hydrolysates. Before prehydrolyzed proteins and their hydrolysates can be consumed safely by sensitive individuals, more research is needed on these and other allied questions (Godfrey, 1996).

Conclusions

Both endogenous and exogenous enzymes have an important effect on the quality of plant foods. Amylases and proteases have long been used in baking whereas use of cell wall degrading enzymes has increased only recently. The baking industry uses proteinases to achieve a number of different functional changes in rheological properties. However, enzymatic enhancement of gelling, whipping and foaming properties are yet to be acheived to a substantial degree. Recent investigations have shown that allergenicity of proteins can be reduced by their careful modification by enzymes. The most common effect of any enzyme modification of protein is to reduce their size either by endo or exo action or a combination of both. The products of enzymic hydrolysis, peptides and amino acids have distinct taste such as sweet sour, salty and bitter. Out of these, the bitter taste has limited the use of protein hydrolysate in food and health care products.

Exogenous enzymes (mainly of microbial origin) have turned out to be good alternative to chemical additives in various food processing and preparations. While exogenous enzymes have large role in starch processing and a smaller but significant role in protein modification, enzymes currently play no role in edible oil and fat processing. It is because necessary enzymes are not available at low costs. Furthermore, discussion on the definition of protein functionality in foods and their measurement remains inconclusive because proteins are never used as pure materials but in complex association with many other substances. Modern enzymology have understood all the ancient processes. A substantial diversity of microbial enzymes has been developed to achieve the desired protein modifications.

REFERENCES

Akers, A.A. and Hoseney, R.C. (1994). Water soluble dextrins from α-amylase treated bread. *Cereals Chem.*, 71: 223-226.

Cheetham, P.S.J. (1996). Applications of enzymes in the food industry. In: *Hand Book of Enzyme Biotechnology,* (Ed.) Wiseman, A. Halsted Press, New York, p. 323-363.

Clemente, A. (2000). Enzymatic protein hydrolysates in human nutrition. *Trends Food Sci. Technol.*, 11: 254-262.

Godfrey, T. (1996). Protein modification. In: *Industrial Enzymology* 2nd ed. (Ed.s) Godfrey, T. and West, S. McMillan Press, p. 302-326

Grassin, C. and Fauquembergue, P. (1996). Fruit juices. In: *Industrial Enzymology,* 2nd ed. (Eds.) Godfrey, T. and West, S. McMillan Press, p. 225-245.

Hamer, R. (1991). Enzymes in baking industry 2nd ed. In: *Enzymes in Food Processing,* (Eds.) Tucker, G.A. and Woods, L.F.J. Blackie, p. 168-193.

Kinsella, J.E. (1976). *Functional Properties of Proteins in Foods: A survey.* CRC Crit. Rev. Food. Ser. Nutr., 4: 219-280.

Lea, A.G.H. (1995). Enzymes in the production of beverages and fruit juices. In: *Enzymes in Food Processing,* 2nd ed., (Eds.) Tucker, G.A. and Woods, L.F.J. Blackie, p. 223-247.

Lim, M.H., Chen, A.O., Pengborn, R.M. and Whitaker, J.R. (1989). Effect of enzymes on aroma changes in vegetables during frozen storage. In: *Trends in Food Biotechnology,* (Eds.) Ghee, A.H., Hen, B.N. and Kong, L.K. Singapore Institute of Food Science and Technology, Singapore, p. 169-176.

Linko, Y.Y., Javanainen, P. and Linko, S. (1997). Biotechnology of bread baking. *Trends Food Sci. Technol.*, 8: 339-344.

Martinez-Anaya, M.A. (1996). Enzymes and bread flavour. *J.Agric. Food Chem.*, 44: 1976-1987.

Motoki, M. and Seguro, K. (1998). Transglutaminase and its use for food processing. *Trends Food Sci. Technol.*, 9: 204-210.

Muller, R. (1997). 1st European Symposium on enzymes in grain processing. *Trends Food Science Technol.*, 8: 92-94.

Noomhorn, A. and Bandola, D.C. (1994). Effect of rice variety: Rice flour concentration and enzyme levels on composite bread quality. *J. Sci. Agric.*, 64: 433-440.

Pawlett, D. and Bruce, G. (1996). Debittering of protein hydrolysates. In: *Industrial Enzymjology*, 2nd ed., (Eds.) Godfrey, T. and West, S. McMillan Press, p. 177-186.

Poutanen, K. (1997). Enzymes: An important tool in the improvement of the quality of cereal foods.*Trends Food Sci. Technol.*, 8: 300-306.

Richardson, T. and Hyslop, D.B. (1985). Enzymes. In: *Food Chemistry*, 2nd Ed., (Ed.) Fennema, O.R. Marcel Dekker, New York, p. 439-535.

Vilasco, P.J., Lim, M.H., Pengborn, R.M. and Whitaker, J.R. (1989). Enzymes responsible for off-flavour and off-aroma in blanched and frozen stored vegetables. *Biotechnol. Appl. Biochem.*, 11: 118-127.

Wasserman, B.P. (1990). Evolution of enzyme technology: Progress and prospects in food technology. 44: 118-122.

West, S. (1996). Olive and other edible oils. In: *Industrial Enzymology*, (Eds.) Godfrey, T. and West, S. McMillan Press, p. 294-301.

Whitaker, J.R. (1990). New and future uses of enzymes in food processing. *Food Biotechnol.*, 4: 669-697.

Whitaker, J.R. (1994). *Principles of Enzymology for Food Sciences*, 2nd ed. Marcel Dekker, New York.

33

Role of Functional Properties in Food Processing

Umaid Singh
Department of Foods and Nutrition,
CCS Haryana Agricultural University, Hisar – 125 004

Man has always selected the types of food crops with some consideration for their nutritional and functional. attributes, although these have not been properly understood. In this context, foods of plant origin have occupied an important place in the daily diets of the people as rich sources of proteins, carbohydrates, minerals and vitamins. The supplementation of cereals with high protein legumes is considered one of the best solutions to the world protein-calorie malnutrition, particularly in the developing countries (Singh *et al.*, 1997).

In view of the increasing utilization of plant foods, their functional properties are assuming greater significance. The functional characteristics provide a set of data which gives information on the fields of application in food formulations (Helmansson, 1979). These can be used as a guideline in product development, especially in wheat-legumes composite flours where proteins are used as major functional ingredients (Fernandez and Beny, 1989). Very often, researchers have emphasized expanding the utilization of grain legumes in the form of meal and flour for use as functional ingredients. The functional properties are provided not only by the proteins of flours, but also by complex carbohydrates and other grain components such as pectins and hemicellulose (Martinez, 1979), and interactions between lipid proteins and carbohydrates. In recent years, functional foods are also assuming greater importance and have attracted the attention of food processors, marketers and consumers (Reilly, 1994).

Definition of Functional Properties

Functional properties are physico-chemical properties, which give information on how a particular ingredient (*e.g.* protein, carbohydrate) will behave in a food system. The functional properties of plant foods are determined by the molecular composition and structure of the individual component and their interactions with one another.

About Functional Properties

The terms most appropriate to functional characterization have been used and described by researchers in the past for the purpose of their being employed for various food formulations. The following functional properties which are generally referred and have found wider applications in food formulations have been fairly discussed in this review.

1. Bulk density
2. Water and oil absorption
3. Flour solubility and swelling power
4. Foam capacity
5. Emulsification
6. Gelation capacity
7. Gel strength
8. Gel consistency
9. Paste viscosity
10. Gelatinization temperature

11. Nitrogen solubility index (NSI) and
12. Protein dispersibility index (POI).

Methods of Determination

A great deal of emphasis is always given to the determination of functional tests. Reliability and reproducibility of analysis are the major considerations in employing the methods of determination of various functional properties. Depending on the availability of facilities and expertise, these characteristics are evaluated by different researchers and food processors. A brief and simplified version of these methods is desirable and described for the benefit of uniformity of determinations by different researchers.

Bulk Density

The simple method described by Narayana and Narasingha Rao (1982) is generally used to determine bulk density. A calibrated centrifuge tube is weighed and the sample is filled to 10 ml mark by constant tapping, until there is no further change in volume. The content is weighed and from the difference in weight, the bulk density of the sample is calculated.

Water and Oil Absorption

The determination of water and oil absorption involves the most simple procedures, which produce reliable and dependable results. Generally, 1 g sample is mixed with 10 ml of distilled water or 15 ml oil for 30 min. The contents are allowed to stand at 30°C in a water bath for 30 min and then centrifuged at 3000–5000 rpm for 20 to 30 min, depending on the availability of facility for this purpose. After centrifuging the content, the volume of the supernatant is recorded and used for determination of water or oil absorption and the results are expressed as g/g sample.

Flour Solubility and Swelling Power

These characteristics are generally determined at temperatures, when the process of gelatinization of starch granules takes place. According to this procedure, 250 g finely ground defatted flour is thoroughly mixed with 15 ml of distilled water and heated to 65°C, being the initial temperature at which gelatinization of starch granules begins. The content is then cooled to room temperature and centrifuged at 5000 rpm for 10 min. The soluble solids content is calculated as percentage of flour soluble in water. At the same time, this can be used to calculate the swelling power, since the sample has been heated at 65°C for 30 min. The starches of some grain legumes may follow two-stage swelling pattern and in such situations, heating at 95°C is also suggested. After heating and centrifuging as described above, the residue is weighted and increase in weight is calculated as the swelling power of the sample at that particular temperature.

Foam Capacity

The foam ability of legume flour is a desirable characteristic for the production of some traditional foods. This is generally determined at pH 7.0. The volume of foam after whipping is expressed as the foam capacity and the volume of the foam over a time period 10–120 min is expressed as the foam stability for the respective time periods. Foam capacity measurements are also made using NaCl solution of 0.2–1.0 M concentration.

Emulsification

The oil-in-water emulsions formed are taken as an index of the emulsification capacity. According to this method, the sample (2 g) is blended in a Waring blender with 100 ml distilled water for 30 sec at high speed. The oil is added in 1 ml or 0.5 ml portions with continuous, as per the requirement. The amount of oil added upto the point where a fine layer of oil is formed, indicating that no more oil is required by the sample suspension. This is recorded as the emulsification value or the emulsification capacity of the sample.

Gelation Capacity

Sample suspensions containing 8 to 12 per cent (w/v) flour in 0.5 per cent increments are prepared in 10 ml of distilled water. Then the test tubes are heated for 1 h in boiling water, rapidly cooled under running cold tap water, and refrigerated for 3 hr at 5°C. The least gelation concentration is that concentration at which the sample does not fall down or slip from an inverted test tube. It may be emphasized that the quantity of sample for preparing suspension would depend on the gelation capacity of the legume flours, some may require higher quantity than the others.

Gel Strength

The gel strength is generally measured using a compression cell (0–5 kg full scale) in an Instron Food Testing Machine (IFTM). The gels are formed first and then compressed. For forming gels, the isolated starch samples (6 per cent starch) are heated at 95°C for 10 min, poured into a round moisture dish (5 cm diameter) and stored at room temperature for 12 h. Thus, uniformly polymerized round starch gel slabs of about 1 cm thickness are removed from the moisture dish and compression force is measured by pressing gel slab between two flat plates considerably larger than the sample slab so that no puncture takes place.

Gel Consistency

The gel consistency is determined by boiling the sample in distilled water and then cooling it to a desirable consistency. The spread of the thus formed gel is recorded as the gel consistency of the sample. According to this method, distilled water (10 ml) is added to a finely ground sample (0.5 g) in a beaker. The suspension is boiled for 20 min and after boiling, the material is transferred in petri dish of known diameter and depth (40 × 10 mm). The material is cooled to room temperature and transferred to a smooth glass surface by inveling the petri dish. The petri dish is slowly removed and the diameter of the gel that spread out on the glass is measured as gel spread and expressed as gel consistency of the sample.

Paste Viscosity

This is also called as hot paste viscosity or peak viscosity. Depending on the concentration of the starch in flour sample, viscosity values are commonly obtained with a Brabender viscoamylograph using 6–8 per cent starch concentration.

Gelatinization Temperature

This is the temperature at which the starch granules get gelatinized and stain with congo-red. The most commonly used method for determining the gelatinization temperature is similar to that followed for cereal starches. The gelatinization temperature of starch is determined by using a light microscope and congo-red (0.2 per cent) is used as a stain. An aqueous solution of flour (starch) is heated using a mini block heater and samples are taken from 60°C onwards at 1°C intervals, until the gelatinization temperature is reached. Starch granules are stained at initial, mid point (50 per cent stained) and final (90 per cent stained). The temperature at which 90 per cent of the starch granules are stained is recorded as the gel temperature of the sample.

Nitrogen Solubility Index (NSI)

The NSI is a function of the pH of the extracting solvent. The higher concentration of nitrogen is soluble in acidic and alkaline pH ranges. The NSI is generally determined at pH 6.8–7.0. The pH of the extracting solvent is adjusted by using dilute acid or alkali and the soluble nitrogen in the supernatant is determined by the Kjeldahl method.

What Does Contribute to Functionality?

The protein and starch are the principal constituents of food crops, since they serve as reserves of storage organs of plant tissues used for food by humans. The most important aspects of vegetable proteins are their functional properties (Hermansson, 1979). Sometimes functional properties are determined in order to predict the behaviour of protein ingredients. Proteins take part in a very complex food structures and hence perform a specific function in foods (Singh, 2001).

Properties of proteins contributing to the functionality of plant foods have occupied an important place and remained a point of focus of attention of food scientists and nutritionists. Total proteins have been reported to be primarily responsible for increased water retention in foods. According to these workers, the protein-water interactions occur at polar amino acid sites on the protein molecules. Most proteins contain numerous polar side chains along their peptide backbone, making them hydrophilic. Hydration of proteins is vital for several of the functional properties such as emulsion capacity, foaming, viscosity and gelation.

In addition, the functional properties are provided by complex carbohydrates and other components such as pectins and mucilages. The starch components are used as valuable ingredients in the food industry because of their versatile functional properties. The water absorption during cooking is a function of the starch granules, which are held together by hydrogen-bonding forces in the form of crystalline bundles, called micelles. When an aqueous suspension of starch granules is heated, these structures are hydrated and eventually sieving takes place, making it an inherited characteristic of starch granules. However, starch is not a good emulsifying agent primarily because it lacks polyelectrolyte character necessary for emulsion stability. This review discusses that the presence of carbohydrates and fibre in legume and cereal flours adversely affects the emulsion capacity. Carbohydrates other than starch also deserve consideration while evaluating the functional properties of various foods. Some people have reported the role of arabinogalactan (a polysaccharide) of black gram in the texture of leavened foods. Starches function as an adhesive binder, encapsulating agent film former, gelling agent, water binder, texturizer and fat sparing agent, with numerous other applications both in the food and non-food areas.

Role of Functional Properties

Since the consumer's choices for foods are expanding, the food industries are increasingly depending on the ingredients, which can provide good functional and nutritional properties to the food being marketed. Particularly, the functional properties are required for the formulation of various value-added food products.

In the first place, water is the most important component of food which plays an important role in the major changes that take place during the cooking of foods. More importantly, water absorption of legume grains particularly of the decorticated ones is directly related to the cooking characteristics and is extremely important in determining all cooking behaviour. Water absorption helps separation of the cells in the cotyledons of legumes. The cell separation, in turn, helps cooking process. The extent to which pulse grains or *dhals* absorb water on soaking is directly related to the cooking time. This characteristic has also a greater application in maintaining the soft texture cookies of cereal-legume composite flours. Proteins are important in maintaining moisture in semi-moist foods and baked cereal foods. In this context, protein functionality of grain legumes could play an important role in value-added bakery products of cereal-legume composite flours. For example, the soft textured cookies could be prepared by using soy flour as an ingredient which enhances the water holding capacity during the baking process, thus giving the finished products with a higher than usual moisture content and a soft texture. Water absorption characteristics represent the ability of a product to associate with water under conditions where water is limiting such as doughs and pastas. Some studies have suggested that the raw and heat-processed legume flour could be useful in food systems such as bakery products, which require hydration to improve handling characteristics. Hydration properties, dispersibility, water absorption, binding, swelling, and viscosity are known to directly influence the characteristics of a food system., In addition to water absorption capacity, the fat absorption capacity is also important, as it improves the mouth feel and retains the flavour.

Protein dispersibility index (PDI) is an important parameter of specification of legume flours for use in bakery as supplements to cereal flours. Utilization of food ingredients, especially high protein materials, depends to a large extent on functional properties. Nitrogen solubility profiles over a pH range are being used increasingly as a guide to protein functionality. Some correlations have been found between functional characteristics of soy flour in various applications and nitrogen solubility index (NSI). Generally, the higher the water dispersible protein in a soy protein concentrates, the better its emulsifying action. Whipping properties of soybean products expressed by foam expansion and foam stability, correlate with water-dispersible nitrogen. Easily dispersible proteins of legumes with those of cereals in a mixed diet could enhance the digestibility of cereal-legume proteins of mixed diets. Among the functional properties of proteins, solubility is probably the most critical because it affects other properties such as emulsification, foaming and gelation. The protein solubility as influenced by heat processing due to denaturation of protein plays a significant role in determining the digestibility of the legume proteins.

The extent to which the functional properties undergo changes as a result of heat treatment could help in estimating the requirement of heat for processing legumes for various food formulations. According to some studies, thick salad dressing-like emulsions were produced by soy flour heated for 30 min and peanut flour heated for 20 min. The viscosity of emulsions produced by soy flours has some applications in meat and bakery products. Viscous emulsions are formed as a result of heat treatment and these could provide a good indicator for the amount of heat required to process the legume flours for human consumption. It was replied by these workers that less heat may be required to process the peanut flour as compared to soy flour because less reduction in emulsion capacity is noticed in the former as a result of heat processing. In other words, higher percentage of reduction in emulsion capacity due to heating would indicate that higher amount of heat would be required to process the flour for human consumption.

The starch granules of pulses have greater stability against the mechanical shear than those of the fragile swollen wheat starches because of the hot paste viscosity that does not show any break down point in pulses. The viscosity patterns of starches are primarily determined by the extent of swelling of the starch granules and by the resistance of the swollen granules to the dissolution by heat or fragmentation by the shear force. Generally, legume starch pastes are more viscous than those of cereal starches, indicating that legume starches have a higher resistance to swelling and rupture than do cereal starches. These properties are very vital for making transparent noodle of legume starches, which are becoming very popular in the orient.

In a very recent investigation, the ability of pre-heated flours to form a viscous paste and to form a gel upon cooling was reported to be due to protein gelation rather than starch gelatinization in legumes. The gelation capacity and gel firmness in both legumes and cereals may also be contributed by protein molecules. A specific functional property may be essential for making a particular food product. It was suggested that cowpea flour was suitable in viscous foods like gravies, soups, baked products but may not be suitable for foods like *idli* and *dossai* where food foaming properties are essential. The foamability of cowpea flour was found to be a desirable characteristic for the production of several traditional cowpea-based food products in Nigeria *e.g.* soups and stews.

The presence of amylose and amylopectin in different proportion in different legumes might attribute to their cooking behaviour and eating quality. Generally, varieties with the highest amylose contents show the lowest peak viscosities in cereals. But this has not been clearly understood in legumes that how does the variation in amylose content would influence viscoamylographic properties. It becomes increasingly apparent that the nonamylose fraction, amylopectin of rice starch may be more influential in cooking behaviour. However, the role of amylose and amylopectin in cooking behaviour of grain legumes needs to examined in greater details.

The preparation of chinese-type noodle with defatted peanut flour and cowpea flour, alone and in combination with wheat was possible because of compatible functional properties. Noodle firmness was affected primarily by the peanut flour. In functional characterization, a set of data is obtained which gives information on the fields of application for a certain protein ingredient. This is especially important for characterization of gels, emulsions and mixtures of structures for the development of sausage formula (Helmansson, 1979). The biscuit dough containing the air-classified proteins of field pea were less sticky than that of the reference doughs and this was attributed to the functional properties of field pea proteins (McWatters, 1980). The ability of cowpea flours to absorb and retain water and oil may help improve binding of the structure, enhance flavour retention, improve mouth feel and reduce moisture and fat losses of extended meat products (McWatters and Heaton, 1979).

The starches used in baking contribute structure, increase batter viscosity and cake volume, control moisture, and increase shelf life (Mauro, 1996). Further, pregelatinized starches offer some distinct advantages in cakes and muffins. The gelatinization characteristics. have been shown to influence the cooking characteristics of rice vatieties. This aspect has not fully examined in grain legumes, particularly in view of the fact that legume flours are good supplements to cereals in making value-added products.

The bulk densities of legume flours play a key role in the formulation of weaning foods. It may be expected that decreased bulk density would be an advantage in the preparation of weaning food formulations.

References

Fernandez, M.L. and Berry, J.W. (1989). Rheological properties and sensory characteristics of wheat bread made with germinated chickpea. *Int. J. Food Sci. Technol.,* 24: 103-110.

Hermansson, A.M. (1979). Methods of studying functional characteristics of vegetable proteins. *J. Am. Oil Chem. Soc.,* 56: 271-278.

Narayan, K. and Narasinga Rao, M.S. (1982). Functional properties of raw and heat processed winged bean flour. *J. Food Sci.*, 47: 1534-1540.

Martinez, W.H. (1979). Functionality of vegetable proteins other than soybean. *J. Am. Oil Chem. Soc.*, 56: 280-286.

Mauro, D.J. (1996). An update on starch. *Cereal Foods World*, 41: 776-780.

McWatters, K.H. and Holmes, M.R. (1979). Influence of moist heat on solubility and emulsion properties of soy and peanut flours. *J. Food Sci.*, 44: 774-778.

Reilly, C. (1994). Functional foods: A challenge for consumers. *Trends Food Sci. Technol.*, 5: 121-123.

Singh, U., Williams, P.C. and Petterson, O.S. (1997). Processing and grain quality of cool season food legumes to meet the market demands. In: *Proceedings International Food Legume Research Conference*, 26-30 September 1997, Adelaide, Australia.

Singh, U. (2001). Functional properties of grain legumes flours. *J. Food Sci. Technol.*, 38: 191-199.

34

Antinutritional and Toxic Factors of Food Crops and their Removal by Processing

Umaid Singh

Department of Foods and Nutrition,
CCS Haryana Agricultural University, Hisar – 125 004

For economic and social reasons, the Indian sub-continent heavily depends on food crops as sources of calories, proteins, minerals and vitamins in the daily diets of the people. It is well recognized that the majority of food crops, particularly the food legume plants including pulses have the capacity to synthesize certain biologically active substances commonly considered to be antinutritional factors since they have been shown to affect animal and human nutrition (Liener 1979). Some of these factors have received little attention because they occur in small amounts and under normal conditions do not pose a serious health hazard. Protease inhibitor amylase inhibitors, oligosaccharides, polyphenols, phytolectins, phytates, cyanogenic glycosides, saponins, lathyrogens, alkaloids and favism factors of varied food crops are the antinutritional and toxic factors that merit attention of food scientists and nutritionists. Food crops will continue to play an important role in human nutrition as valuable sources of calories, proteins, vitamins and minerals in the daily diets of the people in the country. Systematic efforts are needed to remove or reduce the levels of these constituents through research and development activities. However, some of these constituents are of particular significance in food legumes and need special attention (Singh, 1988). The commonly occurring antinutrients in food crops and methods for their removal are discussed in the following sections.

Protease Inhibitors

They are the inhibitors of enzymes involved in the hydrolysis of proteins in the body. These inhibitors combine with the protein hydrolysing enzymes to form an inactive complex and hence reduce the enzyme activity. In addition, some trypsin inhibitors have been found to interfere with the availability of certain limiting essential amino acids of the diet. These are some reports to indicate that human trypsin is essentially uninhibited by the inhibitors of some legume species. Furthermore, it may be pointed out that ovomucoid, the egg white trypsin inhibitor, likewise is incapable of inhibiting human tyrpsin, which perhaps explain why ovomucoid has virtually no effect on the nitrogen balance of human subjects. The trypsin inhibitor and chymotrypsin inhibitor are the two most important forms of inhibitor present in considerable amounts in food crops. Depending on the varieties and growing season, grain legumes contain the highest amount of these inhibitors. Soybeans are well known for their high levels of trypsin inhibitor. These inhibitors have attracted the attention of scientists in many disciplines. Nutritionists are concerned because of their possible adverse affects on the nutritive value of plant proteins. In comparison with soybeans, peas and common beans. Indian pulses including chickpea and pigeonpea offer less problem as far as these factors are concerned (Visitpanich and Batterham, 1982). The beneficial effect of cooking of pulses has been generally attributed to the destruction of a unique class of proteins called protease inhibitors which otherwise have the ability to combine in the intestinal tracts of humans and animal in a specific manner with the trypsin and chymotrypsin enzymes. The low protein digestibility of some varieties of pulses is attributed to their high levels of protease inhibitors. The thermo-labile nature of legume protease inhibitors has long been known (Liener, 1979). The chymotrypsin inhibitors are more heat-stable than trypsin inhibitor when assayed using similar conditions. The inhibitory activities of chickpea and pigeonpea are more heat-labile under acidic conditions and are completely destroyed only when subjected to heat under acidic conditions (Singh, 1988).

Amylase Inhibitors

The proteinaceous alpha amylase inhibitors have received considerable attention from biochemists and nutritionists ever since their presence was revealed in legumes (Jaffe 1973). These inhibitors have been found in wheat, sorghum, and pearl millet. Other cereals and pulses such as mung bean, urad bean and lentil have also been reported to contain some amount of these inhibitors. However, all the pulses show lower amylase, inhibitor activity in comparison with other common beans and peas (Singh 1988). It may be emphasized here that amylase inhibitor of chickpea are more active towards pancreatic amylase than human salivary amylase. Like protein inhibitors, the amylase inhibitors are also heat-labile and these inhibitors could be inactivated by simple boiling of extracts for 10 minutes. Since pulses are usually consumed after boiling, the amylase inhibitor may not be of practical importance except when unheated seeds are eaten, where some inhibition of starch digestion by amylase inhibition may be expected.

Oligosaccharides

Pulses contain flatulence-causing sugars which are called oligosaccharides. The sugars (raffinose, stachyose and verbascose) are not digested by humans. Two enzymes invertase and alpha galactosidase are required for the complete hydrolysis of those sugars. The human GI tract does not possess alpha-galactosidase activity and hence these sugars escape digestion and go to the lower intestinal tract where microflora metabolizes these oligosaccharides to produce flatus. These sugars occur in appreciable amounts in mature seeds and comprise 30–80 per cent of the total soluble sugars depending on the pulse species. Chickpeas are well known for their higher levels of oligosaccharides and also there are large differences among varieties of chickpea.

Polyphenols

In the past decade, the nutritional importance of polyphenolic compounds has been recognized in food legumes. Polyphenols of legumes have been reported to reduce the protein digestibility in animals and humans probably by making protein partially unavailable or by inhibiting digestive enzymes. Pulses contain considerable amounts of polyphenolic compounds that are genotypically variable. The polyphenolic compounds that inhibit the activity of digestive enzymes, trypsin, chymotrypsin and amylase are higher in pulse varieties with dark seed-coat colour (Singh, 1993). Such processing practices as decortication, soaking, germination and cooking have been reported to influence the levels of polyphenolic compounds of pulses. The water-soluble nature of polyphenolic compounds in pulses has been confirmed by several workers. It seems that polyphenols may not pose a serious problem particularly for people in the regions where pulses are consumed after decostication *i.e.* in the forms of dhal and dhal products.

Phytolectins

The phytolectins are toxic factors that interact with glycoprotein on the surface of red blood cells and causing them to agglutinate. Food legumes have long been known to agglutinate the red blood cells. However, hemagglutinating activity has been detected in over 800 different plant species, of which over 600 are in the family leguminosae. Chickpea produced a certain amount of agglutinating activity in cow erythrocytes although not at a toxic level (Singh, 1988). The phytolectins present in legumes are known to exhibit different degrees of specificity depending on the animal species tested. Using human blood cells, some studies of the hemagglutinating properties of some pulses has indicated that pulses have low lectin activity *i.e.* below the toxicity level. Moreover, the phytolectins are highly sensitive to heart treatment and these compounds will not be serious consequences when pulses are consumed after boiling.

Phytates

In the plant kingdom, phosphates are stored in seeds in combination with inositol or as phytates. The phytic acid implies myo-inositol hexakis (dihydrogen phosphate). The Ca-Mg salts of phytic acid are designated as phytin. In most legume seeds, phytate phosphorus accounts for about 80 per cent of the total phosphorus. Most of it is present in the outer aleurone layers of cotyledons. In some legume seeds, phytic acid represent over 90 per cent of the total phytate in the seed. The phytates are largely blamed for complexing dietary essential minerals in legumes and pulses, and thus rendering them poorly available to the systems. Zinc appears to be influenced most by the phytates. Phytates also interact with proteins, resulting in reduced protein solubility and availability. Such a reduction in protein solubility adversely affect its functional properties. The phytates are generally considered to be fairly heat stable. Among the processing methods, germination and fermentation appear to be quite effective in decreasing the phytate concentration.

Cyanogenic Glycosides

The cyanide compounds are widespread throughout the plant kingdom. The cyanogenic compounds in pulses present are cyanogenic glycosides which liberate a carbonyl compound and hydrogen cyanide (HCN). Some pulses *e.g.* pigeonpea and chickpea may contain traces of such compounds below the toxicity level. Among the legumes, the lima beans contain the highest amounts of cyanogens and lima beans with colored testas are likely to have high cyanogenic glycosides.

Lathyrogens

Lathyrism is a paralytic disease affecting the lower limbs in humans and is associated with the consumption of kesari dhal or chick pea. Its outbreaks are mostly confined to some tribal regions in Central India, especially during drought periods, under which this particular pulse species does well. The danger point is apparently reached when daily consumption of kesari dhal exceeds about 300 g per person per day. Human neurolathyrism is attributed to a neurotoxin called β-N oxalyl-diaminopropionic acid. This toxin is also known to occur in other lathyrus spp. as well as in *Vicia* spp. In addition, some varieties of lathyrus contain other compounds that could be lathyrogenic, among them α-γ-diaminoleutyric acid and β-cyamoalanine which are also known to produce neurotoxic effects when injected into animals. These toxins may also be removed by simple procedures such as dehulling, cooking the seeds in excess water and draining, moist heat treatment or roasting at 150°C for 20 minutes. These processing practices remove considerable amount of lathyrogens from the legume spp. being consumed as human food.

Saponins

Saponins are glycosides that occur in a wide variety of plants. They are generally bitter in taste and have the capacity to foam in aqueous solutions. Their ability to haemolyze the red blood cells is also fairly known. Saponins have been reported in soybeans and jack beans, French beans and green peas. But they may not be present in pulses or present in much lower quantities and hence are normally not considered to be a health hazard. Some reports have indicated chickpeas contained some noticeable amount of saponins. However, such results need to be confirmed.

Alkaloids

The alkaloids are not common in legume except in lupin for which they are a major cause of concern. However, lupins are mostly used as animal feed. In recent years, some efforts have been made to process lupin for human consumption. The major alkaloids in lupin are lupanine, and hydroxylupanine. Alkaloids can be removed by soaking the seeds in salt water. In recent years, some efforts have been made to develop lupin varieties with very low level of alkaloids or devoid of alkaloids.

Favism Factors

Favism is a disease characterized by haemolytic anemia. It affects individuals after they have ingested fresh, raw or cooked broad beans (*Vicia faba*). The red blood cells of individuals who are prone to favism exhibit a number of biochemical abnormalities–the most significant of which are reduced levels of glutathione and glucose-β-phosphate dehydrogenase activity.

Recent evidence suggests that the substances responsible for favism are pyrimidines that occur naturally as β-glycosides. These substances known as divicine and isouranil are the aglycone moieties of vicine and convicine, respectively. These compounds decrease the glutathione content of RBC and thus account for the haemolytic effects. Some non-protein amino acids present in some legumes are considered as toxic amino acids. One such amino acid (canavanine) is responsible for favism. Dihydroxyphenylalanine (DOPA) occurs in a number of legumes. DOPA is not directly toxic, but plants containing it turn black as they dry the black compounds turn the seeds less nutritious. DOPA is also thought to be associated with metabolic disorder favism. However, these are few indications that such amino acids are available in food stuffs in amounts large enough to cause some deleterious effects. Some of these factors are highly soluble in water and may be discarded during the cooking process.

References

Jaffe, W.G. (1973). Toxic factor in beans and their practical importance In: *Nutritional Aspects of Common Bean and Other Legumes,* (Ed.) Jaffee, W.G. Arch Latinoamer Nutr, Venezuela, p. 199-210.

Liener, I.E. (1979). Significance for humans of biologically active factors in soybeans and other food legumes. *J. Am. Oil Chem.,* 56: 121-128.

Singh, U. (1988). Antinutritional factors of chickpea and pigeonpea and their removal by processing. *Plant Foods for Hum. Nutr.,* 38: 251-261.

Singh, U. (1999). Cooking quality of pulses. *J. Food Sci. and Technol.,* 36: 1-14.

Singh, U. (1993). Protein quality of pigeonpea as influenced by seed polyphenols and cooking process. *Plant Food Hum. Nutr.,* 43: 171-179.

Visitpanich, T. and Batterham, E.S. (1982). Nutritional evaluation of chickpea and pigeonpea. *Proc. Aust. Soc. Animal Prod.,* 14: 665-670.

35

Food Biopreservatives

A.K. Singh and D.K. Thompkinson
Dairy Technology Division , National Dairy Research Institute, Karnal – 132001

The use of components of naturally occurring antimicrobial system in the preservation of foods has still to enter a period of exponential growth. With the possible exception of Japan, many countries have not invested research efforts in this area. Use of natural antimicrobials along side existing preservatives is an area which needs more investigation particularly in view of the synergistic effects between these natural compounds that could lead to reduction in the levels of preservatives required in the food system. This attains special consideration at a time when general public is concern about food additive per se.

Introduction

The pattern of contemporary life styles necessitates preservation of food stuffs in a safe, economic and effective manner. Biotechnological developments in recent years have paved way for newer approaches for food preservation. Early attempts to preserve foods were based on readily available substances and processes. Earlier approaches involving the use of chemical preservatives to inhibit growth of spoilage bacteria in modern food systems have been recognised as deterrent to health. Today's customer demand product with no chemical preservatives, free from additives, natural with assured safety and better shelf life. This has led to the attention on the possible exploitation of natural substances from plants, animals and micro-organisms as natural preservatives. A great deal of interest has been generated recently, for use of suitable and effective natural biopreservative for the safe usage in various processed foods system.

The availability of natural food biopreservative can be sourced to plants, biological fluids and micro-organisms that are capable of producing substances that act as preservatives. Being natural they have superb stability under processing conditions that make them ideal bio-preservatives for use in foods.

Microbial Metabolites

Starter cultures ensure a high degree of safety in the course of production in modern dairies and milk processing plants. These cultures are protective to human health and are capable of forming substances like bacteriocins, by their metabolism, that inhibit their own species and other micro-organisms. Bacteriocins are low molecular weight proteins produced by many strains of starter bacteria which are bacteriocidal to closely related strains and species (Table 35.1). Many genera of starter bacteria namely *Lactococcus, Leuconostoc, Pediococcus, Lactobacillus, Bifidobacterium* and *Propionibacterium* etc. have been identified as bacteriocin producing stains (Daeschel, 1990; Ray, 1992; Tagg, 1992). In the recent past there has been an explosion of research in the area of bacteriocins from food grade starter bacteria for their future and possible use as food bio-preservative. Being produced from food grade bacteria, their stability under conditions of processing and degradation by proteolytic enzymes of human gastrointestinal tract, without any recognisable problems, make them ideal bio-preservatives for food usage. The bacteriocidal activity of bacteriocins are retained at wider temperature and pH range than normally used in food processing and for relatively long periods under storage conditions.

Table 35.1: Bacteriocin Producing Lactic Acid Bacteria

Microorganism	*Bacteriocin Type*	*Antibacterial Spectrum*
Bifidobacterium	Bifidin	Bifidobacteria, *E. coli*, other pathogens.
Lactococcus lactis subsp.*lactis*	Nisin	*Micrococcus*, *Listeria*, *Staphylococcus*, *Bacillus*, *Clostridium* etc.
Lactococcus lactis subsp.*cremoris*	Diplococcin	Wide range of gram positive
Lactobacillus brevis	Lactobacillin/ lactobracin	*L. brevis* and other related strains of LAB
L. plantarum	Lactolin/Plantericin	*L. brevis*, *L. acidophillus*, *L. viridescens*, *L. xyloses*, *E. faecalis* etc.
L. acidophillus	Lactocidin/Acidolin/ Acidophilin/Lactacin	Narrow spectrum antibiotic- *L. leichmanni*, *E. coli*, *L. helvaticus etc.*
L. helvaticus	Helveticin	*Lactobacillus* ssp. of closely related strains
Leuconostoc mesenteroids	Mesenterocin 5	Enterococcus, *Micrococcus*, *Listeria* spp, *Staphylococcus* etc.
Pediococcus acidolactis	Pediocin A/AcH	*B. cereus*, *Listeria* spp, *Staph. aureus*, *Clostridium perfringes*, *Cl. botulinum*, *Pediococcus*, *Leuconostoc* etc.
Propionibacterium jensenii	Jensenin G	Limited inhibitory effect- *Propionibacterium* and related strains

Nisin

It is the first bacteriocin to be characterised from lactic acid bacteria (LAB) of the genera *Lactococcus*. It is the most thoroughly studied bacteriocin and is the only one approved by regulatory agencies for use in dairy products as bio-preservative. It is commercially produced under the brand name NISAPLIN. It is inhibitory to many strains of gram positive bacterial cells and their spores namely- *Micrococcus, Clostridium, Bacillus, Staphylococcus* etc. Nisin acts as a cationic detergent and produces bacteriocidal effect on cells of gram positive bacteria by destabilising functions of cytoplasmic membrane. In some strains it also causes cell lysis (Ruhr and Sahl, 1985). The level of nisin production differs among LAB strains. The active molecules are present both in medium and on cell surface and the proportion is dependent on pH (Yang *et al.*, 1992). The effectiveness of nisin has been tested against many spoilage and pathogenic bacteria in different food systems. The extent of inhibition have been different in various studies and products (Hurt, 1981; Delves-Broughton,1990). Recent work using nisin in combination with other compounds has revealed that there is a potential to extend its effect to a wide range of gram negative bacteria. Nisin has been used successfully in controlling heat resistant thermophillic spores in commercially sterilised milk and as preservative in both low and high acid canned vegetables (Flower and Gasson, 1991). The future will see a change in nisin technology that is likely to be in the design of modified nisin molecule by genetic methods, which could result in increased microbial spectrum and increased activity in food systems, along with other antimicrobial substances for synergistic effect. Nisin in conjunction with pediocins results in pronounced synergy against various pathogens (Hanlin *et al.*, 1993).

Pediocin

After nisin this is the next most studied bacteriocin from LAB. It has relatively wide host range. Pediocins are produced from species *Pediococcus pentosaccus* and *Pediococcus acidilactis* generally used for food fermentation. Several strains of these two species have been recognised as bacteriocin producers. Notable among them are Pediocin A, Pediocin A1 and Pediocin AcH which have been well characterised. Ruechert (1979) showed that inhibitory substance was a heat stable proteinaceous compound. Pediocin A has wider host range against gram positive bacteria of the strains *Clostrodium, Lactobacillus, Staphylococcus* and *Listeria* (Motlagh *et al.* 1991). The antibacterial activity of pediocin AcH was retained after high heat treatment and at wide pH range. It has been found inhibitory to gram positive bacterial strains and their spores of the genera *Pediococcus, Leuconostoc, Lactobacillus, Micrococcus, Enterococcus, Bacillus, Listeria* and *Clostridium* (Motlagh *et al.*1991; Lewus *et al.* 1992).

The killing through bacteriocidal effect is achieved by absorption at specific receptor on the cell surface before destabilisation of cytoplasmic membrane functions (Bhunia *et al.*,

1991). The effectiveness of pediocin AcH has been studied in sterile and non sterile food systems against gram positive spoilage and pathogenic bacteria by Motlagh *et al.*(1992). They found that there was a rapid drop in viable counts of *Listeria monocytogenes* in all food systems *viz.*, milk, Ice-cream, Cottage cheese, Ground meat and Sausages with largest drop in Cottage cheese. Pediocin A1 is produced in large amounts by *Pediococcus acidilactici* in MRS broth supplemented with 2 per cent yeast extract. The bacteriocin is bacteriocidal to *Listeria monocytogenes* at a temperature range of 4–32°C. The producer strains are resistance to their bacteriocins. The antibacterial activity is retained even after heat treatment at 100°C. Nielsen *et al.*, (1990) reported that pediocin A1 treatment of fresh beef innoculated with *Listeria monocytogenes* was able to reduce the population of pathogen and the level of reduction was directly proportional to the concentration of pediocin used.

Lactacin/Lactococcin/Lacticin

These bacteriocins are produced by various species of LAB from the genera *Lactobacilli* and *Lactococci* namely- *Lactobacillus-lactis* subsp. *lactis; Lactococcus lactis* subsp. *cremoris* and *lactobacillus acidophilus* etc. The bacteriocin have an inhibitory spectrum only against related strains. The bacteriacidal action against sensitive strains is produced by increasing the permeability of the cytoplasmic membrane and dissipating the membrane potential. Having narrow host specificity they have limited but effective usage against food spoilage organisms. Holla *et al.*, (1991) reported that Lactococcin A was inhibitory only to strains of the same genus. However, it inhibits nisin and diplococcin producing strains. Lacticin 481 has been reported (Piard *et al.*,1992) to be inhibitory towards strains of *Lactococcus, Lactobacillus, Leuconostoc* and *Clostridium.* Lactacin A produced from *L. delbrueckii* subsp. *lactis* has a narrow host range and looses its antibacterial activity upon heating to 60°C for 10 minutes. However, Lactocin B produced by *L. acidophilus* is a heat resistant bacteriocidal protein. The Lactocin 27 produced by *L. helvaticus* remained stable after heating to 100°C for 1 hour. It has bacteriostatic effect on sensitive cells and absorbed both by gram positive and gram negative sensitive and resistant cells.

Natamycin

It is also known as pimaricin and is produced by *Streptomyces natalensis*. Natamycin is effective against bacteria, viruses and actinomyces (Luck and Jager, 1996). It acts on candida similar to that of surface active substances *i.e.* on the cell membrane. It forms a complex with ergosterin, and this increases the cell membrane permeability(Oostendrop, 1981). Natamycin is approved as additive which may be applied to the surface of cuts and slices of cheese to inhibit mold growth and related spoilage.

Phyto Metabolites

Herbs and spices are described as dried, aromatic or pungent vegetable or plant substances that contributes flavours and primarily used as seasonings in food rather than nutrition and to which no other legal flavouring principle has been added. In addition to contributing flavour to foods, many spices also exhibit antimicrobial activity (Table 35.2). Spices may be the dried arilla, roots, seeds, fruits, leaves, rhizomes, stigma and styles of the plants or herbs. Some of the common spices and associated antimicrobial substances are given in Table 35.3.

Table 35.2: Spices Containing Antimicrobial Activity

Almonds	Fenugreek	Pepper
Cardamom	Garlic	Peppermint
Celery	Ginger	Spearmint
Cinnamon	Lemon	Tea
Clove	Lime	Turmeric
Cocoa	Mustard	Vanillin
Coffee	Onion	Verbena
Coriender	Oregano	
Fennel	Parsley	

Allicin

The plants of the genus *Allium*, having antimicrobial activity, and widely consumed in the human diet are garlic (*Allium sativum*) and onions (*Allium cepa*). These have been longest and most extensively studied for their antimicrobial activity. Garlic has been used for medicinal purposes for centuries. Cavallito *et al.* (1945) were the first to isolate antimicrobial component from garlic bulbs by steam distillation of ethanolic extract and identified the compound as allicin (2-propenyl-2-propenethiol sulfinate). It is a colour less pungent oil and is the principal odourent and taste of garlic and onion. Extract of *Allium* spp. have been reported to inhibit the growth of several food spoilage and pathogenic bacteria. Salim and Al-Delaimy (1982) demonstrated that aqueous extract from fresh garlic bulbs at concentrations of 5 per cent and 10 per cent inhibit the growth of *Bacillus cereus* by 58.2 and 100 per cent respectively. Garlic sap inhibits the growth of several gram positive and gram negative food spoilage and pathogenic bacteria including *Escherichia, Enterobacter, Salmonella, Shigella and Pseudomonas* etc. (Srivastava *et al.*, 1982). Garlic oil was found to be a potent inhibitor of food spoilage yeasts at concentrations as low as 25µg/ml (Conner and Beuchat,1984). The effect of garlic and onion oils on toxin production by *Clostridium botulinum* in meat slurry indicated that these oils at concentration 1500 ug/g inhibited toxin production (DeWit *et al.*, 1979). The antimycotic effect of garlic and onions have also been documented. Several species of food spoilage yeast and molds are inhibited (Beuchat,1984).

Table 35.3: Antimicrobial Compounds/Actions of Common Spices

Spices	*Antimicrobial Compound*	*Action*	*Microbial Spectrum*
Garlic	Allicin (2-propenyl-2 propenthiol sulfinate)	Bactericidal, Medicinal, Antibacterial	*B. cereus, Staph. aureus, Candida, Torulopsis, Salmonella etc.*
Onion	Allicin	Growth inhibitor, Bactericidal	*E. coli, Shigella, Enterobacter, Pseudomonas* spp. etc.
Oregano	Thymol (5 methyl-2 (1-methylethyl)phenol)	Inhibit growth, germination and toxin production	*S. typhimurium, A. flavous, Pediococcus* etc.
Savory	Thymol	Antifungal, Stimulatory	*Cl. botulinum, Aspergillus, Pediococcus* spp. Pathogenic/non-pathogenic fungi
Cinnamon	Cinnamic aldehyde (3phenyl-2-propenol)	Inhibit mould growth and mycotoxin production	*A. perasiticus, Staphylococcus* and *Bacillus* etc.
Clove	Euganol (2 methoxy-4 (2-propenyl) phenol)	Antimycotic/ Antifungal	*S. typhimurium, B. subtilis, Aspergillus* and *Staphylococcus* spp.
Vanilla	Vanillin (4hydroxy-3-methoxy benzaldehyde	Antifungal and Antimycotic	*Aspergillus* spp., *Bacillus* spp., *Staph. aureus* etc.

Thymol

It is yet another compound that has antimicrobial activity essentially present in the oil fraction of thyme, oregano and savoury seeds. Thymol (5-methyl-2-(1-mehtylethyl)phenol) has a wide spectra of antimicrobial effectiveness that is inhibitory against 25 genera of bacteria. Thymol has been shown to inhibit growth and toxin production of Mycotoxigenic moulds (Akgul and Kivanc,1988). Thymol has been reported to have a strong antifungal activity (Hitokoto *et al.*, 1980). They reported that at a concentration of < 0.4 µg/ml thymol caused complete inhibition of the growth of *Aspergillus flavus* and *Aspergillus versicolor*. Azzouz and Bullerman (1982) reported that a 2 per cent concentration of oregano completely inhibited growth of seven mycotoxigenic moulds and eighteen pathogenic and non-pathogenic fungi. Thyme and oregano may have stimulatory effect on lactic acid production (Zaika *et al.*, 1983).

Cinnamic Aldehyde

It is the major antimicrobial compound in cinnamon, a widely used spice in Indian food cusines. In addition to exhibiting antibacterial activity cinnamic aldehyde (3-phenyl-2-propenal) also inhibits mould growth and mycotoxin production (Dean and Ritchie, 1987). The inhibitory effect of cinnamon on aflatoxingenic moulds was reported by Bullermaan (1974). He reported that 1-2 per cent concentration of ground cinnamon eliminated almost 99 per cent of afflotoxin production.

Eugenol

This is a major constituent of clove oil and present in considerable amounts in the essential oils of all spices. Eugenol (2-methoxy-4-(2-propenyl)phenol) possesses antimicrobial activity and is most effective against *S. typhimurium*, *Staph. aureus* and *V. parahaemolyticus* than thymol, anethol and menthol (Karapinar and Aktug, 1987). Clove oil at concentration of 250 µg/ml has been reported inhibitory to growth and toxin production by several moulds (Bullerman, 1977). Ground clove was observed to inhibit completely the growth of toxigenic *Aspergilli* in culture media (Hitokoto *et al.*, 1980). The rate of germination of *Bacillus subtilis* spores was reported to be adversely effected both by clove powder and eugenol at a concentration of 1200 µg/ml and 200 µg/ml respectivily (Blank *et al.*, 1987).

Caffeine

This is present among the flavour compounds of coffee, tea and cocoa beans. Caffeine (1,3,7-trimehtylxanthyne) is antimycotic as well as antibacterial. Buchman *et al.* (1983) reported inhibition of growth of several mycotoxigenic *Aspergillus* and *Penicillium* spp. at concentrations as low as 1 µg/ml. Caffeine also adversely effected production of afflotoxins, ochratoxin, sterigmatocystin and citrinin by moulds. Various modes of action of caffeine on microbial cell functions has been described. Tortora *et al.* (1985) reported that caffeine uncoupled the regulation of glycolysis and glucogenesis in *S. cerevisiae* and inhibited growth of *A. parasiticus* due to alteration of purine metabolism. Vanos and Bindschedler (1985) reported that *S. aureus; B. cereus; E. coli; Streptococcus faecalis* and *Salmonella* species failed to grow in 2 per cent reconstituted decaffeinated and non-decaffeinated coffee. Caffeine is known to depress bacterial DNA synthesis.

Other Compounds

Medium chain fatty acids: As antimicrobial agents, medium chain fatty acids containing 12-18 carbon atoms are most effective (Jay,1986). The antimicrobial action of fatty acid is primarily static rather than cidal. In general fatty acids are more effective inhibitors of gram positive bacteria and yeast, although some fatty acids also exhibit antimycotic activity (Lindsay,1985). They have observed that fatty acid derivatives reduce both growth and afflotoxin production of *Aspergillus* species. Karaba (1983) reported that lauric; myristic and palmatic acids are effective inhibitors of bacteria and capric and lauric acids are most active against yeasts. The toxicity of fatty acids are determined by their chain length, degree of saturation and geometric configuration. The inhibitory activity of unsaturated fatty acids increases as the number of double bonds increases. With respect to gram positive bacteria, the most effective chain length of saturated fatty acid is 12 carbons, most effective monounsaturated fatty acid is palmitoleic and most effective polyunsaturated fatty acid is linoleic. It is generally accepted that the undissociated form of the fatty acid molecule is responsible for antimicrobial activity. Fatty acids may offer potential as microbial inhibitors in slightly acid foods however, several inherent factors other than acidity and pH of food control the effectiveness of fatty acids. Foods containing antagonists like serum albumin, starch and cholesterol etc. may not be suitable for preservation by fatty acids. Fatty acid esters of sucrose and other polyhydric alcohols also possess antimicrobial activity thus exerting preservative effect in foods. An inhibitory effect of glycerol monolaurate on *V. parahaemolytis* has been observed at concentration as low as 5 μg/ml (Beuchat,1980).

Various mechanism for the antimicrobial activity of fatty acids have been reviewed by Kabara (1983). They pointed that antimicrobial activity is due to inhibition of membrane transport, thus resulting in nutritional deprivation of cells. Gram negative bacteria appears to be less susceptible than gram positive bacteria to inhibition by fatty acids.

Organic Acids

These acids may be present naturally in foods or accumulate as a result of fermentation. Organic acids have multifuctional role in food processing. Apart from providing characteristic flavour to the product, they decrease the pH of the product thus preventing the growth of majority of microbes and they also increase the effectiveness of other additives. There are number of food grade acids (Table 35.4) that are used for texturizing and lowering pH of various processed food systems and to act as preservatives. The undissociated form of acid is responsible for antimicrobial activity (Davidson and Juneja,1990).

Antimicrobial Enzymes

Naturally occurring enzymes also exhibit antimicrobial properties and on the basis of their mechanism are generally grouped into four major classes (Scot,1988).

1. Enzymes which may deprive micro-organisms from the nutrients for their active growth and developments. To this group belongs oxidases; laccases; catalase and aerohydrogenases which can be used for the preservation of fruit juices and beer.
2. There are several enzymes that produce toxic substances lethal to microbes.Under this group comes lipase; lactoperoxidase; myeloperoxidase and xylitol phosphorylase. Some of these enzymes have been used extensively for preservation of raw milk.

3. The third group of enzymes consisting of highly effective antibacterial activity against micro-organisms, by attacking their cell wall and altering cell membrane permeability. Lysozyme; chitonase and mannase etc. comes under this group. They also show pathogenecity against microbes that effect human beings.
4. This class of enzymes are also designated as killer enzymes as they are capable of inhibiting other enzymes that have potential role in microbial metabolism. Proteases and sulfhydryl oxidase are excellent examples of this group.

Table 35.4: Some Common Organic Acids as Food Preservative

Acids	*Occurrence*	*Actions*
Benzoic acid	Raspberries, Canberries, Plum	Antifungal agent, Prevents amino acid uptake by microbial cells
Citric acid	Citrus fruits	Lowers pH, Chelates metal ions, Interfears with glucose utilization in cell wall
Lactic acid	Lactic acid bacteria	Spore inhibitor below pH 5.0, Lowers pH of the system
Malic acid	Apple, Tamarind	Multifunctional additive
Propionic acid	Swiss cheese	Mold inhibitor

Lactoperoxidase

Peroxidases are widespread in nature and there are many types with different redox potentials and substrate specificities. In addition to the enzyme and the substrate hydrogen peroxide, the antimicrobial effect also requires presence of pseudohalogen thiocynate (SCN) as a second substrate. The antimicrobial effect is based on the oxidation of SCN to hypothiocynate (OSCN). The most prominent feature of the system is the combined attack of oxidative and lytic mechanism (Wolff *et al.,* 1986). The LP system have been shown to exert bacteriostatic or bactericidal effects on *Listeria monocytogenes* and *Staphylococcus aureus* (Kamau *et al.,*1990); *Campylobacter* (Borch *et al.,* 1989); *Salmonella* (Wrey and McLaren,1987) as well as other pathogenic bacteria. There have been several attempts to make use of the general antimicrobial effects in food applications. The LP system does not destroy the nutritional quality of milk proteins (Kumar and Mathur,1989), which on occasion is the case with pure hydrogen peroxide treatment of milk. The system has been suggested as preservative in milk and milk products (Earnshaw *et al.*,1989; Kumar and Mathur,1989 and Denis and Ramet,1989). The system has also been exploited as antimicrobial system in infant formulas (Earnshaw *et al.*,1990).

Lactoferrin

This protein that sequester iron have a dual role in biological fluids, restricting bacterial growth and increasing the supply of nutrients. It is present in milk of many species. Lactoferrin is known to have bacteriostatic and bactericidal effects on carcinogenic bacteria (Tenovuo *et al.*,1991). The in vitro antimicrobial effect of lactoferrin have been performed against *E. coli* (Law and Reiter,1977); *Salmonella* and *Shigella* (Batish *et al.*,1988) and *Listeria monocytogenes* (Payne *et al.*,1990) and have been found to be effective against all these bacteria. Ellison *et al.*(1988) demonstrated that lactoferrin damages the outer membrane of gram negative bacteria and alters its permeability resulting in irreversible membrane destruction. These effects are modulated by calcium and magenese ions (Ellison *et al.*,1990).

Lysozyme

It is present in number of different secretions throughout nature ranging from plants to animals and micro-organisms. Hen egg albumin has always been the major source of this enzyme for commercial extraction. Because of its bactericidal activity it has been used as a preservative in food industry and does not cause any harmful effects. Generally gram positive bacteria having simple cell wall are highly susceptible to lysozyme. Whereas, gram negative bacteria tends to be more resistant to lysozyme unless the outer cell barrier are disrupted by physico-chemical treatments. Pellegrini *et al.* (1992) however, demonstrated bactericidal activity of Lysozyme denatured by dithiothreitol against both gram positive and gram negative bacteria. Such activity was thought to be related to the cationic and hydrophobic properties of the molecule rather than residual enzymic activity. The inclusion of lysozyme has been shown to be effective preservative against some strains of *Listeria* in certain foods (Carminati and Carini, 1989). Lysozyme was found to effectively kill high number of organisms in green beans, sweet corn, cabbage, lettuce and carrot stored at 5°C (Hughey *et al.*,1989). Recently a novel lysozyme -dextran conjugate was described by Nakamura *et al.*(1990) which papers to be more effective against gram negative and gram positive bacteria at 50°C.The application of lysozyme ranges from coating surface of fresh vegetables, fruits, fish and meat to preservation of wine.

REFERENCES

Akgul, A. and Kivanc, M. (1988). Inhibitory effect of selected Turkish spices and Oregano components on some food born fungi. Inter. *J. Food Microbiol.*, 6: 263-268.

Azzoz, M.A. and Bullerman, L.B. (1982) Comparative effects of selected herbs, spices, plants compounds and commercial antifungal agents. *J.Food Protection,* 45:1298-1301.

Batish, V.K., Harish, C., Zumdegui, K.C., Bhatia, K.L.and Singh, R.S. (1988) Antibacterial activity of lactoferrin against common food born pathogens. *Aust. J. Dairy Technol.*, 43:16-18.

Beuchat, L.R. and Golden, D.A. (1989) Antimicrobials occurring naturally in foods. *Food Technol.*,43: 134-142.

Bhunia, A.K., Jhonson, M.C., Ray, B and Kalchayanand, N. (1991) Mode of action of pediocin AcH from *Pediococcus acidolactici* on sensitive bacterial strains. *J.App. Bact.*, 70: 25-32.

Blank, G, Al-Khayat, M. And Ismond, M.A.H. (1987) Germination and heat resistance of *Bacillus subtilis* spores produced on clove and eugenol based media. *Food Microbiol.*, 4: 35-42.

Borch, E., Wallentin, C., Rosen, M. and Bjorch, L (1989) Antibacterial effect of lactoperoxidase/thiocynate/hydrogen peroxide system against strains of *Campylobacter* isolated from poultry. *J. Food Prot.*, 52:638-641.

Buchanan, R.L., Joove, D.G. and Jones, S.B. (1983) Caffeine inhibition of aflatoxin production: Mode of action. *App. Envior. Microbiol.*, 46:1193-1200.

Bullerman, L.B. (1974) Inhibition of aflatoxin production by cinnamon. *J. Food Sci.*, 39:1163-1165.

Bullerman, L.B., Lien, F.Y. and Seier, S.A. (1977). Inhibition of growth and afflotoxin production by cinnamon and clove oils, cinnamon aldehyde and eugenol. *J.Food Sci.*, 42:1107-1109.

Carminati, D. and Carini, S. (1989) Antibacterial activity of lysozyme against *Listeria monocytogenes* in milk. *Microbiol., Aliments, Nutri.*, 7:180-187.

Cavallito, C.J., Bailey, J.H. and Buck, J.S.(1945) The antibacterial principle of *Allium sativum* III Its precursors and essential oils of garlic. *J.Am. Chem. Soc.*, 67:1032-1033.

Conner, D.E. and Beuchat, L.R. (1984) Effects of essential oil from plants on the growth of food spiolage yeasts. *J. Food Sci.*, 49:429-434.

Daeshel, M.A. (1990) Application of bactericins in food system. In: *Biotechnology and Food Safety*, (Eds.) Bill, D.D. and Kung, S.D. Butterworth-Heinemann, Boston, USA, p. 91-103.

Davidson, P.M. and Juneja, V.K.(1990) Antimicrobial agents. In: *Food Additives, (*Eds.) A.L. Branen, P.M. Davidson and S.S. Salminen. Marcel Dekker Inc., New York, p. 83-137.

Deans, S.G. and Ritchie, G.(1987) Antibacterial properties of plant essential oils. *Inter. J. Food Microbiol.*, 5: 165-180.

Delves-Broughton, J. (1990) Nisin and its uses as food preservative. *Food Technol.*, 44: 100-110.

Denis, F. and Ramet, J.P. (1989) Antibacterial activity of LP system on *Listeria monocytogenes*. *Microbiol., Aliment, Nutri.*, 7: 25-30.

De-Wit, J.C., Noterman, S., Gorin, N and Kampelmacher, E.H. (1979) Effect of garlic oil or onion oil on toxin production by *Clostridium botulinum* in meat slurry. *J. Food Prot.*, 42: 222-224.

Earnshaw, R.G., Blank, J.G., Fancott, C. and Defrise, D. (1990) Inhibition of *Salmonella typhimurium* and *E.coli* in an infant milk formula by an activated LP system. *J. Food Prot.*, 53:170-172.

Earnshaw, R.G., Blank, J.G., Defrise, D. and Fancott, C. (1989) Preservation of cottage cheese by activated LP system. *Food Microbiol.*, 6:285-289.

Ellison, R.T., Giehl, T.J. and Laforce, F.M. (1988) Damage of outer membrane of enteric gram negative bacteria by lactoferrin and transferrin. *Infec. Immunol.*, 56: 2774-2781.

Ellison, R.T., Laforce, F.M., Giehl, T.J., Boose, D.S. and Dunn, B.E. (1990) Lactoferrin and transferrin damage of the gram negative bacteria outer membrane is modulated by calcium and magnesium ions. *J.Gen.Microbiol.*, 136:1437-1446.

Fowler, G.G. and Gasson, M.J. (1991) Antibiotic-Nisin. In: *Food Preservatives*. Ed. Russel, N.J. and Gould, G.W. Blackie, London, UK. pp 135-153.

Hanlin, M.B., Kalchayaanand, L, Ray, P. and Ray, B.(1993) Bacteriocin of lactic acid bacteria in combination have greater antibacterial activity. *J. Food Prot.*, 56: 252-255.

Hitokoto, H., Morozumi, S., Wauke,T., Sakai, S. and Kurata, H.(1980) Inhibitory effect of spices on growth of toxin production of toxigenic fungi. *App. Envior. Microbiol.*, 39:818-822.

Holla, H., Nilssen, O. and Nes, I.F.(1991) LactococcinA from *Lactococcus lactis spp.cremoris:* isolation and characterization of the protein and its gene. *J.Bact.*, 173:3879-3887.

Hughey, V.L., Wilger, P. and Jhonson, E.A. (1989) Antimicrobial activity of hen egg white lysozyme against *Listeria monocytogenes* in food. *App. Envior. Microbiol.*, 55:631-638.

Hurst, A. (1981) Nisin. Advances in App. Microbiol, 27: 85-123.

Kabara, J. J. (1983) Medium chain fatty acids and esters. In: *Antimicrobials in Foods*. Ed. Branen,A.L. and Davidson, P.M. Marcel Dekker, New York, USA. pp 109-140.

Kumar, S and Mathur, B.N. (1989) Studies on manufacture of yoghurt and mozerella cheese from milk preserved by LP system. *Ind. J. Dairy Sci.*, 42:194-197.

Kumar, S and Mathur, B.N. (1989) Incidence of LP system on milk constituents during preservation of buffalo milk.I. Ion exchange chromatographic and electrophoretic behaviour of casien. *Ind. J. Dairy Sci.*, 42:185-189.

Kamau, D.N., Doores, S. and Pruitt, K.M. (1990) Antibacterial activity of LP system against *Listeria* and *Staphylococcus* in milk. *J. Food Prot.*, 53:1010-1014.

Law, B.A. and Rieter, B. (1977) Isolation of bacteriostatic properties of lactoferrin from bovine milk whey. *J.Dairy Res.*, 44:595-599.

Lewus, C.B., Sun, S.S.and Montville, T.(1992) Production of amylase sensitive bacteriocin by a typical *Leuconostoc paramesenteroide* strain. *App. Envior. Microbiol.*, 58:145-149.

Luck, E.and Jager, M. (1995). *Antimicrobial Food Additives, Characteristics, Uses, Effects*, 2nd edition. Springers, pp. 260.

Motlagh, A.M., Jhonson, M.C. and Ray, B. (1991). Viability loss of food born pathogens by starter culture metabolities. *J. Food Prot.*, 54: 873-878.

Motlagh, A.M., Holla, A., Jhonson, M. C., Ray, B. and Field, R. A. (1992). Inhibition of *Listeria* spp. in sterile food system by pedioccin AcH. *J.Food Prot.*, 55: 337-343.

Nakamura, S., Kato, A. and Kobayashi, K. (1990). Novel bifunctional lysozyme-dextran conjugate that acts both on gram positive and gram negative bacteria. *Agri. Biol. Chem.*, 54: 3057-3059.

Nielson, J.W., Dickson, J.S. and Crouse, J. D. (1990). Use of bactericin produced by *Pediococcus acidolactici* to inhibit *Listeria monocytogenes* associated with fresh meat. *App. Envior. Microbiol.*, 56: 2142-2145.

Oostendrop, J.G. (1981) *Netamycin*. Antonie van Leiwenhock, 47: 170-171.

Payne, K.D., Davidson, P.M., Oliver, S.P. and Christen, G.L. (1990). Influence of bovine lactoferrin on the growth of *Listeria monocytogenes*. *J. Food Prot.*, 53: 468-472.

Pellegrini, A., Thomas, U., von Fellen, R. and Wild, P. (1992). Bactericidal activity of lysozyme and aprotinin against gram positive and gram negative bacteria related to their basic characteristics. *J. App. Bact.*, 72: 18-187.

Piard, J.C., Muriana, P.M., Desmazeaud, M.J. and Klaenhammer, T.R. (1992). Purification and partial characterization of lacticin 481, a lanthionine containing bacteriocin produced by *Lactococcus spp. lactis CNRZ 481*. *App. Envior. Microbiol.*, 58: 279-284.

Ray, B. (1992). Bacteriocins of starter culture bacteria as food preservatives: An overview. In: *Food Biopreservatives of Microbial Origin*, (Eds.) Ray, B. and Daeschel, M.A. CRC press, Boca Roton, Florida, USA, p. 177-205.

Ruechert, P.W. (1979). Studies on the bacteriocin like activity produced by *Pediococcus acidolactici* effective against gram positive organism. *M.Sc. Thesis*, Michigan State University, East Lancing, Michigan, USA.

Ruhr, E. and Sahl, H. (1985). Mode of action of peptide antibiotic nisin and influence on the membrane potential of whole cell and on cytoplasmic and artificial vesicles. *Antimicrobial agents and Chemotherapy,* 27: 841-845.

Saleem, Z.M. and Al-Delaimy, K.S. (1982). Inhibition of *Bacillus cereus* by garlic extract. *J. Food Prot.,* 45: 1007-1009.

Srivastava, K.C., Perera, A.D. and Sarikadis, H.O.(1982). Bacteriostatic effect of garlic sap on gram negative bacteria: An *in vitro* study. *Lebensmitteel wissenschaft und Technologie,* 15:74-76.

Tagg, J. R. (1992). Bacterial BLIS. *Am. Soc. Microbiol. News,* 57: 611.

Tenovuo, J., Lumikari, J. and Soukka, T. (1991). Salivary lysozyme,lactoferrin and peroxidase: antibacterial effects on cariogenic bacteria and clinical application in preventive dentistry. *Proc. Finish Dental. Soc.,* 87: 197-208.

Wary, C. and McLaren, I. (1987). A note on the effect of lactoperoxidase system on *Salmonella in vitro* and *in vivo. J. App. Bact.,* 62: 115-118.

Wolff, S.P., Garner, A. and Dean, R.T. (1989). Free radicles, lipids and protien degradation. *Trends in Biochem. Sci.,* 11: 27-32.

Yahg, R., Jhonson, M.C. and Ray, B. (1992). A novel method of purifying bacteriocins of lactic acid bacteria. *App. Envior. Microbiol.,* 58: 3355-3359.

Zaika, L.L., Kissinger, J.C. and Wasserman, A. E. (1983). Inhibition of lactic acid bacteria by herbs. *J. Food Sci.,* 48: 1455-1459.

36

Use of Antioxidants in Food Preservation

Saleem Siddiqui
Department of Horticulture, CCS Haryana Agricultural University
Hisar – 125 004, India

After microbial spoilage, oxidative degradation leading to overt rancidity is the second most important cause of food spoilage. Certain fruits *e.g.* apple and vegetables like brinjal, on cutting and exposure to air turn brown. This browning has been attributed to air oxidation of phenolic substances present in most plant materials. Enzymes called polyphenoloxidase are present in plants and on exposure to air these are activated and initiate air oxidation and polymerization of phenolic compounds, producing a brown colour. Similarly, freshly extracted juice of certain fruits, turn brown on exposure to air, developing an unpleasant taste. Oils and fats, particularly those rich in polyunsaturated fatty acids, on exposure to air undergo oxidative decomposition, becoming viscous and giving a foul smell. This is referred as rancidity of fats, which makes the fat unfit for human consumption. In case of deep-fried foods *e.g.*

potato chips and other 'namkeens', air contact is increased due to large surface area. The storage life of these is poor and such foods develop stale odour and taste on exposure to air. Hence these must be stored in tightly closed containers and preferably be packed under nitrogen. To a certain extent, these processes can be minimized by the use of suitable antioxidants.

Antioxidants have largely been viewed as tools in the fight against oxidation. With increasing awareness of the importance of antioxidants in health maintenance, their retention through food processing and storage, its use has assumed increasing importance. Antioxidants are a class of chemical compounds that attack 'free radicals' and pro-oxidants.

Generation of Free Radicals

Most of the normal molecules constituting the human body and our surroundings have electrons, constituting the chemical bonds in paired form. and in lower energy level. Unlike these, free radicals have at least one unpaired electron. A free radical is formed when an electron pair bond in a molecule breaks in such a way that each of the fragments of the parent molecule carries an unpaired electron of the original bond. Chemists call it hemolytic fission of a covalent bond.

Free radicals have a tendency to seek an electron from an electron pair bond of another molecule. In this process air creates a new radical and starts a chain reaction. Termination of a free radical chain reaction can result by disproportion reaction or by mutual combination of two free radicals and sometimes due to the formation of a less reactive or a non-reactive radical. Molecular oxygen is paramagnetic and it can be regarded as a weak bis-radical. On activation, it becomes an active free radical. These radicals are formed in response to either direct excitation from incident radiation or secondarily from reactions with oxygen radicals or metastable forms of reduced oxygen, *i.e.*, OH^-, H_2O_2, singlet O_2, and O_2^-. Respiratory oxygen also generates free radicals in a metabolic process. The damage resides in the formation of other organic peroxides propagated through typical radical chain reactions. An antioxidant can quickly terminate such a chain reaction.

Types of Antioxidants

Naturally Occurring Antioxidants

Vitamin-C (Ascorbic acid). vitamin-A (Retinal). vitamin-E (Tocopherol), and melatonin (hormone) occurring in plants or animal tissues are known to act as an antioxidant for the plants in which these occur and for the animals or human beings, using such plants for food. The antioxidant nature of vitamin-C can be realised from the fact that, whereas apple juice turns brown in air. lemon and orange juice, retains the original colour due to the natural presence of the antioxidant vitamin-C. Similarly, vegetable oils containing vitamin-E do not easily become rancid. Sulphur compounds *e.g.* α-lipoic acid and those present in garlics and onions (thiol and mercaptans) are also antioxidants.

The most important group of naturally occurring antioxidants is of phenolic compounds. Thick and dark coloured bark of tropical trees, *e.g.* kher, babool and neem have bioflavonoids and their polymerised products, tannins in their bark, which acting as antioxidants, protect the plant against air oxidation, by oxygen, which gets activated by UV-radiation present in bright and harsh sunlight. Brightly coloured flowers and fruits *e.g.* red grapes, black-berries, blackcurrant and beet root also contain anthocynins and flavones, which are phenolic substances with antioxidant properties. The seed coat of many plants, *e.g.* wheat bran, oilseed and legpme husk are rich in antioxidants (tocopherols as well as bioflavonoids), which protect them from air oxidation when they need to be stored in contact with air for a long period till used as food or germinated to grow into plant.

Human body has its own antioxidant enzymes, some of which have metal ions. copper, zinc, manganese, selenium and iron in their prosthetic groups. These enzymes, acting as redox systems, mop up free radicals under normal conditions. Additionally, the dietary antioxidants are transported through the blood to various tissues and organs of the body. where they act as antioxidants. Human food should contain both water-soluble and lipid soluble antioxidants, which are transported through the blood, and stored in appropriate tissues in the body.

Dietary antioxidants are normally present in human food and these include vitamins, A, C and E besides carotenoids, *e.g.* ß-carotene present in carrot, yellow fruits (pumpkins, mangoes and papayas) and green-vegetables. Carotenoids, a precursor of vitamin-A is even more effective as an antioxidant than vitamin-A. Vitamin-C is present in citrus fruits and green vegetables. Plant pigments other than green and yellow are phenolic compounds, which are collectively known as bioflavonoids (*i.e.* other than chlorophyll and carotenoids), which are antioxidants, widely distributed in plants (tea. coffee, cocoa, gambier leaves, marigold flowers and kutch). Herbal tea and herbal infusion have been used in traditional systems of medicine (Ayurvedic and Yunani decoction). Their healing power has been attributed to the presence of antioxidants, thus providing a scientific basis for their action. Vegetable oils *e.g.* sunflower oil, peanut oil and soybean oil are rich in vitamin-E.

Common sources of dietary antioxidant are shown in Table 36.1.

Table 36.1: Antioxidants and their Source

Antioxidant Type	*Source*
Vitamin-C (Ascorbic acid)	Citrus fruits, green vegetables. strawberries, black currants, guavas
Vitamin-A (Retinal) Vitamin-E	Milk products, eggs, meat. fish, liver vegetable oils (soybean, peanuts. Sun-flower), nuts, peanuts, whole grains, legumes, almonds
Carotenoids β-Carotene	Yellow fruits (mangoes, papayas, pumpkins), carrots
Lycopene	Red peppers, tomatoes
Bioflavanoid	Fruits, lemons, cherries, blackberries. black currants, tea, coffee, cocoa
Minerals Selenium	Shellfish, fish, herring, cod fillet, dairy products, avocado
Manganese	Plant food
Iron	Pigmented fruits, meat, egg, sea food
Copper	Liver, shellfish, cocoa, nuts, mushrooms. vegetables
Zinc	Grains, meat, fish, oysters, eggs, dairy products, vegetables

Synthetic Antioxidants

Some synthetic organic compounds, *e.g.* butylated hydroxytoluene (BHT), butylated hydroxyanisole (BHA). Propyl-gallate and other synthetic phenolic compounds, when added at a very low concentration to a fruit juice or edible oil, can stop their air oxidation, preventing the development of brown colour followed by unpleasant odour and rancidity, respectively. Such substances are referred as 'Synthetic Antioxidants'.

There has been a controversy regarding the use of synthetic antioxidants in food, because some food scientists and medical personals do not regard them as completely safe and non-toxic food additives. However by long term experiments, it is now amply proved that at the permitted concentration of <0.1 per cent, these antioxidants are completely harmless to human beings, and protect against harmful air oxidation of food.

How Do Antioxidants Act to Neutralise Harmful Free Radicals?

Some antioxidants act as sacrificial molecules, *i.e.* they preferentially react with free radicals, thereby getting destroyed and producing harmless byproducts, but protecting vital constituents of the body. Such molecules are themselves reducing agents *e.g.* vitamin-C and anthocynin and bioflavonoids. In other cases, particularly the metal ions acting as antioxidants are constituents of antioxidant enzymes, which are naturally present in the body. These can exist in a variable oxidation-state. After neutralising free radicals by getting oxidised to a higher oxidation state, these are reconverted, by redox systems of the body into the original lower oxidation-state. Thus, these act as catalysts and react in a continuous cycle. Vitamin-C is also regenerated by body redox system and acts as free radical scavenger.

Glutathione is a ubiquitous tripeptide that mediates redox cycling between ascorbate and NADPH. Ascorbate plays a pivotal and clearly defined role in the scavenging of hydrogen peroxide. The carotenoids may function to mediate potential damage from excited states of pigment molecules and singlet oxygen. Protection from photooxidative damage involves the carotenoids in preventing the formation of reactive singlet oxygen. The conversion of violaxanthin to zeaxanthin is mediated by oxygen uptake and cycling back to violaxanthin requires ascorbate. Vitamin E, α-tocopherol, is an exquisite, lipophilic, free radical scavenger which, directly interrupts free radical chain reactions. As with glutathione, mechanisms exist that allow cycling of the quinone (oxidized) and quinols (reduced) using reduced ascorbate or NADPH directly. Finally, various classes of phenylpropanoid derivatives, for example, the flavonoids, lignans, tannins, lignins, serve to minimize the effects of direct photobleaching, particularly that initiated by UV irradiation, and irreversible oxidative damage.

Functions of Food Antioxidants

Packed food when in contact with air, it gets spoiled due to air oxidation. This is purely a chemical process. Toxicity of packaged food can result in a type of food poisoning and to guard against such infection, antioxidants are added to packaged food. Some food processors prefer the use of natural antioxidants, which can be herbal extracts. Herbal extracts can be economical yet safer than synthetic antioxidants. Thus the solvent extracted pigments from tamarind seed coat contain antioxidants suitable for preventing rancidity of oils and improving shelf life of fried foods. Bioflavonoids from marigold flower and catechu (catchin present in 'katha') also contains antioxidants suitable for use in food. Rice and corn bran oils are rich sources of the antioxidant vitamin-E. It has been mentioned that UV-radiation can activate oxygen (single stage), which can damage skin tissues and cause skin diseases (cancer) on prolonged exposure to the sun. To prevent it. cosmetic companies use vitamin-E and other antioxidants in skin-care products.

Quality maintenance and/or improvement through the use of active packaging has received recent attention, with oxygen scavenging techniques being of particular interest. Oxygen absorbent sachets, most commonly containing iron powder or, to a lesser extent, ascorbic acid, may be used to prolong the shelf-life of various foods. Alternative approaches include incorporation of oxygen scavenging materials, such as ascorbic acid, into the packaging itself. Vitamin E has also been incorporated into packaging films, from which it may migrate into the food, so eliminating the need to add antioxidants to the food itself.

Dietary food fibers, which are soluble, non-starch polysaccharides, are depolymerised by free radicals into harmless products (sugars and lower polysaccharides). Thus they act as free radical scavengers or antioxidants. Water-soluble as well as fat-soluble antioxidants are known and an appropriate choice of an antioxidant should be made based on the nature of food.

Maintenance of the structural integrity of foods helps to prevent oxidative changes from occurring. In whole food form, contact of these oxygen sensitive components with oxygen is reduced. Thus, processes that prevent exposure of antioxidants to oxygen to the greatest possible extent have become a requirement. New processing technologies directed towards producing stable foods following minimal treatment are also relevant, not only for the impact these minimal processes will have on product quality, but also on the preservation of antioxidant status.

Future Thrust Area

Antioxidants have largely been viewed as tools in the fight against oxidation. With increasing awareness of the importance of antioxidants in health maintenance, their retention through food processing and storage has assumed increasing importance.

The addition of antioxidants to foods helps to preserve constituents of the food by preventing auto-oxidation. However, genetic enhancement of antioxidant levels is an attractive alternative to fortification and is increasingly feasible. The pathways controlling the biosynthesis of many secondary plant metabolites are well documented. The importance of genetic engineering in crop development programmes has increased in recent years such that techniques to introduce many genes or sets of genes into specific plant species of interest, including major crop species, have now been developed. Additionally, advanced breeding programmes permit new traits to be moved into commercial crops in economically viable time-frames. Thus, there is substantial potential for directed genetic manipulation of crops to enhance productivity in a number of important ways, including maximising antioxidant contents. However, it must be acknowledged that there are important public acceptance issues surrounding the adoption of this technology.

REFERENCES

Alscher, R.G. and Hess, J.L. (1993). *Antioxidants in Higher Plants.* CRC Press, London.

Decker, E.A. (1998). Strategies for manipulating the prooxidative/antioxidative balance of foods to maximize oxidative stability. *Trends in Food Sci. and Technol.,* 9: 241-248.

Lindley, M.G. (1998). The impact of food processing on antioxidants in vegetable oils, fruits and vegetables. *Trends in Food Sci. and Technol.,* 9: 336-340.

Mathur, N.K. and Mathur, V. (2001). Antioxidants: natural ingredients and additives for food. *Beverages and Food World,* p. 13-15.

Part VI
GENERAL

37

Food Processing Industry in India: Emerging Scenario

S.S. Arya
B-31, Kendriya Vihar, Noida (U.P.)

Food processing in India has traditionally been practiced as household or cottage scale activity by small-scale artisans as family profession. Despite lack of basic training in food processing operations, excellent quality sweets, pickles, papads, murabbas fried products and roasted and puffed cereals have been prepared and marketed by these artisans for local consumption. Organised food processing units came in this country to manufacture biscuits, bread, jams, squashes, syrups and fruit confections for British officers and settlers and such units multiplied several fold during the second world war as the number of allied troops increased considerably. These being products needed by westernized elite have all along been held by the government as luxury items and attracted higher tariff. Majorities of these organized units are still in cottage and household sector, which has denied the industry of modern technology vital to enter the world market. Only during eighties it was realized that without establishing organized food processing facilities, it is not possible to sustain adequate growth in agro-horticulture sector which is essential for generating additional employment in rural areas for restricting population migration to cities. Also, it was felt that processed foods provide good opportunity for export. Accordingly, the government had allowed the import of modern technologies like cooker extruders, retorts for pouch processing, I.Q.F., etc., to stimulate growth in food processing and also established a separate Ministry of Food Processing to channelise investment in this sector and to establish linkages with farm sector for ensuring adequate availability of raw materials for processing. The growth in agricultural, horticultural and piscicultural production had improved the raw materials for the food processing industry both qualitatively and quantitatively and many new local and multinational companies had entered the food-processing arena. However, the impact of liberalization policy on the consumption of processed foods in India has not increased to the level expected and the rise in exports is confined to farm products and not to value added processed foods. Only very recently government has removed excise duty on fruits and vegetable products and its effects on exports and employment generation is yet to be seen.

A report released in 1995-96 by the Food and Agriculture Integrated Development Action (FAIDA) popularly known as Mackinsey Report had predicted that India has the potential to be the world's largest food factory and Indian Food Processing is likely to grow three folds by 2005 to be worth 225,000 crores. Such optimistic predictions have been made previously also, but question still remains–is Indian consumer ready to accept value added processed foods in place of home cooked foods and in what form?

Food Processing Industry

The Indian food market is estimated to be of the order of Rs.1,75,000 crores. Of this, the semi-processed food accounts for about 15 per cent and processed food about 10 per cent, whereas fresh food market accounts for 75 per cent. The organized food industry sector consists of about 18,000 factories producing products worth Rs.9,000 crores.

Foodgrains and Oilseeds

Food grains undergo only primary processing involving drying, cleaning and grading. Rice is milled in small and medium capacity mills numbering over 7900 hullers distributed in the rice cultivating areas producing 15,000 crores worth of products. About 2000 units of poha making and puffing, produce Rs.1000 crores worth products. Wheat, on the other hand, mostly reaches the consumers in the grain form and is milled in very small chukkies numbering about 268,000 into atta valued Rs.12,000 crores. About 6 million tonnes (mt) of wheat is milled in roller flour mills into 4.5 mt. of flour, maida, rawa and bran. The maida is mostly used by bakeries producing about 1 mt. bread, 1.9 mt. biscuits and 1.4 mt. other bakery products. There are about 54,000 bakeries and 5,000 pasta units. Legumes reach the consumer in the whole seed, split and dehusked form. Splitting and dehusking is achieved in dal mills numbering about 10,000.

Oilseed processing is a well organized industry carrying out oil extraction through mechanical and solvent extraction processes followed by refining, deodorization and hydrogenation. There exist about 2,22,000 ghanies and expellers and 600 solvent extraction units producing about 75 lakh tonnes of oils worth Rs.8,000 crores. There are 350 vegetable oil refineries and 175 hydrogenation units.

Fruits and Vegetables

India is the second largest producer of fruits and vegetables totaling around 125 mt., but hardly 2 per cent is processed as compared to 70 per cent in developed countries. The installed processing capacity is, however, only 2.0 mt. and even this is not fully utilized. Due to lack of infrastructure and postharvest handling, quick transport and storage facilities about 25-30 per cent of the total harvested produce is wasted. Total fruits and vegetable products worth Rs.270 crores are manufactured in about 5000 factories. Looking at the vast potential for processing and export, the Government of India has abolished excise duty from the current year for the development of this sector. A few multinational companies have entered this sector and private investment of 61,427 crores has been made in this sector. Processing of tomatoes and pulp and juice from mango, orange, guava and apple have been identified as potential areas. Pickles and chutneys are other products receiving greater attention.

The aseptic processing of fruit juices, tomato past, lassis and flavoured milk have been successfully tried. The technology has been assimilated by the industry and there is vast scope in packing juices, juice blends/cocktails that can utilize the diversely cultivated fruits available in the country. Export of fresh and processed fruits and vegetables especially mango, grape, banana, pickles is picking up. During 1996-97, fruits and vegetables worth Rs.460 crores were exported.

Spices

India accounts for 35 per cent of global production of spices. The annual production in 23 lakh hectares is valued at Rs.5200 crores. While 95 per cent of total production is absorbed in domestic market, about Rs.610 crores worth of spices were exported during 1994-95. Of this spice oils, oleoresins and curry powders accounted for Rs.135 crores only. Very strict standards in respect of microorganism's greatest impediment in the export of value added products. Adoption of super-critical fluid extraction technology may provide a boost to the export of oleoresins and spice oils but the process at present is uneconomical in comparison to solvent extraction process.

Frozen Sea Foods

There has been rapid growth in the export of frozen seafoods, which crossed Rs.4700 crores. There are over 300 freezing plants with the estimated capacity of 4000 tpd, processing over 7600 tonnes of shrimp. About 60 IQF units have come up processing 10 per cent of total shrimp export. However, there had been some setbacks to Indian seafood industry by the ban imposed by the European market.

Enforcement of adequate quality assurance programme like HACCP and ISO-9000 in this industry has created confidence in the importing countries. Processing and export of other seafoods like surimi utilizing lesser-priced fish provides excellent potential and shall be tried. Frozen cuttle fish and squid fish are the other products, which are exported in substantial amounts.

Milk

India has emerged as the largest milk producer in world with total milk production touching 75 million tones. While major portion of this is consumed as liquid milk, other processed products like skim and whole milk powder, ghee, cheese, shrikhand, chocolate weaning foods, infant foods, etc., are produced. The milk processing industry is well organized and is able to provide 214 ml of milk per person per day. Besides it has started marketing curd, the traditional Indian products.

Traditional Foods

Indian cuisine has been one of the most accomplished and versatile cuisine based on the understanding of sound scientific principles including nutrition, health and food science. Of late, there has been considerable research interest in developing ready to cook mixes of many food items like Idli, Dosa, Instant Dals, Pulav, Khichdi, Kheer, Halwa, Upma, Cake, etc., but their large scale marketing has yet to take place. Some efforts have also gone into develop ready to eat shelf stable canned and retort pouch foods. Excellent quality sterilized vegetables and mutton pulav, khichdi, potato peas curry and fish curry have been developed by Defense Food Research Laboratory, Mysore. Recently some firms have started marketing these foods. Some efforts have also gone into developing shelf stable ready to serve chapaties, parothas and similar products, but, these have not found wide acceptance in the market mainly because no suitable chapatti making machine is available for mechanized rolling and baking of chapaties.

Even after 50 years of independence it is rather surprising that not a single automatic chapatti rolling and baking plant has been established.

Newly Emerging Process

Increased concerns regarding food safety and conservation of natural flavour and taste are making the consumers look for minimally processed foods with least distortions to the profile of food during processing. Reduced thermal inputs during processing and preservation steps are suggested through new technologies. Several recent developments in the area of food technology suggest that industry can look forward to deployment of viable but acceptable process technologies in the coming years based on current progress achieved so far. A discussion on these will be in order.

High Electric Field Pulses

This technology relies on the lethal effect of strong electric fields for the inactivation of microorganisms. As per the electrical rapture theory, the external electric field induces an electric potential over the membrane, which in turn causes a charge separation in the cell membrane. When the trans-membrane potential exceeds about one volt, the repulsion between charge carrying molecules causes the formation of pores in the cell membrane and thereby the death of cell. The critical trans-membrane potential and consequently the external field strength depends on the type of biological cell, the cell diameter, shape and the growing conditions. By using this technique, the undesirable changes such as protein denaturation and vitamin losses during processing are avoided. The technique is effective against only vegetative microorganisms while spore are not effected.

Oscillating Magnetic Field

Subjecting glass or plastic packed food to one or more pulses of an oscillating magnetic field can lead to destruction of microorganisms. When a large number of magnetic dipoles are present in one molecule, enough energy can be transferred to the molecule to break a covalent bond resulting in the destruction of vital cell constituent like DNA, enzyme, etc., and ultimately the death of the microorganisms. The microorganisms are inactivated when frequencies are in the range of 5-500 KHz. As the exposure time is only a few seconds, no temperature rise occurs.

Intense Light Pulses

Light pulses having a duration of 10^{-6} to 10^{-1} seconds, an energy density of 0.01-50 cm^2, and a wavelength spectrum between 170-2600 nm are reported to be effective in inactivating microorganisms, at the surface of the food materials and within transparent bulk liquids, solids and gases. UV rich light pulses with 30 per cent of the light energy at wave length shorter than 300 nm are considered effective for aseptic packaging materials and air, water and salt solutions. The UV rich pulses inactivate microorganisms and viruses through a combined photo-thermal and photo-chemical mechanisms.

High Pressure Treatments

High pressures in the range of 500-10,000 bar at ambient temperatures have been known to have lethal effect on microorganisms for a long time. However, it was only recently that this basic principle was made use of for the commercial sterilization of foods. High-pressure pasteurization of foods like jams, fruit juices and fruit desserts are already applied commercially in Japan. The technology is not totally effective in destroying bacterial spores and many enzymes. Probably development of combination of high pressure with moderate temperatures may have to be evolved if universal acceptance is desired. Combining high pressure processing with chilled storage and distribution may also be desirable if quality deterioration by enzymes is to be prevented. This technology is conveniently suited to the food industry as it can be helpful in developing foods of high nutrition and sensory properties, novel texture, more convenience and increase shelf-life. The technology is also compatible with the existing range of semi-rigid and flexible packaging materials but in the present form it is not industrially viable.

Ohmic Heating for Aseptic Products

Aseptic processing has emerged as most commonly used method for processing of shelf stable liquid products. The process is based on the sterilization of food in a heat exchanger, followed by filling in paperboard or plastic containers in an aseptic environment. Aseptic processing has not gained wide acceptance for processing of products containing particulate matter. Use of conventional heat transfer technique necessitates over processing of liquid phase to ensure that center of particulate is sterilized. This invariably results in destruction of flavours and nutrients and mechanical damage to particulate. These problems have been overcome with the use of ohmic heating systems. In this method, low frequency alternating current is passed through specially designed electrodes. Since there are no heat transfer surfaces, both solid and liquid phases get heated simultaneously and no scorching or overheating occurs in the food.

Irradiation Processing

Use of γ-irradiation for good preservation has been extensively investigated. While safety of the food processed by using γ-irradiation up to 10kgrey has been well established, its commercial use has yet to pick up. Use of γ-irradiation has also been permitted under PFA for disinfestations in grains and milled products, for sprout inhibition in potato and onions and for disinfestations of spices. γ-irradiation will also be useful for disinfection of poultry and marine products in chilled and frozen state. Because of very large investments needed for establishing γ-irradiation facilities, high throughputs are necessary to ensure viability of the process. Irradiation is likely to find more use as a complementation process alongwith cold storage rather than a sole means of sterilization of foods. In India, use of irradiation is picking up for the disinfestations of spices for export. Irradiation of fresh fruits and vegetables may also prove viable.

Controlled Atmosphere Packaging (CAP)

CAP has been in vogue during the last few years especially for extending the shelf life of fresh produce in the marketing chain. Recently, it has been shown that CAP can also be effective in the preservation of seafoods like fish and bakery products. Packing of fish in an atmosphere of carbon dioxide was found to delay spoilage and provide consumers with a high quality product. Though it is difficult to suggest a standard carbon dioxide concentration in CAP in all varieties of food, carbon dioxide levels varying from 20 to 100 per cent have been suggested. CAP alone is not effective and the product needs to be refrigerated below 3.3°C to retard development of *clostridium botulinum*. Additional treatments such as potassium sorbate and low dose irradiation can further enhance the effects of carbon dioxide. On the other hand modified atmosphere packaging which utilizes suitable packaging films for regulating concentration of carbon dioxide is finding application for preservation of fresh fruits and vegetables because of its similarity and lower cost. However, wide scale use of MAP in fruit and vegetable sector has yet to pick up. Their use will be viable where ambient temperatures are low (below 15°C) and no additional cooling is necessary for wide scale marketing. Modified atmosphere packaging will also eliminate desiccation and moisture loss.

Edible Coatings

Edible coatings are applied to many food products for a variety of aesthetic and protective purposes. Edible coatings serve as invisible packaging and prevent moisture loss, reduce respiration rate in fresh fruits and vegetables, improve handling and reduce spoilage and decay by restricting the growth of microorganisms. The types of materials used include lipids, resins, polysaccharides and proteins. Each group of materials has certain advantages and disadvantages and therefore composite formulations are proving more suitable for various applications. Among the various foods, fruits and vegetables have received largest attention. Many composite patented formulations are marketed for extension of shelf life of fruits and vegetables. There has been considerable research on the suitability of waxes and other materials for the protection of apples, tomatoes, brinjals and other vegetables in India also, but no commercial formulation has emerged from these studies. India being the second largest producer of fruits and vegetables and also having very high percentage of spoilage, it is desirable that this technology be put to use, at commercial level so that spoilage during transportation and marketing is reduced. In majority of the cases coating technology is simple and can be applied at farm level itself. Surprisingly in India, use of edible coating is not permitted under PFA. It is high time that government takes initiative in approving the use of edible coatings and also provide financial incentives in the form of subsidies to encourage the use of coatings which is likely to reduce wastage.

Low Calorie Fat Replacement Aids

In India, average fat consumption is quite low (20 g/mm/day) but consumption pattern is not evenly distributed. Affluent sections of population invariably consume higher level of fats resulting in obesity and hyper-cholesterolemia. Recent awareness of the nutritional implications of high fat intake has led to the search for alternative low calorie fat substitutes. Synthetic fat replacers include sucrose polyesters such as Olestra, Esterified Propoxylates Glycerol (EPG), di-alkyl di-hexydecylmelonate (DDN) and tri-alkoxy citrate. So far, none of these substitutes have been cleared for use in food.

Second approach for formulating fat replacements involves use of hydrocolloids, which thicken or gel in aqueous systems. Oatrim, a maltodextrin, derived from oats is a fat substitute with the ability to lower blood cholesterol levels, both as a replacement of animal fats and through the action of beta-glucan. Maltodextrins with low dextrose equivalent give the most suitable fat substitutes. Oatrim forms fat like gel on heating and cooling 25 per cent dispersion. The calorific value is less than one per gram as compared to 9 per gram for fats. Cellulose, Pectin and egg albumins have also been utilized for preparing fat like replacements. Hydrocolloid based fat substitutes have received good response and are marketed in USA and other western countries.

Biotechnology

Genetic engineering for modified fats: Fatty acids in foods have critical role in deciding about their stability, flavour, texture and nutritional quality. There has been considerable interest in reducing the saturated fatty acid content in foods and to develop vegetable oils with high mono-unsaturated fatty acid contents. There is also considerable interest in increasing the ratio of intake of ω-3 to ω-6 fatty acids. Development in recombinant DNA technology will go a long way in augmenting efforts in breeding and genetics for improving the quality of vegetable oils. Significant success has been achieved in breeding for altered fatty acid composition of vegetable oils to improve their food value in some crops like sunflower and rapeseed. In other cases not much headway has been possible.

All over the world biotechnology is perceived as the panacea for solving the world food problems and intensive efforts are going on to develop crops which are resistant to weeds and insects, which are drought resistant and which are high yielding. Efforts are on to transfer nitrogen fixing genes of Rhizobia into wheat, corn and rice. Genetically improved soybean, corn, cotton, rice, oilseed rape, sugar beet, tomato and alfalfa crops having higher productivity and functional characteristics have been developed. Efforts are also on to evolve caffeine free coffee and fruits that are less susceptible to damage during transport. While there is no doubt that more desirable traits are going to be introduced in various crops through genetic engineering, at the same time we should have a comprehensive protocol for evaluating their safety in the long term consumption. This is necessary to ensure that defective seeds are not passed to the lesser-developed countries in the guise of high yielding crops.

With the rise in population, the pressure on land and water resources is increasing. More intensive agricultural practices are being introduced; the use of pesticides and weedicides is increasingly causing problems of pesticides residues in the food. Therefore, lot of attention is being focused on safety aspects of foods produced by new technologies and practices. Food industry generally considers threat to food safety coming from microbiological spoilage, pesticide residues, mycotoxins, package product interactions, chemical additives and carcinogenic compounds.

Many new techniques now being considered for commercialization on a wide scale such as microwave processing, high pressure treatment, ultra high temperature processing, ohmic heating, all require safety evaluation before consumer can be expected to accept products processed by them. Development of new food ingredients, genetically improved plants and animals and innovative packaging technologies have created exciting possibilities for both the industry and consumers in terms of vast array of new products. All these require greater vigil and safety evaluation efforts before they are cleared for mass consumption. Food scientists and industry have a major responsibility to ensure that new products and process technologies developed by them meet the safety norms.

38

Formulated Foods

G.R. Patil
Division of Dairy Technology, National Dairy Research Institute, Karnal – 132 001

Introduction

Formulated foods/fabricated foods are foods designed and built according to plan from individual components, to yield products having specified physical, chemical, and functional properties (Glicksman, 1971).

The concept of formulated food is not new. These foods have been part of our society since the beginning of recorded history. The earliest known fabricated food is bread. Bread per se does not exist in nature, and while nature made wheat, corn, rye and other cereal grains, it was man who took the harvest and formed it into a food with texture, shape and flavour that made it a joy to eat. Historically, fabricated foods were developed initially not to titillate the palate, but to use available ingredient in a convenient and utilitarian manner.

There are two basic types of formulated foods: those designed to simulate natural counterparts currently referred to as "analogs" such as meat and dairy analogs, and those which have no counterpart in nature but are prepared to give variety and spice to the diet. One of the most common of most popular examples of the latter is jell-o-gelatin dessert. Another popular and universal product is ice cream, designed to please the palate hundreds of years ago and still probably most widely consumed formulated food in the world. Many types of formulated food products are now available.

Why Develop Formulated Foods

1. Lower cost: The cost of certain fabricated foods may be lower while providing about the same nutrition and satiety, *e.g.* margarine.
2. The nutrition of the food may be enhanced: Example fortification of fruit juice with Vit.C, margarine with Vit. A., cereals with iron or other nutrients.
3. Foods can be made more convenient.
4. Foods can be formulated to fill special needs.
5. Formulated foods can provide other meal advantages. They can improve the balance of a meal; they can increase palatability and finally can provide satiety and other advantages.

Principles of Food Formulation

Formulated foods are made by combining the three basic building blocks of all food products–protein, fats and carbohydrates–in ways that provide convenience, texture, flavour and other desirable characteristic. This involves manipulation of these 3 basic components along with water, flavours, preservation, vitamins, etc. to design products with predictable compositions, texture, flavours and storage properties.

In food formulated the aim is to achieve uniformity and high degree of predictability in product attributes which, in tern, is dependent upon complex interrelationship of several aspects of physical chemistry, such as solid-liquid phase equilibrium, multi-component solubility behaviour, polymorphism, nucleation and crystal growth, wetting and emulsification, stability of disperse phases, and mechanical properties of crystal assemblies.

In the design of the formulated product, therefore, it is necessary to take into consideration the complex structures and interactions involved among food components, and whether the final system should constitute an equilibrium state, and if not, what kinetic controls can be built into the system to prevent attainment of equilibrium during normal processing, shelf-life and storage conditions. One or several intermediate steps can, in principle be stabilised by providing for a kinetic barrier. In the case of dispersed systems, such as emulsion, this is often achieved by raising the viscosity of the continuous phases or by incorporating absorbed layers of proteinaceous materials, which act as barrier to coalescence of the disperse particles. Polysaccharides are often used as viscosity modifying agents. The polymer chains of polysaccharides associate to form limited sequences of ordered structures, which in turn make up a random network and give rise to gels of different strength, textures and thermal stabilities. Similarly "edible surfactants" *i.e.* proteins are usually used for stabilizing metastable, highly heterogeneous, disperse systems of complex food products. The properties of polyelectrolyte macromolecules such as proteins or polysaccharides, whose internal conformation can be markedly influenced by alterations of the total charge distribution, are often modified by adjustment of pH and/or type and amount of electrolyte to suit the requirement.

Ingredients of Formulated Foods

The ingredients used in formulated foods must be readily available, economical, safe, and must serve a useful function. Various types of proteins, carbohydrates, fats, etc. used in formulated foods have been discussed here under.

Protein

Plant proteins have played very important role in the growth of fabricated foods. Soy proteins have been greatly used in formulated foods because of its ability to incorporate texture into them. Although the major plant protein ingredient in formulated foods is now soy protein, other sources of plant protein, such as cottonseed, peanuts, rape, sesame, safflower, sunflower, wheat, corn, etc. have been used. In addition to plant proteins as ingredients in formulated foods, other non-conventional protein items such as protein from microbial sources, whey proteins, fish protein and similar products are finding application in formulated foods.

Blends of proteins will be the key to good nutrition in formulated foods. Soy protein, for example, lacks methionine but has more lysine than is needed for proper nutrition. Blending soy protein with cereal products improves the lysine and methionine balance. Animal protein ingredients, such as whey and fish protein, can fill an important role in providing amino acid balance. Mixing protein ingredients may have a secondary benefit. Meats are said to provide "Unknown micronutrients" not present in vegetable protein. This is a problem for presently used foods as well as newly formulated food derived from plant proteins. The best nutritional defence against this unknown micronutrient charge is to strive for greater variety and diversity in our foods and food ingredients. Plant proteins as the primary source of protein in a fabricated food, for example, should be broadened to include proteins from oil seeds, nuts, grains, microbial source and of course animal sources.

Carbohydrates

Carbohydrates in formulated foods are not included primarily for their nutritional values. Carbohydrates are used largely for functional reasons in fabricated foods. The highly soluble common sugars, like sucrose, dextrose and corn syrup, are used mainly as bodying and sweetening agents. Thus they contribute such bodying attributes such as viscosity, texture, density and mouthfeel, among others, while also imparting sweetness and flavour to the organoleptic and sensory aspects of food.

Carbohydrates polymers, more commonly known as gums and starches (hydrocolloids), are used strictly for their functional properties. With the exception of starches, carbohydrate polymers are used at very low levels, generally less than one per cent of the foods stuff. These long chain polymers have many diverse functional properties, but the one major property common to all gums is that of viscosity in aqueous medium. All gums will dissolve or disperse in water to give a thickening or bodying effect. The second major property exhibited by a selected few hydrocolloids is that of setting. Gelation is the phenomenon involving the association or cross linking of the polymer chains to form a 3-dimensional continuous network which traps or immobilises the water within it to form a firm, rigid structure that is resistant to flow under pressure.

In addition to thickening and gelling, gums have many secondary functional properties that are useful in food processing and development ranging from adhesiveness to whippability. In many applications, more than one of these functional properties come into play, and the versatility and uniqueness of these ingredients is sometimes solely responsible for the feasibility and fabricating specific food products.

Fats

For many years, soybean oil has been major component for margarine–a well known formulated food. In addition to soybean oil other sources of vegetable oil such as cottonseed, peanut, sesame, safflower, sunflower; palm, coconut, etc. have been used in formulated foods. The fats are generally tailor made by various processes such as hydrogenation, fractionation, interesterification, etc. to suit to specific requirement.

Flavouring Formulated Foods

While developing a formulated food, the flavour and formulation must be treated as a whole to arrive at organoleptic effects that more effectively mimic what we have accustomed to eating. Therefore, development of synthetic flavours closely parallels to development of formulated foods. As a result several synthetic food flavour, meat flavour, cheese, butter, chocolate flavours, etc. are available in market which can be used to make the formulated products more closely resemble to natural products.

Texture Development in Formulated Food

In developing formulated foods, it is important that desirable textural properties be engineered into the food from the early stages in development. While engineering texture in liquid or semisolid formulated foods is easier, similarly texture of meat like products is more challenging. Most commonly used methods of texturizing proteins are fibre spinning, extrusion and chewy gel formation.

Fibre Spinning

This process is based on changing the protein configuration by solubilizing the protein, unfolding the protein chains and reforming them in roughly parallel fashion thus simulating muscle protein fibre. In addition to predetermined levels of protein, fat and carbohydrate, colours, flavours, vitamins, minerals, amino acids and other constituents could be added as desired. During spinning other textural properties can be imported to the product fibres. Varying pressure in assembling the final extruded fibres can alter the density of texture. Toughness and tenderness can be controlled to a certain degree by the amount of stretch applied to the fibres during the initial spinning steps. Major textural effects can also be produced by additives such as gums or starches, which can effect chewability, stretchability, elasticity, cohesiveness, adhesiveness, moisture retention, rehydratability, freeze-thaw stability and water activity.

Extrusion

Another major technique for fabricating textured vegetable proteins is by extrusion of protein compositions under heat and pressure. It does not produce well-defined fibres but gives fibrous particulates that have good mouthfeel and chewiness similar to meat. The protein in aqueous dispersion is subjected to heat and pressure which causes some realignment of the protein molecules as it is extended from high pressure area into the atmosphere, creating a rapid expansion of the product with a rapid flashing of the water. The mixture is subjected to a pressure above 1000 psi and temperature of 115 to 180°C during processing.

Chewy Gel Formation

A third type of process for creating textural protein foods consists of preparing "chewy gels" under special condition of pH, protein concentration, water content, and heat. The resultant gel has a smooth, uniform hydrated structure that is pleasantly moist in the mouth and yet has enough textural firmness to give proper resistance to bite. Texture and mouth feel can be modified by incorporating hydrocolloids into the gel.

Developments in Formulated Foods

Infant Foods

As the infant grow, mother's milk does not suffice his growth requirements and therefore, wide range of infant formulations have been developed world over to meet the nutritional requirements of different category of infants (Thompkinson, 2000).

Standard Formulae

These formulations supplement of human milk for feeding of normal healthy infants and are based on adjustments of Protein:

Fat: Carbohydrate ratios as a broad compositional simulation to human milk. These formulations are also fortified with general minerals and vitamins.

Soy-based Formulae

Infant formulas based on soy proteins have been developed to aid in the dietary management of infants suspected of being allergic to milk proteins.

Specialized Formulae

This group of formulas is specific products for the dietary management of infants with variety of nutritional and physiological disorders that are associated with the early stages of life. For example, formulation in pre-digested form that is easy to digest or a low lactose diet or lactose free diet. Specialized formulations consisting of pre-digested or hydrolyzed proteins, hydrolyzed lactose, low fat formula, formula for pre-term infant etc fall into this category.

Nutritionally Improved Formulae

Nutritionally superior formulations having chemical and biochemical similarity with human milk (humanized or eternalized formulas) have been developed. Apart from similar gross compositional profile in respect of protein: fat: carbohydrate: minerals, these formulations also incorporate comparable ratios of whey protein: Casein; saturated: polyunsaturated fatty acids; calcium: phosphorus and sodium: potassium. Such formulations closely approach the nutritional qualities of human milk and therefore are easily adaptable by the infantile system without causing great physiological stress on the under-developed organs.

Bifidus-containing Formulae

Notwithstanding the nutritional superiority, the humanized formulations do not offer biological protection against enteric infections as human milk. With the objective to enhance bio protective attributes of ready to use dried infant formula, a process has been developed to incorporate a biomass of Bifidobacteria for its protective role in gut ecology.

Formulations with Bio-protective Attributes

Infant formulations added with bioprotective factors such as lactoferrin, lysozyme, lactoperoxidase, immunoglobulins, etc. have been developed for their direct protective role in the ecological system of the intestinal tract.

Beverages

Current trends indicate that more new beverages than any other class of food are being formulated. These include sports or performance drinks, nutritionally enriched beverages, fruit juices and juices with added proteins. Sports drinks and tea-based beverages grew from a very small quantity to about 500 million litres over the last decade of the twentieth century in Europe. The trend is the same all over the world and the steep hike is expected to continue for at least another two decades. Drinkable yoghurts have almost reached a mature point in their life cycle, with interesting new flavours being added to elongate the top of the curve. Water-type beverages containing minerals are being widely marketed. India seems to have found a very huge market potential for mineral water in the past few years and the trend is likely to stabilize through the early years of this millennium. The role of dietary calcium in preventing osteoporosis being widely understood, recent research has also suggested protective effect of calcium against other health conditions including hypertension, colon cancer and even pre-menstrual syndrome. Calcium-enriched beverages whether dairy-based or otherwise are expected to gain considerable interest and it may be envisaged that products with increasing claims of bioavailability of the mineral will be marketed in the near future.

Dietetic Foods

Slimming Foods

There are a variety of slimming foods on the market, including the following (Kanawjia, 2000):

1. Portion control foods for which the manufacturer calculates and declares the energy content of a complete food as a convenience to the consumer.
2. Energy-reduced foods, which have a minimum of 25 per cent energy reduction of the normal food, or low-energy foods, which have a maximum of 50 kcal (210 kJ) per 100 g or 100 ml of product.
3. Products, which are a sole source of nutrition and are, classified as very-low-calorie diets (VLCDs) at 400-800 kcal

(1680-3360 kJ) per day, and low-calorie diets (LCDs), which contain 800-1200 kcal (3360-5040 kJ) per day. Complete diets below 400 kcal per day are not, at present, classified as foodstuffs and may be taken only under strict medical supervision.

Foods for Special Medical Purposes

Two distinct types are recognized: enteral nutrition, in which foods are administered via the gastrointestinal (GI) tract, either orally or by means of a tube into the stomach or intestine; and parenteral nutrition, in which foods are administered intravenously.

Low Sodium Foods

Processed foods with low or reduced sodium content have been developed for patients with kidney disorders, hypertension, and other medical conditions, which require reduced sodium intake. The EEC draft directive recommended two categories of food: (1) Very-low sodium: the sodium content should not exceed 40 mg per 100 g or 100 ml in the ready-to-eat product, (2) Low-sodium: the sodium content should not exceed 120 mg per 100 g or 100 ml of the ready-to-eat product.

Gluten-free Foods

Gluten-free foods are required for the dietary management of people with coeliac disease, the causative agent being the gluten content of cereals such as wheat, rye, barley and oats. Most of the symptoms result from mal-absorption of nutrients from the intestine caused by damage to the cell wall. Gluten-free foods have been developed utilizing rice and maize flour.

Foods for Intense Performance-Sports Food

To maintain enhanced physical fitness, coupled with intense performance, and the stresses that this places on the body, special formulations have become available such as:

1. Food products specially formulated to supply energy,
2. Food products, tablets, capsules and hydration beverages with a defined content of minerals, trace elements, vitamins, and other substances with nutritional affects, or combinations of these, supporting the physiological performance

3. Food products with a defined content of protein and/or amino acids, especially formulated for intense muscular effort
4. A combination of the products as detailed above.

Food for Diabetics

Formulations based on artificial sweeteners along with bulking agents have been developed for management of hyperglycemia.

Future Trends in Product Formulation

Authentic

Convenience has always been the strongest drive for change. But despite the need for convenience, the foremost requirement for foods is that they should taste, smell and feel right. The number one challenge for the food industry, therefore, is to produce foods that come under the concept of 'authentic', both in terms of flavour and texture. Demographic changes, the top-most among them being large scale immigration, have been instrumental in bringing authentic ethnic flavours into mainstream food processing. Finding a way to reproduce 'authentic' flavour and texture while translating that to more mass-produced items is a considerable challenge.

Healthy

Equally important is the recent transition towards 'healthy' foods. Energy and health have always been derived from food, but what will improve are sophistication in appreciating specific nutritional needs and tailoring foods and their components more closely to the finer detail of bodily absorption rates and special requirements. The close alliance of nutritional knowledge and the delivery of promised benefits in food products offers an unparalleled opportunity for developing a product mix that is the basis of good health. Vitamin and mineral enrichment and the use of dairy fractions including lactoferrin and lactalbumen in dairy as well as functional foods is fairy widespread.

Bio

There is clearly a movement towards the need for food ingredients to be 'bio' and/or 'organic' in origin. The quest for newer prebiotics, probiotics and herbal products being used in foods and supplements is definitely a top trend. Prebiotics including oligosaccharides, fructo-oligosachharides, dietary fibre, and resistant starch fall into the arena of health foods. Probiotic organisms such as lactobacilli and/or bifidobacteria are increasingly being introduced into dairy products such as yoghurts, cheeses and beverages. Special varieties are being developed and patented, either as specific organisms or as part of a food formulation.

Luxury Items

In the western countries, luxury items aimed for the elite society are a fast-developing trend. Most of these are dessert items, pastries and ingredients for these. Since dessert items are more costly than the main course products in these countries, products containing caramel sauce, ice creams in tiny servings, mousse products and custard tarts with extravagant toppings are likely to be expensive. Tremendous opportunities for alternative products and ingredients are envisaged to cater to this segment of the food industry. On a smaller expense-scale, sauces for meats, marinades and specialty-processed fruits will grow in popularity. Alcohol products are acceptable in sauces in the western countries and new technologies have provided ways to add stout red wines or delicate white wines to sauces.

Foods for Demographic Groups

Foods targeted towards specific demographic groups are becoming exceedingly popular, especially in packages that make their usage easier. Ethnic food preferences will continue to be dictated by immigrants. Italian, Mexican, Chinese, Japanese, Indian and Thai foods are becoming exceedingly popular in Europe and America. Although the overall numbers are much smaller than the others, interest in Indian cuisine has doubled in the past five years, thus making it the fastest-growing ethnic category. Fine dining establishments offer a future that blends the cooking from several regions. This fusion cooking might show up as avocado puree topped with an orange sauce or mango chutney over a fish-based item. Asian flavours such as ginger and soy sauce in cooking sauces, salad dressings and snack products are possible examples of such mix-and-match options.

Vegetarianism and Ban Red-Meat Products

Self-described vegetarianism in traditionally non-vegetarian areas of the world is on the increase and is expected to be 20-25 per cent of the population. Soy and other vegetable products are going to be used increasingly in conventional non-vegetarian products such as patties. Propelled by the recent fear of the 'mad-cow' disease in Europe and also the increasing popularity for vegetarianism, poultry and marine products are likely to gain favour the world over. These products will be marinated or seasoned very carefully and are likely to be prepared prior to purchase. The search for alternative sources of gelatin has already yielded results, with fish gelatin and plant-based hydrocolloids being exceedingly in demand.

REFERENCES

Bennett, B. (2000). Riding the next ethnic wave. *Food Processing*, (3): 47-50

Dziezak, J.D. (1989). Ingredients for sweet success. *Food Technol.*, 43: 94-98.

EEC Council (1989). Directive on the Approximation of the Laws of Member States Relating to Foodstuffs intended for particular Nutritional uses. *Official J. The European Communities*, No. L186/ 27.

Glicksman, M. (1971). Fabricated Foods. *Crit. Rev. Food Technol.*, 2(1): 21-43.

Hall, R.L. (1999). Food safety: elusive goal and essential quest. *Food Australia*, 51(12): 601-606.

Kanawjia, S.K. (2000). Dietetic foods and their applications In: *Compendium on Advances in Formulated Foods, Centre of Advanced Studies in Dairy Technology*, NDRI, Karnal, 19 June–10 July 2000.

Katz, F. (1999). Top product development trends in Europe. *Food Technology*, 53(1): 38-38-40, 42.

Lupein, J.R. (2000). Food perspectives and food demand beyond 2000. *Food Australia*. 52(12): 597-600.

Mathur, B.N. (1991). Recent advances in the manufacture of Infant foods. *Indian Food Industry*, 10: 25-29.

Rao, S.V. (1991). Role of specialty foods for meeting nutritional requirements. *Indian Food Industry*, 10: 22-24.

Smith, C.M. 2000. Strategies for growth in the food industry. *Food Australia*, 52(1,2): 48-51.

Sanders, M.E. (1998). Over view of functional foods. *Int. Dairy J,*. 8: 341-349.

Thompkinson, D.K. (2000) Developments in infant food formulations. In: *Compendium on Advances in Formulated Foods,* Centre of Advanced Studies in Dairy Technology, NDRI, Karnal, 19 June–10 July 2000.

39

Thermal Processing and Food Quality

Rajendra Singh

Centre of Food Science and Technology, Chaudhary Charan Singh Haryana Agricultural University, Hisar – 125 004

Almost all the foods receive some sort of heat treatment at one stage or other before consumption. Exception to this may be fruits and some vegetables consumed as raw fresh, juice, salads, chutney or pickles. Thermal treatments are given for achieving one of the following functions of processing:

1. Preservation
2. Cooking
3. Confection

Preservation by thermal processing of foods is achieved by inactivating the causative agent of spoilage at the minimum expense of the nutrients. Causative agent of spoilage may be some intrinsic food enzyme(s), contaminating pathogenic or spoilage organisms and the processes are referred to as blanching, pasteurization or sterilization including a variety of canning processes. These processes are discussed in detail in almost all the textbooks dealing with food processing to the extent that these have become synonyms of thermal processing. This will not be discussed here any more and lecture will be restricted to other two lesser detailed functions of thermal processing.

Cooking include variety of processing methods which include baking, extrusion cooking, roasting, puffing, frying, broiling, grilling and stewing in addition to its common meaning as cooking in hot water or steam. Heat is applied through variety of media:

Hot surface: baking *chapaties,* pan broiling, sauteing, etc.

Pressure explosion: extrusion cooking, puffed snacks

Sand or salt: roasting, puffing

Direct flame: broiling

Hot smoke: smoking,

Hot air: roasting, puffing, baking, grilling, broiling, etc.

Water: simmering, stewing, boiling, etc.

Steam: steaming, pressure cooking,

Oil and Fat: sauteing, frying.

Some times a combination of media is involved in overall process as in case tandoor baking where hot surface generally an earthen pot or structure and direct flame or hot air are utilized. It gives a product with two surfaces with different characters–a shining surface air cooked and the other rough one hot surface cooked.

Confection utilizes the properties of sugar-water system to obtain a range of texture profile in the products. Glass transition temperature with specific temperature programming at different moisture content of sugar with or without other sugars like glucose, invert sugar and starches, gums and gelling agents, fats and proteins or other food solids like milk solids, cream, egg, cereal and pulse flours, cocoa power, cream etc., are responsible for specific texture. Processing temperature and non-sugar constituents determine the color and flavour of confectionery products through varied extent of caramelisation and maillard reaction.

Physico-chemical Changes During Thermal Processing

Foods are composed of various chemical constituents with different physical properties. Thermal processing involves a range of temperature-time treatments from almost just above ambient temperature in cheese cooking to almost 240°C in baking and frying and time ranging from few seconds in HTST, UHT processing to few hours in cooking of pulses and many hours or days in crystallization of sugar. Time temperature interaction with complex system of reactive constituents results in variety of changes in physical and chemical properties of foods.

Loss of Moisture and Solidification

1. Heat treatment under open conditions as those of roasting, baking, jam, jelly, *khoa rabbri* making and *rabbri* allow loss of moisture with concentration of solids.
2. Storage of saturated sugar solution quite above glass transition temperature for long time results in crystallization of pure sugar crystals with high ERR.
3. Fast cooling of low moisture molten sugar below glass transition temperature results in en-mass solidification with crispy textured products.
4. Cooling of high moisture molten sugar mixes at just above glass transition temperature gives a rubbery plastic gummy product with high ERR.

Absorption and Solubilization

Fat and fat soluble components of food absorb dissolve in the frying oil and form colored substances. The foods absorb the frying oil and thus finish the frying oil. Oxy products arising from oxidative changes are preferentially absorbed by frying foods forming hydrogen bonds and covalent bonds with amino groups of proteins and hydroxyl groups of carbohydrates.

Hydration and Solubilization of Polymers

Disruption of Cellular Integrity

At relative low temperatures of 55–60°C, thickening, of cell walls due to hydration of cell wall polymers occurs accompanied by disintegration of plasmalemma leading to loss of cellular integrity. Failure to take up water linked to methyl content of pectin results in dry hard patches in products in case of *'non-soakers'* dry beans. Prolonged heating leads to some breakdown of cellulose and hemicellulose. Cytoplasm forms dense clamps of granules. Solublisation and leaching of pectic substances from middle lamella result in cell separation. Failure of middle lamella to dissolve lead to the condition of *'HTC-hard to cook'* legumes due to partial hydrolysis of phytate and de-esterification of pectin during storage and formation of insoluble Ca-Mg pectate. Activation of hydrolytic enzymes like pectin methyl esterase leads to solublisation of pectin. All these result in softening of tissue texture. Hydrolysis of starch by activating amylases of germinated grains releases fermentable sugars in malted cereals.

Extraction of Pectin

Pectin is extracted from fruit slices by boiling with water for the preparation of fruits jam and jelly. Extraction of pectin in the liquor in during canning causes the defect of gelled brine or syrup and softer texture of the product.

Gelation of Polymers

Gelatinisation and Gelation of Starch

At 50–60°C, amylose molecules migrate out of the starch granules and form gel on cooling the product. Imbibition of water by amylopectin and hydration in granule results in gelatinisation and swelling of starch.

Bursting of Starch Granules

Under limited moisture conditions gelatinisation temperature of starch increases to more than 100°C which results in build up of water vapour pressure inside the grains and starch granules. Fast increase in temperature during roasting causes an explosive release of steam from grains and bursting of grains as in case of puffed cereals. Similarly in extrusion cooking mechanical friction raises the temperature of the mix in side the extruder barrel and keeps it under pressure before release through orifice. A sudden release in pressure on die outlet allow fast release of steam leaving the product puffed.

Draining of Gel

Casein gel set by rennet action and starter culture activity in manufacture of cheese, looses whey when cooked at 35–55°C. Higher the temperature higher is the whey drainage which is assisted by acidity and harder the cheese obtained.

Gelation of Egg Albumin

A heat treatment of above 80°C results in denaturation and coagulation of egg white. Speed of coagulation of egg albumin increases by more than 600 times for every 10°C rise in temperature.

Loss of Nutrients

Loss of Vitamins

Water-soluble vitamins are more prone to heat destruction than fat soluble ones. Loss of thiamin, riboflavin and pyridoxine and vitamin C occur during heating temperatures of cooking foods.

Degradation of Proteins

Proteins loss their tertiary and quaternary structures on heating above ambient temperatures and get denatured. Denaturation of proteins results in loss of their enzymatic activities. Other environmental factors and intrinsic properties play a major role in their denaturation. Sulfur containing–amino-acids and tryptophan are more sensitive to degradation by heat. Release of sulfides from protein degradation results in green discoloration and cooked flavor defects in animal products rich in proteins such as eggs and milk.

Destabilization of Colloids and Emulsion

Destabilization of Protein Structure

Heat treatment of 85–95°C is given to milk to increase the recovery of *paneer* and *chhanna* and improve drainage of whey.

Separation of Phases

Oil-in-water emulsion in milk cream and water-in-oil, emulsion in butter are broken by heat treatment resulting in denaturation of emulsifying proteins and decrease in viscosity of molten milk fat to allow fat separation from the aqueous phase.

Thermo-chemical Decomposition

Maillard Reactions

A reaction sequence is initiated by formation of schiffs base between a carbonyl compound and an amino compound at about 130–150°C. Reducing sugars with free aldehyde or keto groups participate in this reaction along with amino-acids or to a lesser extent free amino group containing amino-acids in the proteins. Through Amadori re-arrangements and enolisation creating further carbonyl group for reaction with amino acids, a series of reactions sets in. This results in degradation of sugars and production of brown colored melanoidin products. Smaller compounds-furfuraldehyde, hydroxymethyl furfuraldehyde, lactic acid, acetic acid, formic acid, carbon di oxide are produced along with complex melanoidins and result in flavor development.

Caramelisation

Burning of sugar at about melting point temperatures results in loss of water molecule and formation of anhydro-sugar. Further reactions lead to polymerization and production of brown colored products known as caramel.

Oxidative Degradation of Fats

Oxygen adds to fat molecule more easily at unsaturation point to form peroxides, which split into hydroxyl and alcoxyl free radicals and sets in a chain reaction. Volatile products like ketones, aldehydes, alcohols acids and hydrocarbons results from oxidative degradation of fats, mainly by splitting of alcoxyl radicals. High temperatures of frying operations accelerate these reactions. Some of the volatile components thus generated are aggressively aromatic in nature though minor in quantitative terms (10–30 ppm). They lead to chemical rancidity of fats and oils.

Hydrolytic Degradation of Fats

Moisture from foods being fried reacts with the fats at high temperatures of more than 100°C and results in their hydrolysis with production of free fatty acids, diglycerides, mono glycerides and glycerol. Glycerol decomposes to volatile compound acrolein, an unsaturated aldehyde responsible for irritation and burning effect in eyes.

Thermal Polymerization

High temperatures of frying operations are conducive to polymerization and cyclization reactions in polyunsaturated fatty acids in fats and oils. Inter-molecular polymer formation results in increased viscosity of oils. Polymerization and cyclization of fatty acids result in loss in their nutritive values.

Thermal Processing

A variety of thermal processes are employed in food processing as indicated above. Conditions and methods employed in these processes are briefly indicated in the following text.

Energy Sources

Heat for processing is generated from four main sources of energy:

1. Solid fuels, *e.g.* coal, coke and wood.
2. Liquid fuel, *e.g.* fuel oils, kerosene and other liquid paraffins.
3. Gaseous fuels, *e.g.* coal gas, natural gas and petroleum gas.
4. Electrical energy generated from above sources as well as from other conventional or non-conventional sources like hydraulic power, nuclear power, wind power, solar radiation etc.
5. Solar energy.

Selection of a particular energy source depends on various considerations such as economics both in terms of running cost of fuel and labour and cost of the equipment and machinery installation and maintenance, effect on food products, flexibility of operation and control and the environmental consideration pollution of air and disposal of generated waste and risk of accidents, etc.

Mode of Heat Application

Heat generated from use of fuel is applied to food for processing either directly or indirectly through heat exchangers or a heated medium such as hot air, hot water or steam at atmospheric pressure or under elevated pressure (super heated steam).

Direct heating by fuel is used in directly fired baking ovens, grilling over fire, kilning of malted grains, fermented coffee beans, cocoa and tea leaves, as well as in drying operations and cooking by solar radiation. Induction, dielectric, microwave and infra-red irradiation heating use the electric energy for processing of foods in direct heating mode. Mechanical energy through friction generation in equipment and food components is another method of direct heat application.

Indirect heating make use of heat exchangers and heated medium. Primary energy sources are used to heat the solid surface or solid powders such as sand, salt or ash or a heat transfer fluid medium such as air, water or other liquids *e.g.* mineral oil, o-dichlorobenzene, etc. which is used for heat transfer directly or through an exchanger to the food.

Processing Methods

Thermal processing preservation methods is used to modify raw materials to develop texture, flavour, colour, to inactivate anti-nutritional factors and to make the nutrients digestible and food palatable. Different method used are listed below with brief description.

Cooking

Heating the food either in water or without added water at temperature ranging from just above ambient to above 100°C for variety of purposes in open or closed containers by direct heating in hot water or by injection of steam in the container.

Frying

Using a pan or an oil bath for shallow or deep frying, various food items are fried with a frying bath ratio (ratio between the contact surface (in dm^2) and weight (in kg) of oil) varying from 50 to 200 and 1 to 2, giving an oil film of 0.5–1 mm and depth of 5–10 cm, respectively. The temperature for shallow frying varies from 120 to 230°C and time from few minutes to 10–20 min. For deep frying temperature of oil is generally maintained at 150–200°C and operation is carried for several hours a day with several batches of frying each taking a few min. Heat is applied to the oil by burning fuels or by using electrical resistance heating devices.

Baking

Bread, biscuits and cakes are baked in batch or continuous process in direct fired baking ovens or in electrically heated ovens with temperature varying from 150 to 250°C. The ovens have natural or forced convection or top and bottom radiation heating systems for heat transfer. Time of baking depends upon the type of system and product and varies from few minutes to several min. Indian unleavened bread, *chapaties* and *roties* are baked on hot plate and in direct flame while *idllis* in the steam chamber.

Roasting and Puffing

Cereal grains and pulses are roasted by application of heat at 150–180°C in air using drum roasters or solid powder-sand or salt in open hearth pans in batch process. Vegetables are roasted in hot air or in hot ash. Roasting of some grains types at low moisture results in their rupturing and puffed texture.

Extrusion Cooking

Coarsely ground grains and pulses mixtures with added spices and salt at a moisture content of 15–20 per cent are cooked by simultaneous grinding and pasting under pressure and under the influence of frictional as well as applied heat and puffed on release of pressure at the outlet of die. The temperature applied is about 120°C and residence time about 2–5 min.

Concentration and Eutectic Boiling

Concentration of milk with application of heat to obtain condensed milk, *khoa, rabbri* and solidification of sugar-water is used in preparation of variety sweets and confectionery products.

Proper control of thermal processing in terms of temperature, time and where applicable the pressure is an essential requirement of quality production in processed foods. Any deviation from the required control drastically affects the quality of finished products which have already had irreversible changes in their chemical composition and physical states.

40

Processing and Packaging of Honey

H.D. Kaushik and Ombir
Department of Entomology, CCS Haryana Agricultural University
Hisar – 125 004

Honey is the sweet and purest substance stored in combs and produced by honeybees from nectar of flowers or from secretions of living parts of plants, which they collect, transform and combine with certain specific matters. Honey consists of various sugars mainly of glucose and fructose. It also contains proteins, amino acids, enzymes, organic acids, minerals, pollen and other substances as well as yeast. The colour of honey varies from nearly colourless to dark brown. Consistency may be fluid, viscous or partly to wholly crystallized. The flavour and aroma vary with its floral origin from which it is obtained and are important characteristics available in the comb. During handling and processing, it should be our aim to preserve them.

Honey is a good preservative and can be kept for a long time without deterioration. In the present time, honey is produced in large amount, handled at various levels, *i.e.* from production, extraction, collection, processing, storage and consumption and stored in different conditions, which determines its extent of deterioration.

Categories of Honey

Honeys obtained through out the world are categorized as:

Apiary or Table Honey

Centrifugally machine extracted from beehive with optimal moisture content and superior quality.

Squeezed Honey

Obtained by squeezing or pressing the honeycombs of wildbees (*Apis dorsata* or *A.florea*) with high moisture content liable to fermentation.

On the basis of moisture content of honeys, Bureau of Indian Standards (BIS) has grouped them as:

1. Special grade (less than 20 per cent)
2. 'A' grade (less than 22 per cent)
3. Standard grade (less than 25 per cent)

Honey with more than 25 per cent moisture content is rejected for consumption purposes. The coastal area and rockbee honeys have thinner consistency with 28 per cent or more moisture content and liable to ferment, while north and central India honeys are thicker having moisture content from 17-20 per cent.

Honey is hygroscopic *i.e.* exposed to the air it will attract moisture from atmosphere. In a very dry, warm atmosphere, honey will lose water and quality will improve.

What is Processing?

It means indirect heating of honey at controlled temperatures and duration just sufficient to kill all the yeast cells in honey, which is alike the pasteurization of milk. The processing leads to reduction/ prevention of granulation and fermentation of honey.

"Least processing is the best processing" is the thumb rule. The keys to good honey processing are care, hygiene and understanding in most of the western countries as they extract the honey hygienically with less than 18 per cent moisture content. Therefore, their processing is limited to controlled indirect heating and straining for ease in its subsequent handling. On the contrary in India and other tropical and sub-tropical countries, concept of processing is entirely different with an ideal of indirect heating to liquefy it, destroy all yeast cells to prevent fermentation and delay granulation, reduce excess water content with retention of its flavour, colour and taste. It is generally overheated in open vessels without proper temperature and duration controls to delay crystallization, which adversely affects its valuable ingredients. This process increases production of hydroxy methyl furfural (HMF) and L: D ratio which may result in to charring of honey.

Man's effort to convert honey to his use through any kind of processing result in some quality deterioration, though minimum processing becomes essential with following steps:

Heat Treatment

1. To increase the flowability of raw or granulated honey, heat treatment is essential to liquefy and reduce the viscosity at 40°C for 30 minutes. However, flowability, of honey is a problem in cooler climates as compared to tropical or sub-tropical climates.
2. To separate wax, honey is heated at 45°C in pre-heating tank provided with a coarse filter at the top and stirrer and strained through 80 μ strainer.
3. To destroy yeast cells mainly sugar tolerant species of, *Zygosaccharomyces* and *Saccharomyces*, rapid heating and cooling is done at 63°C for 7-8 min., 69°C for/min or 71°C. These cells are responsible for fermentation of honey with high moisture content.
3. To dissolve glucose crystals and remove incorporated air, honey is processed by heating to 60°C for 5-7 min. or 71°C for 2-3 min. Honey itself is supersaturated solution of granulating sugars but granulated honey is not acceptable to common consumer.
4. To reduce moisture content in honey, it is heated at 60-65°C by either sprinkling on hot revolving drum or in open vessel before straining and filtration.

Straining

To remove foreign particles like dust, beeparts, etc., the honey having temperature of 35°C is passed through screens of different meshes for rapid straining with reduced viscosity of honey. Before straining, heating of honey may not always be required under Indian conditions. Honey passes from one screen to next to have final passing through finest one. BIS specified to use 180 μ strainer in the first of the three settling tanks. Honey processing plant as specified by BIS consists of:

1. A plate form
2. Hot water bath
3. Processing unit, and
4. Settling tanks.

Filtration

Sometimes to remove pollen, colloids and small air bubbles left in honey after straining, it is filtered under pressure. The filtered honey remains in liquid form for longer period as compared to unfiltered one. This process is useful for granulated honey, which is not acceptable to the users. In USA, filters are made of diatomous earth.

The processing and handling loss for various honeys ranges from 2-8 per cent. Moisture reduction of 8-10 per cent, 100 per cent destruction of yeast cells, 60 per cent reduction in pollen and elimination of foreign particles enhances the quality and shelf life of honeys to considerable level. Although such processing increases slightly the HMF content with in the permissible limits.

Storage of Honey

After processing, honey may be stored preferably in clean, sterilized dry large containers made of stainless steel having airtight lids to avoid chemical reaction and outside contamination. Honey should be stored at as low a temperature as possible. Damage to honey for long storage at 30°C is same as that of heating at 70°C for few hours. Yeast will not grow at temperatures below 11°C preventing fermentation. Increase of HMF and darkening of honey above 10°C is rapid. Do not store unprocessed honey with high water content which may lead to fermentation and even explosion. Therefore, airtight containers with air valve on top to release CO_2 liberated during fermentation should be used. Storage place should be clean, cool and dry. It is advisable to use few large containers rather than large number of small ones to save space and avoids losses in transferring honey.

Packaging of Honey

Honey processing and packing is a highly technical field, require professional competence, proper organization, machinery, etc. At present, this is being done through network of cooperative societies especially in states of Himachal Pradesh, Karnataka, Kerala, Maharashtra, Orissa, Punjab, Tamil Nadu and Uttar Pradesh.

Bottling of honey in retail containers on large scale is done with automatic equipments. Townsend (1978) recommends bottling at 57 to 60°C to dissolve any small crystals. Filled bottles must be cooled before they are stacked. Honey is usually sold with quantities specified by weight and labelled with following information:

1. Name and Address of Supplier
2. Weight (gms or kgs)
3. Description of content (flora, etc.)
4. Statement about origin, if imported

Honey is sold in glass or food grade plastic or steel jars/ containers. Glass jars are considered the best for packaging honey but because of weight, fragility and transparency, cardboard and plastic containers are preferred. Packaging may be in pails of 20 kg to sachets of 10 g or in 500 g to 1 kg jars.

A good container should be hermetically sealed so that product retains its taste. Never heat honey on fire. A consumer can soften honey in glass jars by slight warming in water bath.

REFERENCES

Crane, E. (1990). *Bees and Beekeeping: Science Practice and World Resources.* Heinemann Newnes, Oxford, U.K., p. 614.

Jean Prost, P. and Medori, P. (1987). *Apiculture: Know the Bee, Manage the Apiary*. Oxford and IBH Publishing Co. Pvt. Ltd., New Delhi. pp. 659.

Goyal, N.P. and Sharma, R.K. (eds.) (1993). *First National Conference on Beekeeping*. Organized by NHB, Ministry of Agriculture, GOI, New Delhi. pp.117.

Wakhale, D.M. (1995). Quality testing and equipments: Perspective and Planning for Honey and Other Bee Products. Lecture delivered at *National Workshop on Trainers Training and Appropriate Area for Applied Research and Policy Changes*. Held at Y.S.Parmar University of Horticulture and Forestry, Nauni, Solan from 2-3 June, 1995.

41

Nutritional and Medicinal Value of Honey

Ombir and H.D. Kaushik
Department of Entomology, CCS Haryana Agricultural University, Hisar – 125 004

Honey is the wonderful gift of Nature and God to the mankind. It is sweet, viscous, liquid substance produced by honeybees from the nectar of blossoms or from secretions of extra floral nactaries, which they collect, transform and combine with specific substances and store in honey comb. It was the only sweetening agent known to early man and it found its way into traditions, customs and in food of every religion and nation. It has been also acclaimed as a restorative healer and health provider since ancient times as it cures wounds, burns and ulcers and enhances resistance against diseases. Honey is a desirable and delicious variety in our diet. It is twice stolen sweet, which is changed from an easily spoiled, thin sweet liquid to a stable high density and energy food by honeybees.

Honey contains about 80 different compounds, predominantly sugars (fructose, glucose, maltose and sucrose). It also contains minerals, amino acids, proteins, vitamins and enzymes. Honey has different colours, aromas and flavours. It is almost a complete natural food and is held in high esteem next to *Amrit*. In combination with milk, honey makes almost complete food for mankind. It is a very good source of energy, which provides about 3000 cal/kg. It not only provides energy but also provide mineral and vitamins for human body. The following components are present in honey:

Moisture

It is most important characteristics of honey. Honey contains 15-25 per cent moisture, which influence the keeping quality and granulation. Higher the moisture content earlier the fermentation of honey.

Sugars

Honey is excellent source of carbohydrates, which are its major components (70-80 per cent). Sugars contribute 95 to 99 per cent of total solids. It has 13 plant sugars predominantly fructose and glucose (almost in equal proportion), that are easily digestible and provide instant energy. The other common sugars are maltose and sucrose.

Table 41.1: Composition of Honey

Moisture (per cent)	15-25
Total sugar (per cent)	70-80
Fructose (per cent)	38.0
Glucose (per cent)	32.0
Sucrose (per cent)	2.0
Maltose (per cent)	7.0
Other sugars (per cent)	1.5
Minerals (per cent)	0.5
Protein (per cent)	0.2
Acid (per cent)	0.2
Enzymes	Traces

Minerals

Generally, minerals are referred to as *ash*, which varies from 0.02 to 1.0 per cent. About 11 minerals and 17 trace elements are available in normal honey. The dark coloured honey contains more mineral than light coloured honey. The minerals, which are available in honey are potassium, calcium, sodium, phosphorus, iron, magnesium, manganese, sulphur, copper, chlorine and silicon.

Vitamins

Honey contains several vitamins. At least six vitamins are present in honey but due to their small quantity these have a little nutritional value. Riboflavin, thiamine, pantothenic acid, niacin, pyridoxine and ascorbic acid are available in honey in variable amount depending upon source of nectar.

Enzymes

The enzymes present in honey are invertase, glucose oxidase, diastase and to lesser degree catalase and acid phosphatase. Almost all these are added to honey by honeybees.

Proteins and Amino Acids

A little quantity of proteins is found in honey, which varies with nectar sources. About eighteen free amino acids are present in honey in small amount.

Aroma and Flavour

The aroma and flavour of honey are important characteristics from consumer point of view. There are many different honey aroma and flavour as many plant species, from which nectar is collected. The flavour and aroma is sensitive to heat and storage.

Colour

The colour of honey ranges from whitish to dark brown depending upon the plant source from where nectar is collected.

After analysing the honey, it has been confirmed that honey has high nutritive values and easily digestible nutrients. In India, honey is licked at the tongue of a baby at the time of birth. Dr. Luttinger recommended the use of honey in infant feeding because of rapid absorption; it complements the iron deficiency in human and cow's milk. It increases appetite and it has soothing effect, which reduce fretfulness. Honey is also advised for athletes, sports persons, mountaineers, deep-sea divers, patients in hospital and workers in factories. Honey quickly replenishes the energy lost in these activities. So honey is a nutritive food for all the ages, castes, and religions of people.

Table 41.2: Specification of Honey

Characteristics	IS:4941 Grade Special	IS:4941 Grade A	IS:4941 Grade Standard	PFA A07.03	Agmark 1986
Moisture (max. %)	20.00	22.0	25.00	25.0	20-25
Specific gravity	1.41	1.39	1.37	–	1.35
Reducing sugar (min. %)	70.0	65.0	65.0	65.0	65.0
Sucrose (max. %)	5.0	5.0	5.00	5.0	5.0
Fructose/glucose ratio (min.)	1.00	1.0	1.0	0.90	0.95
Ash (max. %)	0.50	0.50	0.50	0.50	0.50
Hydroxy methyl furfural (ppm max.)	30	40	40	40	40
Acidity (max %)	0.2	0.2	0.2	0.2	0.2

Honey is used in hundred of food products around the world. It is predominantly used in cereals, bakeries and confectionery products. Honey may be used as honey spread on bread, biscuits, etc. for breakfast. It may also be used in some squashes such as lemon, mango and orange squash. Honey can also be used in limejuice or water, which is a good energizer.

Medicinal Value of Honey

Honey is valued not only as a food but also for some of its therapeutic attributes. It has been used in medicine throughout historical times and probably earlier also. The earliest recorded medicinal prescription that includes honey is dated back to about 2000 B.C. Seven hundred prescriptions with honey for both internal and external uses have been recorded in Ebers Papyrus about 1500 B.C. (Crane, 1996). Honey is used most commonly as a vehicle or one of the ingredients of medicine in Ayurveda, which enhances the effect of the medicine. Today honey is used as one of the component in many cough syrups. Several scientists observed that honey contains active principles of great therapeutic value, which is an efficient mean of combating diseases and prolonging life span. Honey enhances the resistance against diseases by improving the biological process of organs and systems. Its medicinal properties neutralize fatigue, compensatory hypotonia when used in the preparation of beverages (Mihailesu *et al.*, 1974). Honey also has some antibiotic properties.

Antibacterial

Several scientists reported bactericidal effects of honey. Sackett (1919) observed certain bacteria quickly perished in heat-sterilized honey. Dold *et al.* (1937) first time reported that antibacterial effect of honey was due to a material, which they called *inhibine*. The *inhibine* effect was due to hydrogen peroxide produced by glucose oxidase to form gluconolactone from honey glucose (White *et al.*, 1963). The inhibitory action of honey was more on gram-positive bacteria than gram-negative bacteria. Honey has bactericidal effect on many pathogenic bacteria including entropathogens such *Solmonella* spp. *Shiegilla* spp. and *Escherichia coli* and other gram negative organisms (Ibrahim, 1981; Jeddler *et al.*, 1985).

Healer

Honey acts as a good wound healer and is an ideal topical dressing material for wounds, burns and ulcers. Honey cleared 60 per cent septic wounds as compared to *Savlon* (36 per cent) in the same period observed by Hamdy *et al.* (1989). Efem *et al.* (1992) observed that honey inhibited most of the fungi and bacteria responsible for wounds and post surgical infections. Topical honey dressing is found to be more effective than ampicillin or ampicillin honey mixture (Gupta *et al.*, 1992). This is because the honey also enhances the healing of wounds. In some cases that had failed to respond to conventional treatments, wounds became sterile within one week of topical application of honey. Honey also heals burn's wounds. Efem (1988) observed that early burn's wounds healed quickly after honey application. The gangrene patients received systemic antibiotics and topical application of unprocessed honey showed distinct advantage over the orthodox method (Efem, 1993).

Honey also has some Common Time Tested Therapeutic Uses

1. A good appetizer.
2. Improves digestion.
3. Acts as laxative when taken with warm water, milk or coffee or with lemon.
4. Acts as blood purifier.
5. Healing agent for wounds and burns.
6. Cures mouth ulcers.

7. Strengthens the heart and circulatory system.
8. Good for anaemic patients.
9. Helps in reducing extra fat.
10. Enhances children physical fitness if given 5-10 g honey a day.
11. Cures the sore throat.
12. Beneficial to bronchitis patients.
13. Good for jaundice patients.
14. Helps typhoid patients in early recovery.
15. A mixture of *Aonla* juice and honey in equal proportion soothes breathing problem.
16. Honey with ginger and *Tulsi* cures cold and cough.
17. Use liquid and well-filtered honey as eye drop.
18. Honey taken with 1/4 teaspoon of *Jaiphal* power relieves the symptoms of hiccup, gas, nausea and vomiting.
19. Sleeplessness can be checked by taking honey with 1/2 gm. of *Jaiphal* powder.

The consumption of honey as a food is very low in India due to food habits, and its utilization is mainly for medicinal purposes. Since long honey has been recognized as a health promoting food. Hence, it is desirable to promote the use of honey either directly or through value added products.

REFERENCES

Chaudhary, O.P. and Taori, K. (1993). Honey: The most nutritious food. *Khadi Gramodyog*, 34(11&12): 742-746.

Crane, E. (1975). *Honey: A Comprehensive Survey*. Willium Heinemann Ltd., London.

Crane, E. (1996). The past and present importance of bee products to man. In: *Bee Products; Properties, Application and Apitherpy*, (Eds.) Mizarati and Lensky. Plenum Press, New York, p. 1-13.

Dadant and Sons. (1984) (Ed.). *The Hive and The Honeybee*. Dadant and Sons, Illinois.

Goyal, N.P. and Sharma, R.K. (1993). *Proceedings of First National Conference on Beekeeping*, organized by NHB, Ministry of Agriculture, GOI, New Delhi.

Gulati, R. and Sihag, R.C. (1996). Medicinal Value of Honey: A Review. *Indian Bee J.*, 58(3): 139-143.

Iannuzzi, J. (1998). Health From The Hive: You are What You Eat. *American Bee J.*, 138(8): 597-599.

Sharma, T.R. and Jyothi, Y.S. (1999). Honey: It is More Than Just Sweet. *Indian Food Industry*, 18(6) 349-357.

42

Utilization of Food Industries Wastes

S. Sehgal

Professor, Food and Nutrition, College of Home Sciences, Chaudhary Charan Singh Haryana Agricultural University, Hisar – 125 004

The food and agro industry is central to the economic and development policies of most of the countries. A country may be theoretically self sufficient in food, economic and logistical constraints have often made it difficult for the poorer segment of the population to acquire adequate food pricing and pricing support and control of foodstuffs have, therefore, become very important consideration in decision making process of a country.

Conceptually, recycling and residue utilization have the object to optimize the economic value of residue and prevent the formation of waste resulting from all forms of productive activity and/or consumption. Utilization of the residues is synonymous with sound environmental management and helps to overcome pollution problem. Waste can pose environmental problems if not solely handled and disposed of.

The wastes which are of interest are those emerging from food processing industries and major aim should be utilization rather than disposal. The raw material for different industries is different and varied. Not only may the nature of waste differ from site to site but it also vary, on one site, from one time of the year to another. Additionally the volume and concentration of the waste material is not be constant. So the first problem is to ensure that a process which can work consistently inspite of the major fluctuations in the nature and quantity of raw material. In practice, the usual response has been to construct relatively large units of plants, which incorporate substantial 'holding capacity' in order to even out the fluctuations and introduce some stability.

Which product is to be made by utilization of waste? Here there seem to be two basic choices, namely to modify the waste to produce an upgraded material of higher value that the waste itself or to produce something quite distinct by using the waste.

Dairy Waste

Wastes from modern dairy plant constitutes only 1–4 per cent of milk output. General dairy wastage arises largely from drainage and washing of tankers and equipment plus spillages. An immediate aim is, therefore, to reduce these direct wastages by fairly obvious means of good housekeeping and management.

Whey is the fluid separated from milk or skimmed milk in the process of making cheese or casein. World cheese production is increasing fairly rapidly with a corresponding increase in whey production. Whey contains about 4.7 per cent of lactose, 0.7 per cent of protein and 0.5 per cent of salts and the lactose content of dried whey product may be as high as 73 per cent. Utilization of whey is mostly done by conventional technology, including drying of whey to powder, lactose extraction and direct feeding to animals. Most of the whey powder is being used in animal feed in many of the countries whereas in USA the use of dried whey for human foods includes uses in baked goods, ice-cream, processed meat and processed cheese. Large quantities of whey is utilized in feeding directly to animals especially pigs in U.K.

Lactose extraction is another well-established use for whey. Uses for lactose which include pharmaceutical are however, fairly limited. Whey-based beverages have achieved modest popularity in a few countries.

The unconventional processes of gel filtration, ion exchange and ultrafiltration aim at extracting the protein from whey in undenatured form. True protein amount is about 0.6 to 0.7 per cent in whey. It has an excellent amino acid spectrum. In the undenatured form proteins also possess functional properties of high solubility, aeration, heat gelation and stability in acid solution. These functional properties indicate use in the flour and sugar confectionery industries in the soft drink industry and in respect of the excellent nutritional value, for example, baby foods.

Anaerobic fermentation of lactose permeate produces methane which can be used in the steam-raising plant of a cheese creamery.

Wastes from Meat and Poultry Plants

Animals are grown and slaughtered to provide nutritious meat for humans. The economic of the world's meat industry demand that animal by-product be utilized so that the livestock industry can stay economically competitive with vegetable protein sources. If animal by-products are not effectively utilized, of course, a valuable source of potential revenue is lost and the added and increasing cost of disposal of these products is incurred by the industry. The cost of live animals often exceeds the selling price of its carcass therefore, the value of by-product must pay the expense of slaughter and generate the profit for the meat slaughtering operation.

Wastes effluents from meat, poultry and fish processing plants contain large amounts of protein and fat and usually have much higher BOD levels than town sewage. It has been estimated that between 2 per cent and 5 per cent of the total carcass protein is lost in the effluent from the processing plant. The protein recovery system of these factories waste involves two stages, the first stage being a flocculation pretreatment whereby a proportion of the soluble protein is rendered insoluble and is recovered by air flotation or sedimentation followed by dewatering and drying. Residual soluble protein is extracted from the effluent by the passage through a bed of newly developed ion-exchange resin.

Fat content is recovered prior to protein recovery by air flotation. By such treatment it is possible to produce a colourless effluent with a relatively low BOD value.

Protein recovery from effluents has been evaluated nutritionally by amino acid analysis and by feeding trials in pigs and poultry. Good growth rates were obtained with no evidence of toxic effects.

Meat processing waste consists of solid and liquid portions. Solid include manure and paunch manure (contents of slaughtered animal's digestive tract) and bits of animal tissues. The liquid portion includes blood and other body fluids. In a practical sense, most of the small tissue solids and often the manure, paunch solids become mixed with water used for carcass rinsing or for clean up and are drained to the sewer as liquid waste. Blood is usually collected, especially in large plants and dried for animal feed.

There is a long list of animal by-products which includes variety of meat for human consumption, edible fats for shortening, margarine, sweets and chewing gum, livestock feed, pet food and feed for aquatic farming, gelatin for confectionery items, ice-cream and jellied food products, rennin used in cheese making etc.

Cane Sugar Industry

The two main by-product of cane sugar industry are bagasse and molasses. Bagasse is composed of cellulose, pentosans and lignin. Essentially it can be said that the paper can be obtained from cellulose, furfural from pentosans and plastics from lignin.

Molasses, being rich in sugar is used as a chemical raw material and as a nutrient for micro-organisms producing either, through their own proliferation, yeast or through their specific catalytic action, a number of organic compounds such as ethanol, butanol, acetic acid, citric acid, glutamic acid etc.

Wastes of Cereal and Pulse Industry

Byproduct of wheat milling are mill feed, germ, wheat shorts and bran Mill feed consists of mixture of bran, germ and shorts. Germ can be separated and used for production of wheat germ oil. The residual solvent extracted from wheat germ is rich in protein and B-vitamins and can be used in the preparation of weaning foods. Bran is used mainly as poultry and cattle feed. The by-product of rice milling are husk, bran or polishing and rice bran oil. Rice husk is used in manufacture of insulation boards.

Seed coats and brans which are the byproducts of shellers and dal mills are discarded as a waste and these can be used effectively as a potent source of dietary fibre in the diet. Efforts are being made to develop some of the value added products utilizing cereal brans and seed coats of various pulses upto 30 per cent.

Fruit and Vegetable Processing Industry

During processing more than 50 per cent of fruit weight goes as waste. Thus the waste utilization is both a necessity and a challenge. There are more than 2000 licensed fruit and vegetable processing units in India which generate more than 1,50,000 tones of processed products. At present most of the solid wastes from these factories is either thrown in garbage or disposed off by spending about Rs. 25–30 per tonne.

The waste of fruit industry have greater economic significance. Wastes from these industries are mainly solid wastes having high biological oxygen demand (BOD). Wastes from fruits contain cellulose, hemicellulose, starches, pectin, vitamin, mineral and other nutrients. In the recent years, there has been an increased interest all over the world in almost all the industries especially food processing to utilize waste and byproducts for oil, animal feed, fuel, fertilizers, manure and sometime value added products like fermented and pectins, tartaric acid, citric acid etc.

Fruit and vegetable processing wastes are available in bulk quantities. They are highly seasonal and during the season, considerably large quantities sometimes several hundred tonnes of these wastes become available in one factory and their utilization and disposal become a difficult and serious problem. Generally waste is classified as field waste, market and storage waste and consumer and processing factories waste. These wastes can be utilized by producing either an upgraded material of higher value or new products. The available technology for the preparation of products from waste materials are not being fully utilized except for the recovery of pectins, peel oil from citrus and use of orange peels in marmalade.

During the processing of guava into juice and jelly considerable amount of pomace is left after separating the clear extract. This left over pomace can be used in preparation of guave cheese, jelly, toffee etc. Some work has been done to utilize guava waste for production of ethanol, oil from seeds, pectin from pomace and animal feed from guava pomace and seeds.

In preparation of grape juice and wine, grape stems and grape pomace are obtained as waste products whereas raisins seeds and stems are the by-product of raisins manufacture from grapes. These wastes obtained from processing of grapes can be processed into a variety of by-products such as cream of tartar, seed oil, tannin, alcoholic beverages sugar syrup, jelly, chutney, pectin, oil cake, fertilizer, food colourant etc.

Approx. 75–80 per cent of all citrus wastes are converted into by products. Citrus peel products include dried citrus pulp, citrus pellets, citrus molasses, citrus peel oil, essence oil. Citrus seed oil can be used as food oil or working oil, as a salad. Pectin can be produced from lemons, limes, grape fruit and orange wastes. This pectin can be further used for jam, jelly, marmalade and as a thickener and stabilizer in foods such as puddings and processed milk products. Peels can be used in marmalade and as a seasoning in cakes, cookies and sherbat mixes.

Mango stone can be used in production of edible fat, mango flour, animal feed, manure etc. Mango pulp/peel can be used in beverages, peel wax, pectin etc.

Wastes ranges from 60–70 per cent in case of apple processing. It is in the form of pomace wastes, cores, cull fruit and by-products are pectins, cider, beer, vinegar, soft drinks, jelly etc.

Wastes ranges from 40–50 per cent in case of pineapple processing. By-products prepared are vinegar, syrup, citric acid, candy, cattle feed etc.

In processing of Daflana for canning and dehydration, the banana peel is a waste product. The pulpy portion can be scraped from thick peel and can be utilized for preparation of banana cheese. Pseudo-stem of the banana plant can be utilized as raw material for preparation of paper pulp.

The waste left over in the drying of vegetables like potato, cabbage, cauliflower etc. can be used in making cattle feed. It may also be possible to utilize the leafy wastes of vegetables for preparation of some value added products.

The vines and pea hulls obtained after processing of peas can be dehydrated and used in the preparation of stock feed by suitable blending with other materials.

In the canning of vegetables and the manufacture of tomato products, a considerable proportion of the vegetable is discarded as or is often only partially utilized. The trimmings of tomatoes left over after processing can be used for the preparation of tomato juice, puree or ketchup of fairly good quality. When tomato juice is made by the cold press method, the seeds are still good for raising tomato plants. The seeds can be separated from the skin after drying the material and utilized for an extraction of edible oil. Tomato seeds are also reported to be a good source of proteins and amino acids.

When planning and implementing waste utilization, careful consideration should be given to identify the most appropriate alternative. Basic factors to be considered include the quantity availability, seasonality and characteristics of the waste availability of a sustained market for the resultant product, economics and ability of the utilization, technology to fit in within available resources, people, capital, equipment and existing system.

43

Hurdle Technology for Shelf Stable Food Products

Joginder Singh Berwal
Government of India, Ministry of Agriculture, New Delhi

The Paleolithic man started learning the preservation techniques, which have been improved upon by man in different eras by employing empirical methods. With the advent of scientific knowledge, he had tried to establish the principles behind these methods. This led to evolving of concepts and technologies being used by today's food industry.

Origin of Hurdle Technology

After almost a pan global review of certain well established traditional food products, emerged the hurdle technology. This concept was evolved at "Federal Center for Meat Research". Kulmbach, Germany by Leistner and Rodel in 1976. According to them, "the processes used for preservation provide shelf stability to a food product by laying hurdles which are difficult to over come by the microorganisms causing food spoilage and poisoning" (Berwal, 1994). Since the formation of the concept, the hurdles have been improved, modified and developed to produce shelf stable products (SSPs). Hurdle technology is variably called combined processes, combined methods, combination preservation, combination techniques or barrier technology. Food preservation by hurdle technology employs the intelligent use of combination difference, preservation factors or techniques to achieve multi target, mild but reliable preservation effects. Many promising hurdles have been identified so far.

The spoilage and poisoning of foods by microorganisms is a problem that is increasing world wide despite the wide range of preservation techniques employed. In fact the current consumer demand for food that appears more for natural and fresh food has prompted the manufacturers to use more mild preservation techniques. Thus for the benefit of these food manufacturer's there is an urgent need to improve old and new methods that allow further production of safe and stable foods. The concept of hurdle technology addresses this need.

Introduction to Hurdle Technology

Hurdle technology deliberately combines existing and new preservation techniques to establish a series of preservation factors (hurdles) that microorganisms in question are unable to over come. Hurdle in a food are substances or processes inhabiting deteriorating processes. The parameters like water activity (a_w), low temperature (t), acidification (pH), redox potential (Eh) and preservatives (pres.) are called hurdles.

Hurdle technology is of fundamental importance for the preservation of intermediate and high moisture foods because the hurdles in a stable products control microbial spoilage and food poisoning as well as desired fermentation processes. The application of this concept proves to be very effective, since an intelligent combination of hurdles secures the microbial stability and safety as well as the sensory, nutritive and economic property of a food. For the economy of a good item it is important how much water in a product is compatible with the microbial stability of this food. The sensory quality is determined by a number of positive and negative hurdles because in many cases the prevention of deteriorative processes and maintenance of sensory quality are opposing. The intensity of processes and preservatives has to be kept as low as possible and is possible through balancing and combining the hurdles. The concept of hurdle is therefore, important since the balancing of hurdles or combination of processes produces stable food products by controlling microbial spoilage, food poisoning and in some instances desired fermentation processes in refrigerated storage, because hurdles applied did not allow growth (Leistner, 1994).

Significance of Hurdle Technology

An intelligent application of hurdles contributes to both improvement in product quality and stability. An adjustment and calculated application of hurdles could result into saving of money, energy and several other resources. Each hurdle contributes to the safety not against the microorganisms. Since the self-stable products (SSPs) do not require refrigeration, they travel long distances for finding place in the market to reach the consumers. This distance is even more in case of defense force as remote as Siachin Glacier in North. Reinforcing strong combination of hurdles can reduce the Perishability of hurdles (Berwal, 1994).

Shelf Stable Products (SSPs)

SSPs are food products that have been mildly heated (70 –110°C core temperature) in containers, but sufficient to inactivate all but speculated bacteria. Growth of surviving force of *bacilli* and *clostridia* are inhibited by sufficient decrease of a_w, pH and Eh. No refrigeration is required for their storage.

"Hurdle technology" foods can also be defined as products whose shelf life and the microbiological safety is extended by use of several factors none of which individually would be totally lethal towards spoilage or pathogenic microbes.

Hurdles

Up to now about 50 different hurdles have been identified in food preservation and they have been very well documented by Bogh Sorenson (1994) as below:

Physical Hurdles

Heat Processing, Sterilization, Pasteurization, Blanching, Storage temperature (Chill temperature, Freezer Temperatures), Radiation, Ultraviolet radiation (UV), Ionizing radiation, Electromagnetic energy (EME), Microwave energy, Radio frequency energy, Oscillating magnetic field pulses, High electric field pulses, Photodynamic inactivation, Ultra high pressure, Ultrasonication, Packaging (vacuum packaging, "active" packaging, or modified atmosphere packaging (MAP), Aseptic packaging).

Physico-chemical Hurdles

Water activity (a_w), pH, Redox potential(Eh), Preservatives *e.g.* Nitrate ($NaNo_2$) in combination with nitrite (or nitrate), Carbon dioxide (CO_2), Oxygen (O_2), Ozone, Lactic acid, lactate, Acetic acid, Sulphite or SO_2 (sources of SO_2 are dissolved salts, mainly sodium metabisulphite, or gaseous sulphur dioxide), Smoking, Phosphates, Glucono-d-lactone (GDL) Glucono-d-lactone (GDL), Phenols, BHA (butylated hydroxy anisole), BHT (butylated hydroxytoluene), TBHQ (2-tertiary butylhydroquinone), Chelators (citrates, lactates, pyrophosphates and EDTA (ethylene diamine tetra-acetic acid)), Surface treatment like Diphenyl, O-Phenyl phenol, Thiabendazole, Ethanol, Propylene glycol, Maillard reaction products, Spices and herbs (*e.g.* allicin in garlic, allyl isothiocyanate in mustard, etc.), Lactoperoxidase, Lysozyme,

Antibiotics

Pimaricin (natamycin).

The above mentioned hurdles should be used in combinations which are best suited to requirements of stability of any product. Single hurdles used alone have not been successful as indicated below:

Hurdle (Used Alone)	*Undesirable Effect*
Refrigeration	Chilling injuries, weight loss.
Freezing	Nutritional, texture loses, discoloration, aesthetic change, enzymatic browning.
Controlled atmosphere	Softening, discoloration, retarded spoilage
Pasteurization	Nutritional losses, sensory losses
Sterilization	Nutritional losses, sensory losses
Drying	Discoloration, flavour changes, mould growth
Preservatives	Consumer aversion resistance

To avoid the above undesirable effects the intensity of the hurdles has to be decreased. If we decrease the intensity the shelf life will also decrease. In many cases prevention of deteriorative processes and maintenance of quality are opposing actions. Therefore, to maintain optimal quality the intensity of hurdles has to be kept as low as possible and the same can be achieved by combination of various hurdles. A balanced and combined application of hurdles could have synergistic effect.

Different hurdles in food could not just have additive effect on stability but may act synergistically. A synergistic effect should become true if hurdles in a food hit, at the same time, different targets (*e.g.* cell membrane, DNA, pH, a_w, enzyme systems) within the microbial cell and thus disturb the homeostasis of microorganisms present in several respect. Homeostasis is the tendency to uniformity or stability in the normal status of organisms. If the homeostasis of microorganisms, that is their internal equilibrium is disturbed by preservative factors (hurdles) in foods they will not multiply, remain in lag phase or even die before homeostasis is re-established (repaired). Thus food preservation is achieved by disturbing the homeostasis of microorganisms in a food temporarily or permanently (Leistner, 1994).

Basis of Food Preservation Using Hurdle Technology

Food preservation involves inhibiting growth of microorganisms, shorten their survival, or cause their death by exposing them to a hostile environment. The response of the microorganisms to the changed conditions decides the ultimate result. A better understanding of homeostasis, metabolic exhaustion, and stress reactions of microorganisms is vital for selecting hurdles and further evolving new ones.

Homeostasis

The desire to live is the crux of homeostasis in other words the tendency to uniformity or stability by the organisms under the changed conditions. Maintenance of different physical and chemical parameters within certain ranges is critical for both the microorganisms and bigger organisms. Gould (1998) has observed that during evolution a wide range of more or less rapidly acting mechanisms (*e.g.*, osmoregulation to counterbalance a hostile water activity in food) have developed in microorganisms that act to keep important physiological systems operating, in balance, and unperturbed even when the environment around them is greatly perturbed. This implies that the most effective method of food preservation is one which leads to collapse of the homeostatic mechanism. To revive the damaged homeostasis tremendous energy is required and the restriction of energy supply scuttles the repair mechanism.

Metabolic Exhaustion

Another phenomenon of practical importance is the metabolic exhaustion of microorganisms, which could lead to "autosterilization" of foods (Leistner and Karan-Djurdjic 1970). Mildly heated (95°C core temperature) liver sausage was adjusted to different water activities by the addition of salt and fat, and the product was inoculated with *Clostridium sporogenes* PA 3679 and stored at 37°C. Clostridial spores that survived the heat treatment vanished in the product during storage if the products were stable. Later this type of behavior for both *Clostridium* und *Bacillus* spores was regularly observed during storage of shelf-stable meat products (SSP), especially F-SSP (Leistner 1994). The most likely explanation is that bacterial spores that survive the heat treatment are able to germinate in these foods under less favorable conditions than those under which vegetative bacteria are able to multiply. Therefore, during storage of these products some viable spores germinate, but the germinated spores or vegetative cells derived from these spores die. Thus, the spore counts in stable hurdle-technology foods actually decrease during storage, especially in unrefrigerated foods.

Apparently, microorganisms in stable hurdle-technology foods strain every possible repair mechanism for their homeostasis in order to overcome the hostile environment. By doing this they completely use up their energy and die. This leads eventually to autosterilization of such foods.

Stress Reactions

Stress reactions of microorganisms could prove to be limiting for hurdle technology. Some bacteria become more resistant (*e.g.*, toward heat) or even more virulent under stress, when they generate stress shock proteins. The synthesis of protective stress shock proteins is induced by heat, pH, a_w, ethanol, etc. as well as by starvation. These responses of microorganisms under stress might hamper food preservation and could turn out to be problematic for the application of hurdle technology. On the other hand, the activation of genes for the synthesis of stress shock proteins, which help organisms to cope with stress situations, should be more difficult if different stresses are received at the same time. Simultaneous exposure to different stresses will require energy-consuming synthesis of several or at least much more protective stress shock proteins, which in turn may cause the microorganisms to become metabolically exhausted (Leistner 1996).

Keeping in view various intricacies involved in producing shelf stable food the multi-target preservation approach should be adopted for shelf stable foods. Employing different hurdles simultaneously in the preservation of a particular food should achieve optimal microbial stability.

User Guide for Making a Shelf Stable Food Product Using Hurdle Technology

Define the sensory attributes and expected shelf life the novel or modified food product frame a processing technology to be followed

↓

Manufacture the food product

↓

Analyse the a_w, pH, preservatives (levels), processing temperature, storage temperature, shelf life etc.

↓

Predictive microbiology should be used for testing preliminary stability

↓

Challenge test the food product with food poisoning and spoilage microorganisms (use higher/doses of inoculate and storage temperatures than normal so as to take care of abusive conditions)

↓

Depending upon necessity as indicated by the findings of the challenge testing the hurdles be modified (increased/decreased, added or deleted) but the sensory quality should not be compromised

↓

Modified food product made

↓

Modified food product is against challenge tested

↓

If necessary hurdles are modified again

↓

Predictive microbiology is done to assess the safety of the food

↓

Established hurdles of the modified or novel food are exactly defined, including tolerances

↓

Pilot plant studies carried out

↓

Scale up process must be validated

↓

For the industrial process the critical control points (CCPs) and their monitoring be established

↓

Manufacturing process should be controlled by HACCP

All above steps although involve experts from various

disciplines but it is recommended that they should work as a team. The success lies in the combination of the processes by a combined team of experts.

44

Sensory Evaluation of Foods

G.R. Patil and Dharam Pal
Dairy Technology Division, N.D.R.I., Karnal

Introduction

In a modern food-processing unit, a number of quality assurance procedures are used to examine and maintain quality of a product at different stages starting from reception of raw materials to surveillance of the finished products. These tests are physical, chemical, microbiological and sensory. Amongst all these methods, sensory evaluation is of paramount significance. In fact, with the tremendous growth of economy, and competition, the high cost of R and D and too high failure rates in new food products developments, the research workers and technologists has started turning to sensory responses, as validating tests for better prediction of success. The sensory measurement offer following advantages:

1. It identifies the presence or absence of perceptible differences.
2. The important sensory characteristics are measured in a fast and quantifiable manner.

3. Identifies particular problem that cannot be detected by other analytical and instrumental techniques.
4. It helps in ensuring that the consumers get consistent, non-defective and enjoyable product.
5. Used for establishing correlation with chemical, physical and instrumental measurements.

The sensory evaluation procedures have been studied in considerable details with the result that this scientific discipline has come to be recognised as of fairly objective in nature (Learmond, 1987). The inherent variability of sensory evaluation by human subjects can be greatly overcome through appropriate selection and training procedures, coupled with application of statistical methods so as to take full advantage of the high sensitivity of human sense organs that even today surpass the most sophisticated instrumental means for flavour, texture and colour examination (Ogden, 1993).

Sensory Evaluation Procedure

A successful implementation of sensory evaluation programme requires proper laboratory facilities, trained sensory panels and adoption of pre-use sensory methods. A sensory panelists works as an analytical instrument, hence should be carefully selected and rigorously trained for a particular product so as to obviate inconsistency (Pal *et al.*, 1995).

Laboratory Facilities

Nakayama and Wessman (1979) observed that sensory evaluation facilities at the manufacturers levels should be as elaborate or as large as those in R and D Centre. The emphasis is on conducting tests that produce valid, reliable and reproducible results. They recommended that the facilities be divided into two distinct areas–one for sample preparation and another for evaluation- with appropriate lighting, ventilation and temperature controls so as to create an environment conducive to preparation and evaluation. Stone and Sidel (1993) emphasised that the laboratory facility should be flexible enough to handle current and future testing activities as well as to provide a workable environment for the staff.

Test Room

In general, test room should meet following requirements:

1. Walls and ceilings should preferably be of light (off-white or light neutral gray) and mat colours. Unnecessary decorations should be avoided.
2. For seated assessments there should be dividers between the places for each assessor.
3. Tabletops and dividers should have a mat, light neutral gray colour.
4. The lighting in the test room should be from lights having a colour temperature of 6500 K. The intensity of the lighting should be constant and uniform.
5. The temperature in the test room should be constant.
6. The test room should be kept free from odours.
7. The noise level should be kept to a minimum during the assessment.
8. In assessments where it is necessary with sample preparation, the preparation area should be sheltered from assessors.
9. In addition the room should be equipped in a manner providing the maximum convenience to the assessors, with special reference to temperature and humidity.

Recruitment, Selection, Training and Monitoring of Assessors

1. The trainee assessor should have satisfactory personal qualifications for assessments. Trainee assessors should be pre-selected by the use of screening tests, which assess candidate's ability to discriminate sensory attributes.
2. The trainee assessor should know sensory evaluation in general. It is essential that the trainee assessor is familiar with the products to be evaluated. This implicates a training programme during an appropriate period. The programme should contain:
   (a) Theory about factors of practical importance for sensory evaluation

(b) General training on methods, scales and description of sensory impressions

(c) General training on detection and recognition of sensory attributes and specific sensory effects,

(d) Relevant theory about the products,

(e) Thorough training in sensory evaluation of the products to be evaluated.

3. The trainee should not be accepted as selected or expert assessor of the products before he is able to reach results, which are sufficiently in accordance with the results of a panel of experienced assessors. He/she should be able to duplicate the scores and the use of nomenclature.

4. Selected or expert assessors of the products should participate in a minimum number of sensory evaluations every year. The assessment for each assessor should be continuously submitted to a statistical treatment. There should be defined specific demands when new training or evaluation of fitness of assessors is necessary.

Requirements for the Panel

1. The number of assessors for the dairy products in the panel should be uneven, the minimum number being 3.

2. Several factors should be taken into consideration for optimum performance of assessors before assessing:

   (a) The assessors should not be suffering from any illness that may affect their performance of performance. Under such circumstances another assessor should enter the panel,

   (b) The assessors should be in time for the assessment to make sure that they have spread enough time for the assessment,

   (c) The assessors should not use strongly scented perfume, aftershave, and deodorant or eat highly flavoured (spiced) food, etc.

   (d) The assessors should not smoke, eat or drink anything other than water during the last 30 min before the assessment.

Documents/Equipment

1. Necessary documents for the sensory evaluation of the various products should be available *i.e.*:
   (a) Recommended methods,
   (b) Product specifications,
   (c) Nomenclature of product defects.
2. The equipment should not affect the samples or the assessors in an undesired manner.

Sensory Tests

A large number of the sensory tests are available and new methods continue to develop. Broadly, these tests can be classified into three groups, as follows (Stone and Sidel, 1993).

1. Discriminative (difference testing)
2. Descriptive
3. Affective (acceptance/preference)

Discriminative testing is one of the most useful analytical tools available to the sensory professionals. It is on the basis of a perceived difference between two products that one can justify proceeding to a descriptive test in order to identify the basis for the difference. Within this general class are a variety of specific methods, *e.g.*

1. Paired comparison test
2. Duo-trio test
3. Triangle test
4. Multiple sample test, and
5. Other test methods such as dual and multiple standard tests.

The main objective for all these methods is to answer a simple question: is the stored product different from the fresh one? Obviously the response to this question can have major consequences. If the conclusions from a discrimination test are to be accepted by management as reliable, valid and believable, hence it is important that each test be conducted with proper consideration for all aspects of the test design, product preparation and handling, implementation, data analysis and interpretation. Failure to appreciate all the subtleties of a test increases the risk of data misinterpretation.

Descriptive analysis has been defined as a sensory methodology that provides quantitative descriptions of products based on the perceptions of a group of qualified subjects (Stone and Sidel, 1993). It is a total sensory description taking into account all sensations that are perceived: visual, auditory, olfactory, kinesthetic and so on, when the product is evaluated. Descriptive analysis results provide complete sensory descriptions of an array of products and provide a basis for determining those sensory attributes that are important to acceptance. The results enable one to relate specific process variables to specific changes in some of the sensory attributes of a product. From the product shelf life viewpoint, descriptive-information is essential in finding out the types and extent of changes that have taken place during storage from which one can establish cause and effect relationship.

A descriptive test involves relatively few subjects whose training is primarily focused on development of descriptive language, which is used as a basis for scoring the product. Some of the useful descriptive methods recommended for food analysis are: flavour profile, texture profile, quantitative descriptive analysis and scoring.

Affective or acceptance testing, a valuable and necessary component of every sensory program, is performed keeping in view the consumer's levels. It refers to measuring liking or preference for a product. In that respect, this method is most relevant to the development of new and formulated foods. Preference can be measured directly by comparison of two or more products including control, if available with each other, that is, which one of two or more products is preferred. Indirect measurement of preference is achieved by determining which product is scored significantly higher than another product in a multiproduct test or which product significantly more people score higher than another. The two methods most frequently used to directly measure preference and acceptance is the paired comparison test and the 9 point Hedonic scale.

Assessment

1. Each attribute is to be evaluated separately. The attributes are defined as follows:

(a) Appearance is the assessment of the visual properties, both internal and external, of a dairy product, including for instance shape, colour, openings etc.

(b) Consistency is evaluated by feel, especially by the skin and muscles of the mouth (mouth feel) and by the fingers (finger feel).

(c) Flavour (olfactory and gustatory sensations during tasting) and odour is estimated by smelling the products and by chewing and salivating the product in the mouth.

2. Samples should preferably be scored by the assessors in random order.
3. The number of samples should not be too large to ensure that the assessment carried out consistently.
4. Highly flavoured samples and/or samples with high fat content should be served after less highly flavoured samples and/or samples with lower fat content.
5. The presentation of samples should be done in such a way that the assessors do not mix up the samples.
6. The assessors should be instructed to do the technical performance of the sensory evaluation in a correct way (size of the sample in the mouth, time of chewing etc.).
7. Deviation on one sample should not affect the sensory evaluation of the next sample. A suitable remedy to prevent any affect from one sample to another should be used where necessary (*e.g.* mouth rinse).
8. In order to calibrate the assessors a minimum of 2 calibration samples should be assessed and the results discussed, before starting the assessment. The calibration samples should also be available during assessment.
9. The assessment should be carried out without intercommunication.
10. A description of the defects, if any, should be given. In doing so, the assessor should use the nomenclature of defects compiled separately for each product or group of products. The defects should be listed according to the significance for the quality of the product.

Sensory Evaluation for Quality Assurance

Though many food manufacturing units in India are conducting some type of sensory testing as a quality assurance programme, particularly in respect of inspection of incoming raw materials and outgoing processed products, in most of the cases there is no systematic and scientific application. The precise role of sensory evaluation to attain sound quality at different stages of production and processing is discussed in following sections.

Inspection of Raw Materials

The final quality of the food product undoubtedly depends on the initial quality of raw material. Concerted testing efforts must, therefore, be made at the reception point of raw material. Highly trained and experienced sensory monitor should be involved to critically examine each lot of raw materials for smell, taste and sight. In addition to these, other ingredients for example, sugar, stabilizers, emulsifiers, food colours, flavours and preservatives, coagulants, additives etc. should also be carefully examined for the absence of sensory defects. These ingredients should also conform to various quality standards.

Packaging Materials Inspection

Containers, labels and cartons should also be inspected by sensory evaluation technique, primarily through sight checks. The use of colour tolerance standards, verification of the correctness and scanability of the universal products code, and detection of appearance defects in glass or plastic containers, errors in label copy, and texture differences in the surface finish of closures and containers-all these utilize basic sensory evaluation techniques (Reece, 1979). The interaction between the packaging material and the product is of great significance and mainly determined by the sensory tests.

Process Controls

Sensory evaluation plays a most important role during in-process phase of the quality assurance programme because it is a quick and reliable screening check. It is at this point that appropriate, trained panelists in this area must reach the judgment as to whether the appearance, texture and flavour of the product are satisfactory for packaging. Errors of omission or even commission can have serious consequences and lead to product recalls or product-liability situations, both of which are extremely costly and can have adverse effects on the manufacturers image.

Final Product Inspection

Sensory evaluation checks on the finished product, primarily by panels, are important and should be performed with the same degree of diligence as earlier in the product-cycle. This is the last chance to catch non-conformance to the specifications and desirable characteristics before the product is shipped to the consumer marketplace, and it also serves as a routine audit of the effectiveness of the incoming inspection and in-process control parts of the total quality control programme. The inspection of final product also includes the packaging style, neatness, fullness and proper sealing in addition to the overall quality of product.

Product Surveillance During Storage and Marketing

All products are to be stored and marketed under the conditions commensurate with the type of product. If the storage conditions, that is, the temperature, humidity, sanitation profile of storage room is incompatible with the product, serious losses will take place. A manufacturer should therefore monitor these conditions as a part of total quality assurance programme. Sensory evaluation is adopted as a most fast and reliable test to examine the nature and extent of deteriorative changes in product quality, thus help in selecting the best storage conditions (Pal *et al.*, 1998).

Shelf Life Determination

Shelf life determination is an important aspect of any product development programme or quality assurance protocol. Now it has also become mandatory to indicate the expiry date *i.e.* "use before date" on all the food products. It implies that the expected shelf life of product has to be determined in shortest possible time by adopting sensory, chemical, microbiological and instrumental methods. Aided by state of the art approaches, including usage of computer, sensory evaluation can be considered an objective method of product-quality assessment. Different sensory tests coupled with relevant statistical methodologies make sensory evaluation an important tool for shelf life determination. Several successful attempts have been made to develop shelf life prediction models for different food products (Arora *et al.*, 1995) each relaying heavily on a sound sensory evaluation technique.

Sensory Evaluation for Product Development

In the formulation of entirely new foods, or modification of the existing products, sensory tests can provide helpful input in many ways (Erhardt, 1978, Baker *et al.*, 1994). These are discussed as under:

1. *Confirm intentional/unintentional changes in product/ ingredient formulation*: Sensory evaluation can provide data to confirm that changes have been made in the direction indicated by the consumer tests. They can also be used to determine whether unintentional changes have occurred in the product during reformulation. The evaluation provides assurance that the desired changes are perceptible in an objective test situation. Sensory evaluation can also be used to determine whether additions of a certain ingredient affect the flavour of a product; *e.g.* whether vitamin fortification replacement of fat with vegetable or mimic results in the perception of "off" or undesirable notes.
2. *Determine, if optimization has been achieved*: In new product development, sensory tests provide the developer with correcting information and guidelines for product improvement.
3. *Observe changes during processing/storage*: Sensory panels can be used to detect and identify the differences between the desired product and production. Ageing effects can be studied through sensory evaluation.
4. *Identify areas for improvement*: The effects of various packaging materials on product characteristics, possible changes in the product over time, changes in the product characteristics during processing can be determined by sensory tests.
5. *Monitor product quality*: Tests can be designed to monitor the quality of the manufactured product and to determine whether products slowly drift away from the intended quality over a period of time.
6. *Substantiate advertising claims*: Sensory tests can also be used to provide data to substantiate advertising claims and statements. If the product is billed as "better tasting", "shiner", "brighter", or "more comfortable", panelists can pinpoint products characteristics and their responses can be measured.

REFERENCES

Amerine, M.A., Pangborn, R.M., Roessler, E.B. (1965). *Principles of Sensory Evaluation of Food*. Academic Press, NY, USA.

Arora, K.L., Pal, D., Verma B.B., Rajorhia, G.S., Garg, F.C. and Patil G.R. (1995). Shelf life prediction model for rasogolla manufactured in Bikaner. *Indian J. Dairy Sci.*, 48: 50-56.

Baker, R.C., Hahn, P.W. and Robbins, K.R. (1994) Fundamentals of New Food Product Development, Elsevier Science B.V., The Netherlands.

Billmeyer, B.A. and Wyman, G. (1991) Computerization Sensory Evaluation System. *Food Technol.*, 45(7): 100-101.

Dethmers, A.E. (1979) Utilizing Sensory evaluation to determine product shelf life. *Food. Technol.*, 33(9): 40-42.

Erhardt, J.P. (1978) The Role of Sensory Analyst in Product Development. *Food Technol.*, 32(10): 7-59.

Lamond, E. (1987) Sensory Evaluation can be objective. In: *Objective Methods in Food Quality Assessment*, (Ed.) J.G. Kapsalis. CRC Press, Inc. Boca Raton, USA, p.3-14.

Ogden, L.V. (1993) Sensory Evaluation of Dairy products. In: *Dairy Sci. Technology Handbook*, (Ed.) Y.H. Hui. VCH Publ., N.Y, p. 157-276.

Pal, D., Sachdeva, S. and Singh, S. (1995). Methods for Determination of Sensory Quality of Foods. A critical appraisal. *J. Food Sci. and Tech.*, 32: 357-367.

Pal, D., Patel, A.A. and Gupta, S.K. (1998) Methods and significance of sensory evaluation of foods for shelf life determination. In: *Modern Food Packaging*, Indian Institute of Packaging, Mumbai p. 56-62.

Nakayama, M. and Wessman, C. (1979). Application of sensory evaluation to the routine maintenance of product quality. *Food Tech.*, 33(9): 38-39.

Reece, R.N. (1979). A quality assurance perspective of sensory evaluation. *Food Tech.*, 33(9): 37.

Stone, H. and Sidel. J. (1993). *Sensory Evaluation Practices*. Academic Press Inc., London.

45

Nutritional Considerations of Processed Foods

Umesh Kapil and Monika Obrah
Department of Human Nutrition, All India Institute of Medical Sciences, Ansari Nagar, New Delhi – 110 029

Introduction

Good nutrition is the fundamental basic requirement for positive health, functional efficiency and productivity. A proper diet is essential from the early stages of life for growth, development and active life. For maintaining good health and physical efficiency the diet should provide adequate amounts of all nutrients. Carbohydrates, fats and proteins are macronutrients, which are needed in large amounts. Vitamins and minerals constitute the micronutrients and are required in small amounts.These nutrients are necessary for physiological and biochemical processes by which the human body acquires, assimilates and utilizes food to maintain health and activity (NIN, 1998).

Changing Scenario in Food Habits

We are today, carrying the double burden of nutritional deficiencies among the poor section of the population on the one hand and problems such as overweight and obesity among the affluent sections attributable to wrong diets and unhealthy life-styles, on the other. For a nation to be healthy, strong and productive, the nutritional status of its people must be good (NIN, 1998).

Population explosion, demographic changes, rapid urbanization and alterations in traditional habits contribute to the development of certain unhealthy dietary practices leading to diet-related chronic diseases. The application of modern science and technology has effectively increased the production of processed foods and ready-to-eat foods. Today, there is an increased demand for processed, ready-to-eat foods due to changes in the lifestyle. Rapid urbanization has increased the intake and demand for processed foods however, there is a trend towards replacing traditionally cooked foods with processed foods, variety of foods, which are available and are within the reach of the common man, can be selected judiciously to formulate nutritionally adequate diets.

Although taste is still the king, consumers have been paying more and more attention to what they eat and how it affects their health. With health professionals advising consumers what to eat and what to avoid, the media highlighting the latest scientific and medical evidence about substances in foods that pose health risks.The government is also taking a direct interest in food labeling and health claims.The consumer recently has become highly aware of what goes into food and what should be taken out.

This heightened awareness has influenced the development of attitudes, preferences and demands for certain processed foods. Today's consumers want products that promote health by providing the extra bonus of nutrition and protective influences, Providing safe and nutritious food–what consumers demand–is the task of all involved in the food sector. Although, there is a plethora of food laws as well as national and international standards and regulations, industrially processed foods are often perceived as being less healthy than fresh products (Kroll, 2001).

What are Processed Foods ?

Foods that are subjected to technological modifications either for preservation or for converting into ready-to-use/eat foods, eliminating laborious household procedures are called 'Processed Foods'. Manufacture of processed foods is on the increase in the unorganized, small-scale and cottage-industry sectors and these products may not always be in conformity with food standards. Some of the examples are ready mixes, dehydrated foods,pasta products, canned foods, confectionaries, bakery, dairy products and breakfast-foods (Kroll, 2001). There are many reasons why foods should be processed, like

1. To increase productivity in order to feed large amounts of people.
2. To maintain shelf-life of a product.
3. To conserve quality of a food product.
4. To make unpalatable foods more palatable.
5. For convenience
6. To conserve the texture of a food product.
7. To prevent microbial growth (additives are often used for this reason).

Methods of Food Processing

Most food is processed and packed to stop it from becoming inedible.

Some of the methods of preservations are as following:

Drying

It is an ancient way of preserving foods. Dried food is hard on the outside so microbes find it hard to penetrate. Drying does destroy most of the vitamins, but can lead to higher taste concentration.

Salting, Pickling and Smoking

These are also the traditional ways used all over the world to preserve food. Meat is covered in salt then hung to dry. This is called curing.The salt draws a lot of moisture out of the meat and also protects against the microbes. Sugar is used to preserve fruits. Thus,we have jams and sweet pickles.The sugar stops the microbes from growing after the preserve is boiled.Vinegar is used to pickle vegetables.The vinegar retards microbe growth because it is acidic and stops exposure to oxygen, which the microbes need to survive.

Smoking involves hanging meat over wood fires and allowing the hot smoke to cure the meat. This also gives it a distinctive flavour. More recently all these procedures have been replaced by other methods of preservation such as pasteurization, sterilization, irradiation, freezing, canning and the use of other chemical preservatives (Swaminathan, 1993).

Nutritional Limitations of Processed Foods

Processed foods are generally consumed either as a part of a meal, or as a snack item. Their contribution in terms of essential nutrients depends on the type of processing and fortification, frequency of use and the quality consumed.

Processed foods are generally refined and a majority of them are:

1. Rich in fat
2. Rich in salt/sugar
3. Calorie dense
4. Lack dietary fibre
5. Lack micronutrients

According to a Scitific report processed foods can lead to development of certain degenerative health conditions like (Arthritis, Allergies, Diabetes) (Pottinger).

Thus, caution needs to be exercised when processed foods constitute a major part of the meal. Breakfast cereals are increasingly being used in urban areas. They are expensive. In contrast, traditional breakfast items like idli, dosa, upma and roti which are rich sources of nutrients. Puffed and parched rice products besides being crisp and tasty, are easily digestible. Consumption of sugar based processed foods should also be limited because they provide empty calories, apart from being high in fat and prepared from refined cereals. For prevention of diet related chronic diseases, sugars and refined cereals should be used sparingly.

Processing takes a lot of nutrients out of the nutrient rich foods. For this reason, processed foods are fortified with synthetics. Most of the pantothenic acid, folic acid, biotin, Choline, inositol, vitamin B-6 and E are discarded in the milling of breads and the refining of flour for cereals. To make up for the loss of nutrients during processing, synthetic vitamins and minerals are added. Although helpful, these synthetics are not as healthy as their natural counterparts.

Harmful Effects of Processed Foods

The major harmful effects of processed foods are as follows:

1. Slower digestion leading to toxin production in the body. Modern diets consist of over-processed foods and lifeless foods, which produce little enzyme activity and contain little or no fibrous material. We eat devitalized food that is not fresh, with fewer nutrients. The entire natural fiber, such as grain peels, are removed in the refining stages. Because of this the diet is poor in fiber, digestion is slow and the waste materials stay in the body longer. This sets up the perfect condition for formation of toxic materials to form.
2. Food additives/artificial colours loads the body with foreign chemicals: Processed foods can contain huge amounts of chemicals and toxins which were added in the processing. Daily we eat lots of chemicals like coloring agents, artificial flavors and preservatives that can make the food appear more attractive and taste fresher. Even small amounts of different processed foods, can substantially contribute to the burden of foreign chemicals in the body.

Thus, consumption of processed foods may not only affect intake of nutrients, but also in addition, increase the risk of exposure to various chemical additives (World Cancer Research Fund and American Institute for Cancer Research, 1997)

A number of food additives in processed foods and beverages have been identified as carcinogens in experimental studies.Artificial sweetners such as Saccharin, Aspartame and Cyclamates have been found to be carcinogenic in nature (NAS, 1992). Limonene, an artificial flavour has been shown to cause renal cancer in rats(7). However, no judgement can be made on the cumulative lifetime effect on the cancer risk of additives that are mutagenic or carcinogenic as further research needs to be conducted in this field.

Steps to Improve Nutritional Value of Processed Foods

Fortification of Processed Foods

One way is to fortify products with substances that provide protective influences against certain diseases and conditions.

These protective substances are included in two groups. One group includes currently marketed fortifiers that are accepted by health professionals as protective influences, such as vitamins, minerals, proteins, fiber and omega-3 oils. The other group includes substances still in the research stage, such as extracts of garlic, certain spices, caffeine and tea, to determine if they can be extracted for later incorporation as protective influences into so-called "Designer foods".

Elimination of Harmful Agents

Reduction or removal of those ingredients that are related to health risks. Fat, cholesterol, sugar and salt are ingredients linked with such conditions as cancer, heart disease, obesity and related problems, and hypertension. In order for consumers to lower their intake of these ingredients, manufacturers must develop "light" versions of the products containing them, and this requires the use of various types of replacers and substitutes (Flamm, 1991) .

Nutritional Considerations During Food Processing Methods

Canned products should be canned at the time of peak nutritional value. With canning, the product is protected from heat, air and light. Optimal amounts of nutrients are usually obtained with the exception of Vitamin C and to some degree Vitamin B. When heat and oxygen are used for drying purposes, there may be losses of vitamin A, C and beta-carotene. When sulphurdioxe is used to dry products, there may be substantial losses of vitamin B1. During the drying process, exposure to light and air will affect vitamin C. Thus, care should be taken so as to have minimal exposure to light and air.

With freezing, the main nutrient lost is vitamin C. Since water expands when frozen, there may be damage to vegetable cells leading to a loss of fluid and nutrients from the cells. With milling and refinement of products such as bread, there will be losses in B complex vitamins, fibre and micronutrients such as magnesium, chromium and zinc.

Food Safety and Quality

Despite the progress made in medicine and in food science and technology, illness caused by food-borne pathogens represents, in Europe, an average of 120 per 100.000 annual cases of illness. In some European countries, there are at least 30.000 cases of acute gastero-enteritis per 100.000–many of which are thought to be food-related.

Processed foods have an excellent safety record and most of the food-borne intoxications are linked to inappropriate storage and preparation (Stalder and Marne, 2001). Processing food can improve nutritional quality and safety, for example, by increasing digestibility of certain foods and by increasing the availability of some nutrients. Food processing can also eliminate or neutralise certain toxins. All fresh foods lose vitamins and other nutritive qualities during transport and storage. A wide variety of new, non-thermal processes such as irradiation and high pressure as well as the use of lactic acid bacteria, biotechnology and microbiological intervention will enhance the quality and safety of foods. From a nutritional point of view, these techniques are promising since they are likely to reduce loss of nutrients to a greater extent than traditional food processing methods.

Undesirable compounds such as heterocyclic aromatic amines (HAA) can be formed during processing. Carefully controlled industrial processing can reduce the quantity of HAAs produced when compared with home-cooking. Transfatty acids (TFA), which can raise LDL cholesterol levels in the blood, are formed during industrial hydrogenation of vegetable oils and fats. It has been speculated that these compounds could be responsible for between 30.000 and 150.000 deaths a year–although this view has been challenged. New methods in fat technology are enabling the production of fats and oils with a low TFA content.

Minimal Use of Food Aditives

Processed foods containing food additives must declare them on the label either by name or number. Most approved food additives have been allocated a number in the Food Standards Code which allows the specific additive to be identified. For example, the designation on a product's label of 'colour (102)' means that the food contains the yellow colouring tartrazine. Avoiding toxins in the diet is an important initial step toward enhancing the health and lowering the risk of disease. Avoidance and discrimination of processed foods are crucial proactive steps in most natural health care programs.

Some of the key additives to avoid and their associated health risks:

1. Hydrogenated Fats-cardiovascular disease, obesity.
2. Artificial Food Colors-allergies, asthma, hyperactivity; possible carcinogen.
3. Nitrites and Nitrates-these substances can develop into nitrosamines in body, which can be carcinogenic.
4. Sulfites (sulfur dioxide, metabisulfites, and others)-allergic and asthmatic reactions.
5. Sugar and Sweeteners-obesity, dental cavities, diabetes and hypoglycemia, increased triglycerides (blood fats) or candida (yeast).
6. Artificial Sweeteners (Aspartame, Acesulfame K and Saccharin)-behavioral problems, hyperactivity, allergies, and possibly carcinogenic. The government cautions against the use of any artificial sweetener by children and pregnant women. Anyone with PKU (phenylketonuria-a problem of phenylalanine, an amino acid, metabolism) should not use aspartame (Nutrasweet).
7. MSG (monosodium glutamate)-common allergic and behavioral reactions, including headaches, dizziness, chest pains, depression and mood swings; also a possible neurotoxin.
8. Preservatives (BHA, BHT, EDTA, etc.)-allergic reactions, hyperactivity, possibly cancer-causing; BHT may be toxic to the nervous system and the liver.

9. Artificial Flavors-allergic or behavioral reactions.
10. Refined Flour-low-nutrient calories, carbohydrate imbalances, altered insulin production.
11. Salt (excessive)-fluid retention and blood pressure increases.
12. Olestra (an artificial fat)-diarrhoea and digestive disturbances (Elson, 2000).

Conclusion

In coming years, with larger constraints on time in the home, demand for processed foods is going to increase. Care is necessary to see that a nutritionally sound diet is not compromised through unwise intake of various processed foods.

Processed vegetables and fruits, which are currently available, cannot be equated with fresh vegetables and fruits.

During the last decade, science and technology have contributed at an accelerated pace to the introduction of new products. While in the past, the introduction of "novel" foods and processes did not require formal safety evaluations, but today they must address to toxicological, microbiological and nutritional considerations. Therefore, the safety of "novel" foods is now more carefully assessed than for any new food product in the past. In addition to their safety, "novel" foods will provide new products that will not only improve the quality but also contribute to a varied, nutritious diet.

REFERENCES

National Institute of Nutrition (1998). *ICMR Dietary Guidelines for Indians: A Manual*, 1: 58-62.

Kroll, D. (2001). Changing needs and preferences of health consious consumers. Project report, www.healthy.net.

Swaminathan, M. (1993). *Advanced Textbook on Food and Nutrition*. 2: 405-425.

Pottinger, D. Effect of processed foods on health-an experimental study. Processed foods. www.mcvitamins.com/proc-foods.

World Cancer Research Fund and American Institute for Cancer Research. (1997). *Food, Nutrition and the Prevention of Cancer: A Global Perspective*, 1: 472-485.

NAS. (1992). *Diet, Nutrition and Cancer*. Washington DC: National Academy Press.

Flamm, W.G., Lehinan-MC Keeman L. (1991). The human relevance of the tumor-inducing potential of δ-limonene in male rats:implications for risk assessment. *Regal Toxicol. Pharmacol.*, 13: 70-86.

Stalder, R. Nestle and Marne La Vallee. (2001). *Report on Safety of Processed Foods*, France.

Elson, M. Haas. (2000). Staying healthy with Nutrition, and the new staying healthy shoper's guide: Feed your family safely.

46

Nutritional and Medicinal Value of Mushrooms

R.L. Madaan
Institute of Agricultural Technology, Training and Education,
Directorate of Extension Education,
CCS Haryana Agricultural University, Hisar – 125 004

Nutritional Value

Mushrooms with their flavour, texture, nutritional value, very high productivity per unit area and time, less dependence on land and ability to grow on a variety of residual agricultural wastes, have rightly been identified as a food source to fight malnutrition in developing countries. It has been estimated that the annual production of crop residues in India is more than 300 million tones. Though, the statistics about the ways in which these crop residues are being utilized are not available, but a sizeable portion is being either burnt or left to rot. Burning of crop residues, besides posing environmental problems, is a waste of energy and poor utilization of crop residues. These crop residues could be converted into food, feed and fuel by mushroom cultivation.

Mushrooms have been recognized by FAO as food contributing to the protein nutrition of the countries depending largely on cereals. Mushrooms like other vegetables, contain about 90 per cent moisture and are basically low calorie food (about 30 calories/100 gm. fresh weight). Total carbohydrate content is 4-5 per cent, which consist of chitin, hemi cellulose and glycogen. Very little amount of free sugars (0.5 per cent) is present. Starch is absent. The fat content is very low (0.3 per cent), but it is rich in linoleic acid (70 per cent). Cholesterol, the dreaded sterol for heart patient, is absent and in its place ergosterol is present, which can be converted into Vitamin D by human body. Fibre content is high, which is helpful in excretion of waste and prevention of constipation.

Vitamin A, D, E, K are almost absent in mushrooms. It is the Vitamin of B-complex group and Vitamin C, which are present in significant quantities, which include thiamine, riboflavin, niacin, biotin and pantothenic acid. Folic acid and Vitamin B-12, which are normally absent in vegetable foods, are present in mushrooms. They are also fairly good source of Vitamin C (4-8 mg/100 g). Phosphorous, sodium and potassium are present in high quantities. They are comparatively deficient in calcium and iron, but the latter is present in available form. High K : Na ratio present in mushrooms is desirable for the patients with hypertension (Table 46.1).

Table 46.1: Average Food Value of Mushroom as Raw or Lighly Cooked

Vitamins	*Per 100 gram*
B1 (Thaimine)	0.1 mg
B2 (Riboflavin)	0.4 mg
Niacin	0.5 mg
Pantothenic Acid	6.2 mg
B12	0.05 mg
C (Ascorbic Acid)	2.0 mg
Minerals	
Phosphorus	75.0 mg
Potassium	620 mg
Iron	1.0 mg
Copper	1.0 mg
Salt	6.0 mg

Contd...

Table 46.1–Contd...

Vitamins	*Per 100 gram*
OTHERS	
Fibre	1.0 g
Protein	3.0 g
Fat	Traces
Sugar	Absent
Starch	Absent
Cholestrol	Absent
Calories	30

Protein malnutrition particularly in infants, children and pregnant and lactating women is the biggest nutritional problem in our country. It is the quality of protein, which is more important than quantity and is in fact the requirement of 8 amino acids (10 in case of children), which the human body is incapable of synthesizing in desired quantities. Deficiency of any of the essential amino acids limits the utilization the total protein in direct proportion of the deficit of those particular amino acids. Based on the contents of these amino acids, protein has been classified as good, bad and intermediate quality. In general, as a single source, vegetable proteins (cereal, pulses and vegetables) are of poorer quality than the animal proteins (milk egg, meat, etc.). It has been recommended that 1/3 of the total protein intake should be animal origin to make up for the deficiencies in vegetable proteins. Deficient intake of protein, especially the good quality protein, is then main dietary defect in India. Cereals the staple constituents of Indian diet are deficient in two essential amino acids namely lysine and tryptophan. Mushrooms are very rich in lysine and tryptophan, which can effectively supplement the cereals in our diet. Mushrooms contain 20-35 per cent protein (about 3 per cent on fresh weight basis), which are higher than most of the vegetables (Table 46.2).

Quality of mushroom protein is superior to that of vegetable proteins and is as good as animal proteins. Digestibility of mushroom protein is also high. Due to low calories, low fat (rich in linoleic acid and lacking in cholesterol) and high K: Na ratio, mushrooms are the dieticians choice for those with obesity, hypertension and heart disease. With low calorie, no sugar and starch, and high protein, they are said to be the 'delight of diabetic'. Alkaline ash content coupled with high fibre present in mushrooms is suited to those with hyperacidity (Table 46.3).

Table 46.2: Composition of Common Edible Mushroom (gram/100 gram)

Mushroom/ Vegetable	*Mois- ture*	*Protein*	*Fat*	*Carbo- hydrate*	*Fibre*	*Ash*	*Calories*
Agaricus bisporus	90.1	2.9	0.3	5.0	0.9	0.8	36
Volvariella volvacea	90.1	2.1	1.0	4.7	1.1	1.0	36
Pleurotus sajor caju	90.2	2.5	0.2	5.2	1.3	0.6	35
Cabbage	91.9	1.8	0.1	4.6	1.0	0.6	27
Cauliflower	90.8	12.6	0.4	4.0	1.2	1.0	30
Potato	74.7	1.6	0.1	22.6	0.4	0.6	97

Source: Gopalan *et al.* (1971). *Nutritive Value of Indian Foods.* FAO (1972).

Table 46.3: Medicinal Value of Mushroom at a Glance (Health-wise)

Composition	*Activity*
☆ Low calorie and high protein diet	Diabetic patients
☆ Almost no starch and sugar	
☆ High K : Na ratio Low	Obsesity
☆ Low calorie and low fat (lacking cholesterol)	Hypertension Arthrosclerosis
☆ Alkaline ash and high fibre content	Hyperacidity, Constipation
☆ Rich in Folic acid and available Iron	Pernicious Anaemia
☆ Anti viral effect	Recent reports of effective against AIDS virus
☆ Anti bacterial effect	
☆ Anti tumour effect	
☆ Renal effect	Extend life span of chronic renal failure patients

Medicinal Value

Mushrooms have long been considered to have medicinal value. In fact, the early herbalists were more interested in medicinal properties in mushroom than in their basic value as a source of food. In 1991, according to an estimate the medicinal products of mushrooms generated 1.2 billion dollars. In Japan, a top selling cancer drug, 'PSK' from mushroom (*Coriolus versicolour*) in 1987 has annual sales, totaling 358 million dollars and accounted for 25 per cent of the country's total sales of anti-cancer drugs. Chang and Buswell (1996) have coined the term 'mushroom nutricuticals' for the medicinal preparations from mushrooms.

Due to unique chemical composition, mushrooms are suited to specific groups suffering with some ailments/disorders. As a low calorie high protein diet with almost no starch and sugars, mushrooms are the 'delight of the diabetic'. Due to high K: Na ratio, few calories and low fat (rich in linoleic acid and lacking cholesterol), mushrooms are suited to persons suffering from obesity, hypertension and arthrosclerosis. Alkaline ash and high fibre contents make them suitable for those suffering from hyper acidity and constipation. Besides this, mushrooms are rich in vitamins; folic acid and available iron that reduce the degree of pernicious anaemia. Water-soluble polysaccharides of the several mushroom species have been shown to display anti-tumour activity. Some of the mushroom species display anti-bacterial and anti-viral activities (Table 46.4).

Table 46.4: Medicinal Properties of Some Mushrooms

Sl.No.	*Mushroom Species*	*Activity*
1.	*Pleurotius ostreatus*	Anti-tumour
2.	*Pleurotius griseus*	Antibiotic
3.	*Pleurotius japonicus*	Anti-leukemia
4.	*Lentinus edodes*	Inhibits growth of subcutaneous sarcoma. Decrease hypertension and Plasma cholesterol level
5.	*Auricularia auricula*	Anti ulcer effect
6.	*Flammulina velutipes*	Anti hypertensive and reduce plasma cholesterol
7.	*Coriolus versicolor*	Anti-tumerous
8.	*Schizophyllum commane*	Effective against tumours and carcinoma

Some mushrooms produce toxins, which destroy nematodes. It also serves as therapies for hyperlipemia and diabetes and also selective inhibitor of myxovirus multiplication. Thus, mushroom bears a number of useful functions for human health, beside their contribution to human nutrition. Though, few commercial mushroom products for medicinal purpose are available, but a day is not far to claim further increase in market value of mushrooms.

Products of Mushrooms in the Market

Concord Sunchih

Concord International Trading Pvt. Ltd., Australia from fresh cultivated *Gandoderma* fruitbodies.

Claim

Improves immunity, anti-hypertension, hepatitis, bronchitis, piles, asthma, diabetes. Dosage: 1 to 2 caps twice daily for health and 3 caps thrice daily for illness.

Griferon

Maitake products, Inc. New Jersey, USA, from *Grifola frondosa*.

Claim

It combines well with conventional chemotherapeutic drugs against cancer. Dosages: 5-6 drops TDS for general health and for therapeutic dose 0.5-1 mg./kg body weight TDS.

Prospects in India

Prospects for medicinal mushrooms are immense in India as this is the land of the 'Ayurveda' and appreciates the healing power of the nature. Already India has made headway in production and marketing of herbal products and very little effort will be needed to put the mushrooms in the chain. What are needed are indigenous development of technology for production of medicinal mushroom and also the medicinal preparation from these mushrooms. One of the areas, which have been neglected, is that mushroom flora available in India and also the commercially edible mushrooms have not been subjected to pharmacological scrutiny.

47

Requirements for Pesticide Residues Analysis in Foods

K.S. Sandhu and Usha Bajwa
Punjab Horticultural Postharvest Technology Centre,
PAU Campus, Ludhiana – 141 004

There are two types of quality requirements for foods, *i.e.* general and mandatory quality requirements. The first type includes appearance, texture, flavour, nutrients etc. The second category requires the absence of harmful substances like pesticides, toxic heavy metals, microbial toxins and any other substance which is harmful for human health. The scope of present lecture is limited to detection or estimation of pesticides residues in foods. The declaration of pesticides qualitatively or quantitatively has become inevitable in the WTO regime for export and import. About 176 pesticides are listed which come into foods from their intentional or unintentional use on the crops or from treatment of animals or as contaminants. The maximum residue limits vary with the type of commodity. Use of many pesticides is restricted by law. The maximum residue limits are fixed by national (PF A in India) and international (Codex Alimentarius Commission, 1993) agencies.

The identification and quantitation of pesticide residues in foods is very cumbersome job and requires skilled lab staff, separate laboratory, standard procedures and specific instruments. The sampling should be done according to pre-defined procedure. The basic steps for analysis of pesticides are given below:

Sampling

↓

Sample preparation

↓

Extraction of pesticides

↓

Partitioning

↓

Cleanup

↓

Concentration of extract

↓

Dilution with appropriate solvent

↓

Identification and estimation of pesticides

**(GC, HPLC, Mass Spectrometer, GC-MS,
UV-VIS Spectrometer, TLC etc.)**

The general requirements for residue analysis are discussed briefly in this lecture to provide basic information to the participants.

Sampling

The sampling procedure varies with the commodity and type of sample. The numbers of units or quantity for sampling also vary with the accuracy of detection (per cent violation incidence), type of sample and purpose of analysis. For details the readers are referred to Cochran (1963). The sole objective is t6 draw a sample as uniform as possible from whole lot of produce.

Guidelines on Good Laboratory Practice in Pesticide Residue Analysis

The ultimate goal in fair practice in international trade depends, among other things, on the reliability of analytical results. This in turn particularly in pesticide residues analysis, depends not only on the availability of reliable analytical methods, but also on the experience of the analyst and on the maintenance of "good laboratory practice in the analysis of pesticides". These guidelines define such good analytical practice and may be considered in three inter-related parts:

1. The Analyst
2. Basic resources, and
3. The Analysis.

The Analyst

Residue analysis consists of a chain of procedures, most of which are known, or readily understood, by a trained chemist, but because the acceptable margin for error is smaller than in most types of analysis and any mistake can invalidate the whole analysis, attention to detail is must. The analyst incharge should have an appropriate professional qualification and be experienced and competent in residue analysis. At anyone time the majority of staff need to be trained in the correct use of apparatus and basic laboratory skills and the principles of residue analysis. They must understand the purpose of each stage in the method being used and the importance of following the method exactly as described and of noting any unavoidable deviations. A clear understanding of the terminology involved is also essential.

Basic Resources

The Laboratory

In ideal circumstances the laboratory and its fittings should be designed to allow tasks to be allocated to well defined areas where maximum safety and minimum chance of contamination of samples prevail. Fittings should be of materials resistant to attack by chemicals likely to be used in the area. Thus, under such ideal conditions, separate rooms would be designated for sample receipt and storage, for sample preparation, for extraction and clean up and for instrumentation used in the determinative step.

No smoking, eating, drinking or application of cosmetics should be permitted in the working area. Only small volumes of solvents should be held in the working area and the bulk of the solvents stored separately, away from the main working area. The use of highly or chronically toxic solvents and reagents should be avoided whenever possible. All waste solvent should be stored safely and disposed of frequently.

The main working area should be treated as a solvent laboratory and all equipment such as lights, macerators and refrigerators should be "spark free" or "explosion proof'. Extractions, clean-up and concentration steps should be carried out in a well ventilated area, preferably in fume cupboards. Fume hoods or canopies above the bench are generally ineffective in removing vapours at the working level.

Safety screens should be used when glassware is used under vacuum or pressure. There should be an ample supply of safety glasses, gloves and other protective clothing, emergency washing facilities and spillage treatments kits. Adequate emergency fire extinguishing equipment must be provided. The staff should be given periodic medical checks.

Equipment and Supplies

The laboratory will require adequate supplies of electricity and water and various gases, either piped or from gas cylinders, of proven quality. Adequate supplies of reagents, solvents, glassware, stationary phases, etc., are essential.

All laboratories require an adequate range of reference standard pesticide of known and reasonably high purity. The range should cover all parent species for which the laboratory is monitoring samples as well as those metabolites, which are included in MRLs.

The Analysis

Avoidance of Contamination

One of the major areas in which pesticide residue analysis differs significantly from macro-analysis is that of the problem of contamination. Trace amounts of contamination in the final samples used for the determination stage of the methods can give rise to errors such as false positive results and to a loss of sensitivity that may prevent the residue analyst from achieving the necessary limits of determination. Contamination may arise from either construction material, from the environment or from the procedure.

Bench polish, barriers creams, soaps containing germicides, fly sprays, perfumes and cosmetics are all commodities that give rise to laboratory contamination and are especially significant when an electron-capture detector is being used. There is no real solution to the problem other than to ban their use in the laboratory.

Greases, plasticizers, rubber bungs and tubing, thimbles, filter papers and cotton-wool can also give rise to contamination of the final test resolution.

Pesticide reference standards should always be stored in a room separate from the main residue laboratory, Field samples, sample preparation and formulation analysis should also be kept separate from the main residue laboratory.

Contamination of glassware, syringes and gas chromatographic columns can arise from previous samples. All glassware should be cleaned with detergent, rinsed thoroughly and then rinsed with the solvent to be used. There must be separate stock of glassware for pesticide residue work.

Chemical reagents, adsorbents and general laboratory solvents may contain components that interfere in the analysis. It may be necessary to purify reagents and absorbents by heating and it is generally necessary to use redistilled solvents. De-ionized water is often suspect and redistilled water is preferable. In many instances tap water or well water may be satisfactory. All glassware, reagents, organic solvents and water should be checked for possible contaminants before use.

Apparatus containing polyvinylchloride should be suspect and if shown to be a source of contamination, should not be allowed in the residue laboratory. Other materials containing plasticizers are suspect but silicon rubbers are usually acceptable. Sample storage containers can cause contamination and glass bottles with ground glass stoppers should always be used. Instrumentation should always be housed in a separate room. The nature and importance of contamination can vary according to the type of determination technique used and the level of pesticide residue to be determined. These contamination problems, which are important with methods based on gas chromatography or high performance liquid chromatography, may well be less significant if a spectrophotometric determination is used, and vice-versa. For relatively high levels of residues the background interference from solvents and other materials may be insignificant in comparison with the amount of residue present. Also, many problems can be solved by the use of specific detectors. Furthermore, if the contaminant does not interfere with the residue being sought, its presence may be acceptable.

Avoidance of Losses

In an ideal situation samples should be stored at chill (3–5°C) temperature, away from direct sunlight, and analyzed within a few days. However, in many instances samples may require storage for an extended period (6–9 months) before analysis and the following precautions should be observed:

1. Storage temperature should be approximate – 20°C, at which temperature degradation of residues of pesticides by enzyme action is extremely slow.
2. All samples should be re-homogenized after freezing as there is a tendency for water to distil out and to collect as ice crystals, which, if discarded, will affect the analytical result.
3. Neither the containers used for storage nor their caps or stoppers should allow migration of the chemical being sought into the container. The containers must not leak. All samples should be labeled clearly with permanent labels and recorded in a sample book.
4. The extracts and final test solutions should not be exposed to direct sunlight.

Standard Operating Procedures

Written, adequate, standard operating procedures must be available for use by analytical staff. Any deviations from procedures must be recorded and authorized by the analyst incharge.

Validation of Methods

The amount of effort allocated to the validation of methods will vary considerably. In a routine laboratory monitoring for compliance with Codex MRLs or national tolerances, standardized methods will be used in most instances. These should be validated initially and thereafter checked periodically.

In all laboratories, regular checks should be made on the effects of variation in sources of supply of chemicals, solvents, etc. The performance of the method should be checked by, for example, the recovery of standards, added at appropriate levels, taken through the method both alone and in the presence of each new substrate.

Maintenance of Overall Analytical Performance

In all laboratories engaged in pesticide residue analysis there is a need for regular assessment of the methods in use, both at the tolerance level and at the lower limit of determination. Recovery of pesticides from "spiked" samples is commonly used as a measure of efficiency of extraction, but it must be recognized that such studies are of limited value. Where possible, emphasis should be placed on checking recoveries where residues are in a "real" state *e.g.* in field treated samples, since a method that gives adequate recoveries from samples spiked with parent compounds may be inadequate for the efficient extraction of the parent compounds and their metabolites from the "real" samples. The evaluation of a method should include, where possible, the extraction of radio-labelled compounds or a comparative extraction with a method of known efficiency. Recoveries should be within the range 70–110 per cent with a mean of greater than 80 per cent.

Regular analyses of substrates known to be free of pesticide residues is necessary in order to check that contamination is not occurring. Care should be taken that standard solutions of pesticides are not decomposed by the effect of light or heat during storage or become more concentrated owing to solvent evaporation. Equal care must be taken to ensure the stability of reference standard compounds. Regular injection of standards during gas chromatographic analysis of a series of samples allows the performance of the determinative step to be checked.

A means of monitoring the performance of a method (or any analyst) is to introduce check samples at regular intervals. These check samples should be introduced as routine samples without any indications being given as to their special nature.

Various national and international organizations now organize collaborative studies on particular methods and/or check sample programmes. The latter present an ideal way for laboratories to assess their own performance. If possible, check samples should be introduced as routine samples so that the analyst concerned does not attempt to "make a special effort", which would invalidate the samples as a test of laboratory performance.

Confirmatory Test

When analyses are done for regulatory purposes, it is especially important that confirmatory tests are carried out before reporting adversely on samples containing residues of pesticides not normally associated with that commodity or where MRLs appear to have been exceeded.

Contamination of samples with non-pesticidal chemicals occurs from time to time atld in some chromatographic methods. These compounds may have similar properties to pesticides and may therefore be miss-identified as such. Examples in gas chromatography include the responses of electron-capture detectors to phthalate esters and of phosphorus–specific detectors to compounds containing sulfur.

Confirmatory tests can be divided into two types. Quantitative tests are necessary when MRLs appear to be exceeded whilst confirmation of identity is also needed in these cases, and when a typical residues are encountered. Qualitative tests may involve chemical reactions or separations where some loss of the residue occurs. Particular problems occur in confirmation when MRLs are set at or about the limit of analytical determination.

In quantitative confirmation at least one alternative procedure should be used and the lower result reported. In qualitative confirmation, an alternative technique using different physico-chemical properties is desirable.

A number of instrumental procedures for estimation/confirmation are discussed in the following paragraphs.

Gas Chromatography (GC)

GC is the commonly used instrument for qualitative and quantitative determination of pesticide residues. The specific columns and detectors are used for different groups of pesticides. The results obtained in the primary analysis should always be quantitatively and qualitatively confirmed using at least one alternative column involving a stationary phase of different polarity. The quantitative results obtained should be within 20 per cent of the primary analysis and the lower figure should be reported, since the higher figure may have been enhanced by interference from co-extracted material. Further quantitative confirmation is required if the results differ by more than 20 per cent, except when the MRL is set "at or about the limit of determination" when a variation of up to 100 per cent would be acceptable. In choosing the alternative column material, consideration should be given to separating any pesticidal or interfering compounds known to have retention times on the primary column identical to that of the residue detected. The alternative column may be a packed column or, preferably, a capillary column whose greater resolving power may be utilized. Whilst the use of an alternative gas-chromatographic column may not always give positive confirmation, it will often quickly disprove a suspected identity. In either case further confirmation is required to identify the residue.

When pesticides containing several chemical elements are present, detectors showing enhanced response to these elements may be used. Detectors such as flame photometric (sulfur, phosphorus and tin), alkali flame ionization (phosphorus and nitrogen) and coulometric/conductivity (nitrogen, sulfur and halogens) can give valuable additional information on residues. The sulfur/phosphorus response ratio obtained by using aflame photometric detector can give useful information in the case of phosphorothioates.

Thin Layer Chromatography (TLC)

In some instances, confirmation of gas-chromatographic findings is most conveniently achieved by TLC. Identification is based on two criteria, Rf value and visualization reaction. The scientific literature contains numerous references to the technique, the IUPAC Report on Pesticides (Batora, 1982) reviews the technique and serves as a convenient introduction. The advantages of thin-layer chromatography are speed, low cost and applicability to heat sensitive materials; disadvantages include (usually) lower sensitivity than gas-liquid chromatography and frequent need for more efficient clean-up. In some countries problems may be encountered when high humidity or temperature cause lack of repeatability.

Higher Performance Liquid Chromatography (HPLC)

HPLC can be used advantageously for the confirmation of residues initially found by gas chromatography or by other techniques and may be in certain circumstances the preferred quantitative technique. Post- or pre-column derivatization, and/or use of different detectors, are further options available to the analyst, especially when heat sensitivity or low volatility make the compound to be analyzed amendable to gas chromatography.

Mass Spectrometry

Results obtained using mass spectrometry present the most definitive evidence for confirmation/identification purposes. Where the apparatus is available it is usually the confirmatory technique of choice.

There are two principal methods of introducing samples into the instrument. The preferred method utilizes gas-chromatographic separation prior to introduction into the mass spectrometer. This allows full mass spectral analysis of the peak observed during the primary analysis. Alternatively, samples can be introduced using the direct insertion probe technique. This method can be used in conjunction with TLC and HPLC when these have been used as initial confirmatory procedures. Residues separated into these techniques are isolated and subjected to mass spectrometry.

To increase sensitivity, particularly with fast scanning quadrupole instruments, techniques known as single and multiple ion detection have been used. A sufficient number of fragment ions must be selected to ensure unambiguous identifications. Increased sensitivity with respect to the molecular ion may be obtained by using chemical ionization in place of electron-impact. As mass spectrometers are generally sensitive at the nanogram level some extracts from primary gas chromatographic analysis may require concentration before mass spectrometric analysis, particularly when electron-capture detectors have been used for quantification. In some cases additional clean up may be necessary, particularly if full spectra are to be obtained.

Problems can be encountered with heat-sensitive compounds during mass spectrometry and particular care has to be taken when coupling gas chromatographs to mass spectrometers. As there is almost no differential response to compounds in mass spectrometry, complications can arise in the presence of co-eluting contaminants. A useful introduction of the various techniques is given by Hutzinger and Safe (1973) and Waller and Dormer (1980).

Spectral Measurements

At present little use is made of infra-red. Ráman or nuclear magnetic resonance spectroscopy in pesticide residue analysis. Instrumental techniques using multiple reflection cells, microcells, microphobes, laser light, Fourier Transformation etc. are being developed. These improve the quality of spectra and enhance the sensitivity and may enlarge the application of these techniques as post column detection methods for identification of compounds isolated by chromatographic techniques.

Criteria for the Selection of Analytical Methods

Whenever possible, the following criteria were used when selecting analytical methods for inclusion in the list:

1. Published in books, manuals or open literature; collaboratively studied or known to have been validated in a large number of laboratories.
2. Capable of determining more than one residue, *i.e.* multi-residue methods.
3. Suitable for as many commodities as possible at or below the specified MRLs.
4. Applicable in a regulatory laboratories equipped with routine analytical instrumentation.

References

Batora, V., Vitorovic, S.Y., Their, H.P. and Klisenko, M.A. (1982). IUPAC report on pesticides. *Pure and Appl. Chem.*, 53: 1039-1049.

Cochran, W.G. (1963). *Sampling Techniques*, 2nd Ed., John Wiley and Sons, Inc., p. 74-75.

Codex Alimentarius Commission (1993). *Codex Alimentarius Vol II: Pesticide Residues in Food*. Food and Agriculture Organization of United Nations and World Health Organization, Rome.

Hutzinger, O. and Safe, F. (1973). *Mass Spectrometry of Pesticides and Pollutants*. CRC Press.

Waller, O.R. and Dormer, O.C. (1980). *Biochemical Applications of Mass Spectrometry*. John Wiley and Sons, New York.

48

Integrated Pest Management in Stored Grains

Shiv K. Singal
Department of Entomology
CCS Haryana Agricultural University, Hisar – 125 004

Agriculture contributing 32.7 per cent of the Gross Domestic Product (GDP), is the backbone of Indian economy. Over 2 per cent average annual growth in yield production has been maintained. The total foodgrains production in India has increased to over 205 million tonnes in 1999-2000 against 51 million tonnes in 1950-51. In order to meet the demand of our ever increasing population growing at the rate of 1.8 per cent, we would need 260-264 million tonnes of food grams in 2030.

The produced foodgrains have to be stored either for short or long duration for utilization in the coming times for food, feed and seed. In storage, these suffer 9.33 per cent losses due to various agencies such as rodents, insects, microorganisms and moisture in India in comparison with other underdeveloped countries where losses are as high as 30-50 per cent. Therefore, with an urge to produce more there is still urgent need to protect the produce in the storage to meet the challenge. As a result of improved storage technology, these losses in Government Warehouse could be reduced to minimum but at farmers' level where 60-70 per cent of the produce is retained the losses are still too high.

The increased emphasis on adopting Integrated Pest Management (IPM) strategies for stored product insects is primarily due to consumers' demand for food free of insecticidal residues (Subramanyam and Hagstrum, 1995). Therefore, the current scenario of stored grain protection in India depend totally on timely and judicious use of IPM strategies. The IPM in storage can be broadly classified as (1) Preventive and (2) Curative, for dealing with insect-pests of stored grains (Table 48.1).

Preventive

For storage insects "Prevention is better than cure" is most appropriate. this can be achieved through sanitation and legal aspects.

Sanitation

Clean the receptacles/stores thoroughly. Remove the old loose mud-plaster and apply new one to fill cracks and crevices. Bury deep or burn the cleanings.

Disinfest the stores with some malathion 0.5 per cent, primiphos methyl 0.5 per cent or pyrethrin 0.02 per cent @ 31/100 m^2 with frequency of treatment 15 days.

1. Use new gunny bags
2. Treat old gunny bags by dipping in cypermethrin 0.01 per cent or fenvalerate 0.01 per cent or malathion 0.1 per cent for 10 minutes and dry in shade before filling the grains (Source: Package and Practices, HAU, Hisar).
3. Dry the grains thoroughly to bring down moisture below 10 per cent before storage.

Legal Method

Entry of an insect which is not found in the particular area can be prevented by the imposition of Destructive Insect Pests Act, 1914.

Table 48.1: Insect Pests of Stored Grain

Sl.No.	*Scientific Name*	*Common Name*	*Order*	*Family*	*Commodities Attacked*
A.	**Primary pests**				
1.	*Sitophilus oryzae* (L.)	Rice weevil (Sursuri-sundwali)	Coleoptera	Curculionidae	Wheat, rice, maize, jowar and paddy
2.	*S. zeamais* (Mots)	Maize weevil (Makka ki sursuri)	Coleoptera	Curculionidae	
3.	*S. granarius*	Granary weevil (Vedeshi sursuri/ Pahari sursuri)	Coleoptera	Curculionidae	
4.	*Rhizopertha dominica* (Fab.)	Lesser grain borer (Ghun)	Coleoptera	Bostrychidae	Paddy, rice, wheat and maize
5.	*Trogoderma granarium* (Everts)	Khapra (Pai)	Coleoptera	Dermestidae	Wheat, maize, jowar, rice, pulses, and oilseeds
6.	*Sitotroga cerealella* (Olivier)	Grain moth (Anaj ki titali)	Lepidoptera	Gelechidae	Paddy, maize, jowar and barley
7.	*Corcyra cephalonica* (Stainton)	Rice moth (Chawla Ka Patanga)	Lepidoptera	Pyralidae	Processed products of cereals, pulses, oilseeds and milled species
8.	*Plodia interpunctella* (Gubner)	Indian Meal moth (Makka ka Patanga)	Lepidoptera	Pyralidae	Maize, cereals, dry fruits
9.	*Ephestia cautella* (Walker)	Almond or Fig moth	Lepidoptera	Pyralidae	Wheat, ice, maize, jowar, groundnut species

Contd...

Table 48.1–Contd...

Sl.No.	*Scientific Name*	*Common Name*	*Order*	*Family*	*Commodities Attacked*
10.	*Callorobruchus chinensis* (L.)	Pulse Beetle Dhora (Bhunbhuna)	Celeoptera	Bruchidae	Pulses
11.	*C. maculatus* (F.)	-do-	-do-	-do-	-do-
12.	*C. analis*	-do-	-do-	-do-	-do-
B.	**Secondary pests**				
13.	*Tribolium castaneum* (Herbst)	Red rust flour beetle (Atta ka keet/pata)	Coleoptera	Tenebrionidae	Broken grains/mechanically damaged grains, germ portion and milled products
14.	*Tribolium confusum* (Jacq)	Confused flour beetle	Coleoptera	Tenebrionidae	
15.	*Latheticus oryzae* (Waterhouse)	Long headed flour beetle (Atte ke chooti sursuri)	Coleoptera	Tenebrionidae	Cereal flours, packaged food, rice and rice products
16.	*Oryzaephilus surinamensis* (L.)	Saw toothed grain beetle (Chawal ki sursuri)	Coleoptera	Silvanidae	Rice, wheat, maize cereal products, oilseeds and dry fruits
17.	*Cryptolestes ferrugineus* (Olivier)	Rust red grain beetle (Moonchwali sursuri)	Coleoptera	Cucujidae	Rice, maize, wheat, different flour and groundnut
18.	*Cryptolestes pusillus* (Schon.)Curative	Flat grain beetle	Coleoptera	Cucujidae	-do-

Curative

Non-chemical Approach

Ecologial Control

Insect population in stores can be checked through temperature and grain moisture regulation.

Low Temperature

Temperature below 14°C results in death particularly of immature stages of almost all insect-pests. Death occurs rapidly at freezing point (-10 to -20°C). Low temperatures are usually obtained by aeration with cool ambient air but refrigeration systems can also cool bulk grains (Banks and Fields, 1995).

Some species are more cold hardy than others, in general *Tribolium casteneum* and *T. confusum* and *Oryzaephilus mercator* are most cold susceptible, whereas *Trogoderma granarium, S. granarius, Ephestia elutella* and *Plodia interpunctella* are most cold tolerant species (Fields, 1992).

In most field applications of cold temperature to control stored product pests, the insects are exposed to 10 to 20°C for some days before being exposed to lethal cold temperature. Exceptions to this rule are commodities that are rapidly cooled in freezers to control insect population. When insects are exposed to cool temperature (20 to 10°C), they become acclimated and their cold hardiness increases by two to ten folds (Smith, 1970 and Evans, 1983).

High Temperature

Most of the stored grain insects die at 50 to 60°C with a period of 10 to 20 minutes. Exposures to temperatures only about 5°C above the optimum for population will stop the development. The exposure of 2h at 50°C eliminated almost all pests but studies on heat tolerant species especially pupal stage of *Rhizopertha dominica* are lacking. Grain heating is done by hot air fluidized bed, infra-red, high frequency dielectric and microwave heating for achieving uniform grain temperature throughout the storage structure.

Grain Moisture

Grain moisture regulates biological activity of insects in storage. Atmospheric humidity and grain moisture are closely related factors. Even though stored grains are dried to safe moisture level before storage, moisture content in due course comes to equilibrium with humidity in the air. It is usually necessary to dry the grains to bring down its moisture content to a safe level to minimize or avoid losses caused by insects. Paddy at 13 per cent moisture does not allow insects to multiply in storage (Hall, 1980). Grains stored at around 10 per cent moisture content escape from the attack of insects (except '*khapra*' beetle).

Controlled Atmosphere

Controlled atmosphere (CA) disinfestation technology involves the alteration of natural storage gases, *i.e.* CO_2, O_2 and N_2 so as to prevent multiplication of insects, mould growth and quality deterioration of the foodgrains. The controlled storage atmosphere may be achieved in several ways (Benks and Annis, 1990). About 9.0 to 9.5 per cent CO_2 in air is lethal to all insects. Such an atmosphere are called 'high CO_2'. Low oxygen atmosphere (2 to 4 per cent) generated by purging the pure N_2 and from other sources is called 'Low O_2'. Allow metabolic processes within a storage to remove O2, usually associated with release of CO_2, *i.e.* in storage O_2 is consumed by grains and insects during respiration and CO_2 is produced. Thus, O_2 level will be reduced below 1 per cent and CO_2 will automatically increase, which will be lethal to all stages of insects. Such atmosphere is referred as 'hermetic storage'. Controlled atmosphere technology for control of grain pests has been extensively used in the field.

Inert Dusts

These are non-toxic and commonly used by farmers. There are four basic types of inert dusts, *i.e.* earth, diatomaceous earth, silica aerogels and non-silica. The principal mode of action for inert dusts is that they cause insects to desiccate. Insects die when they lose about 60 per cent of their water or about 30 per cent of their total body weight. Silica aerogels can absorb as much as three times their own weight in oils. It is believed that as insects move through the grains, the silica aerogels absorb waxes from the insects' cuticle. Because the stored product insects live in very dry environments with limited access free water, water retention is crucial to their survival. Also, since insects are very small they have a large surface area with respect to their body weight and therefore have greater problems for retaining water than larger animals.

Insects protect against desiccation in many ways but the waxy cuticle, which inert dusts destroy, is one of the main mechanisms in maintaining water balance. Food grains stored with inert dusts are to be cleaned and washed before consumption.

Use of Edible Oils

Mustard and groundnut oils @ 7.5 ml/kg of stored grains are effective for protection of stored pulse grains against pulse beetle 'Dhora' upto 9 months (Singal, 1995).

Plant and Other Non-toxic Grain Protectants

Environmental friendly biopesticides such as 'neem' *Azardiracta indica* (*e.g.* leaf, seed kernel powder, oil and crude extract), possess repellent, antifeedent and feeding deterrent properties against storage insects (Savitri and Rao, 1976). 'Neem' seed kernel powder 4.0 per cent 'neem' seed oil 1 per cent and 'mahua' oil 1 per cent proved repulsive, potent oviposition inhibitor in checking damage by pulse beetle, *Callosobruchus chinensis* upto 8 months in pigeonpea (Singal and Chauhan, 1995).

Wheat grains mixed with *neem* and *dharak* (*Melia azadirachta*) leaves @ 4 per cent (w/w) were infested least by the insects (J.C. Mehla–Personal communication). The rice grains treated with turmeric powder, *Curcuma longa* @ 3.25 per cent (w/w) were found to be less damaged by rice weevil, *Sitophilus oryzae* as compared to other treatments and control.

Biological Control

Bacteria such as *Bacillus thurengiensis* (Bt) is currently registered as protectant for use on grains U.S.A. The larvae of Indian meal moth, *Plodia interpunctella* and almond moth, *Epheslia cautella* showed high levels of susceptibility to Bt (Nawanze *et al.*, 1975). NPV, GV and cytoplasmic polyhedrosis virus isolated mainly from lepidopterous insects have demonstrated in several cases and may be included to provide long term protection. *Dinarmus* spp. is a larval-pupal parasitoid of *Callosobruchus* spp., *Bruchus spp., Bruchidius atrolineatus* and *Acanthoscelides* obtectus in legume seeds. Natural population of *dinarmus basalis and Eupelmus* sp. were found parasitizing larvae of *C. maculatus* and *B. atrolineatus* in Niger in both field and storage. Eggs of grain moth, *S. cerealella* and *E. cautella* are attacked by egg parasitoid such as *Trichogramma* spp. *Uscana mukerjii* attacks eggs of *C. maculaltus* and *U. lariophaga* attacks eggs of *B. atrolineatus* (Kapila and Aggarwal, 1991; Van Haus *et al.*, 1991).

Uscana semifumipennis resulted in eggs parasitization upto 90 per cent of *Caryedon serratus*, a serious pest of groundnut. Biological control has a limited scope in the pest management of stored grains but is becoming an increasingly important part of IPM approach.

Cultural Control

Pulse beetle attacks whole pulses only. Split pulses escape the attack of *Callosobruchus* spp. and as such most suitable for storage.

Use of Improved Storage Structures

The traditional receptacles used for storage of foodgrains are not proof against insects, rodents, fire hazards, moisture and sub-soil water, etc. Secondly, the traditional structures are also not suitable for fumigation/disinfestation during storage.

1. Improved traditional storage structures: Some of the improved structures developed by different agencies are: *Pusa* Bin, *'Puri'*, *'Gade'*, *'Patara'*, *'Kothi'*, *'Pucca Kothi'*.
2. Modern storage structures: Food storage receptacles constructed on scientific basis and techniques can store foodgrains in safe and sound conditions for long periods. Such scientific structures are:

Indoor Bins

1. Domestic Designs
2. *'Gharelu Thekka'*
3. High Moisture Hollow Bamboo Bins

Outdoor Bins

1. Flat Bottom Metal Bins
2. Hopper Bottom Metal Bins–Module I to IV
3. Composite Bins
4. R.B. Bins
5. High Moisture Grains Storage Bins.

Urban Bins

Circular or square in shape. Polyethylene lined gunny bags or polyethylene bags of 300-400 gauge thickness also protect paddy from cross infestation of insects for upto 2 years as well as stop moisture migration to the paddy stored in these bags. However, *R. dominica* and *S. oryzae* can penetrate polyethylene lining. Sometimes the loss of seed viability was also greater in polyethylene lined bags than in jute bags without polytheylene.

Chemical Control

Inspite of regular prophylactic measures, infestation develops due to hidden infestation or cross infestation, etc., then some curative measures have to be adopted and the application of chemical is done.

Knockdown Chemicals

Pyrethrum spay, lindane smoke generator or fumigant strips are aimed against flying insects but may kill the insects at surfaces as well as in cracks and crevices.

Grain Protectants

Mixing of chemical dusts with the food grains meant for human consumption is not recommended. However, the grain meant for seed can be protected by mixing insecticides like pyrethrum dust or malathion 5 per cent @ 250g per quintal of seed.

Deltamethrin 2.8 EC at 0.14 and 0.18, cypermethrin 10 EC at 0.06, fenvalerate 20 EC at 0.03 and nimbecidine at 10 and 15 ml/kg seed were effective in controlling the infestation of pulse beetle for 135 days of storage (Dhillon *et al.*, 1998).

Fumigants

Aluminium phosphide (Phostoxin and Calphos). 7-10 tablets of 3g each 1000^3 ft space with an exposure period of 7 days.

Ethylene dibromide (EDB) available in ampules, its prescribed dosage is 3 ml/q for wheat and pulses but 5 ml/q for rice and paddy. It is also applied @ 1.7 litres for 1000^3 ft space. Such stocks are left undisturbed for 7 days before consumption, the grains are to be aerated.

Ethylene dichloride carbon tetrachloride (EDCT) is used in mixture in the 3:1 v/v ratio for large scale food grain fumigation @ 30-40 kg/1003 m in large scale storage. For small storage use 55 ml/q on food grains irrespective of bulk or bag storage.

REFERENCES

Benks, H.J. and Annis, P.C. (1990). Comparative advantages of high CO_2 and low O_2 types of controlled atmospheres for grain storage. In: *Food Preservation by Modified Amospheres*, (Eds.) M. Calderon and R. Barkai Golan. CRC Press, Roca Baton, FL, p. 93-122.

Evan, D.E. (1983). The influence of relative humidity and thermal acclimation on the survival of adult gram bulles in cooled gram. *J. Stored Prod. Res.*, 19: 173-180.

Fields, P.G. (1992). The control of stored products insects and mules with extremes temperatures. *J. Stored Prod. Res.*, 28: 89-118.

Flinn, P.W. and Hagstrum, D.W. (1990). Simultaneously comparing the effectiveness of various stored grain management practices used to control *R. dominica*. *Environ. Ent.*, 19: 725-729.

Savitri, P. and Rao, S.S. (1976). Studies on the admixture of neem kernel powder with paddy in control of important storage pests of paddy. *Andhra Agric. J.*, 23: 137-143.

Singal, Shiv K. (1995). Testing some vegetable oils for protection of gram seed during storage against *C. chinensis*. *J. Insect Sci.*, Ludhiana, 8(2): 210-216.

Singal, Shiv, K. (1998). Effect of different nontoxic and insecticidal dusts as protectants against *Callorobruchus chinensis* infesting stored chickpea. *Proc. Intl. Conf. on Food Security and Crop Science*, CCSHAU, Hisar, India, pp. 252.

Singal, Shiv and Chauhan, R. (1995). Inhibitory effect of biopesticides and other materials on development of pulse beetle, *C. chinensis* on pigeonpea, *Cajanus cajan*. *Proc. 82nd Sessions of Indian Science Congress Assoc., Section Zoology, Entomology and Fishereis*, Calcutta, Part III, p. 126.

Subramanyam, Bhadriraju and David, W. Hagstrum. (1995). *Integrated Management of Insects in Stored Products.* Marcel Dekker, Inc., New York, p. 426.

Singal, Shiv and Chauhan, R. (1995). Inhibitory effect of biopesticides and other materials on development of pulse beetle, *C. chinensis* on pigeonpea, *Cajanus cajan*. *Proc. 62nd Sessions of Indian Science Congress Assoc., Section Zoology, Entomology and Fishereis*, Calcutta, Part III, pp. 128.

49

Management of Food Processing Units

Atul Dhingra
Department of Business Management,
CCS Haryana Agricultural University, Hisar, India

Introduction

Management is nothing but planning, implementation, and control of activities for optimum utilisation of resources. Management of any plant or factory calls for decisions on the following vital aspects:

1. Production systems
2. Product selection
3. Process selection
4. Plant location
5. Plant layout
6. Demand forecasting
7. Capacity planning

8. Production planning and control
9. Quality control
10. Work study
11. Inventory control
12. Maintenance
13. Waste management
14. Value engineering
15. Research and Development

Besides all these activities the following are important related areas:

1. Marketing
    - (a) Product
    - (b) Pricing
    - (c) Promotion
    - (d) Distribution
2. Finance
    - (a) Procurement of Funds
    - (b) Utilisation of Funds
    - (c) Accounting
    - (d) Taxation
    - (e) Auditing
3. Human resource management
    - (a) Labour Legislation
    - (b) Social security
    - (c) Recruitment and Selection
    - (d) Training and Development
    - (e) Motivation
    - (f) Performance appraisal

Plant Location

Before starting the production we have to decide about the site where we want to establish our factory. Many factors are considered while selecting the site. They are:

1. Availability of raw material
2. Nearness to potential markets
3. Nearness to fuel and power
4. Supply of labour
5. Infrastructure facilities
6. Transportation facilities
7. Climatic conditions
8. Political situation
9. Taxes
10. Local attitude towards industry

All these factors are considered before the selection of suitable site for food processing plants.

Production Systems

Basically there are two types of production systems, namely continuous and intermittent systems.

Continuous System

In this system production is carried out throughout the year without any stoppages. Usually items are produced for stock and same product is produced. In this more and more automation can be used.

Intermittent System

In this system items are produced for order. Production process is stopped after completion of a job or batch. This type of production system is most suited for food processing units.

Product Selection

It involves the selection of the products and its variants we want to prepare. First of all a product range is identified which can be produced on available facilities and then 2-3 products are finalised for production. Usually brainstorming sessions amongst the experts is used for finalising what products should be produced ?

Process Selection

After the decisions regarding products to be produced is decided, the process of manufacturing them come to limelight. One can use automation, others may use manual process. Basically the type of product being manufactured, technology available, and money in hand help us in deciding about the process to be used.

Plant Layout

It is the arrangement of machines in the plant. Computers, Models, sting diagrams are the tools for deciding about the layout of the machines in the available space. Three main types of layouts are available:

1. Product layout
2. Process layout
3. Stationary layout

Demand Forecasting

Before starting the production of goods their demand is determined. Both statistical and non-statistical methods are used for demand forecasting.

Non Statistical Methods

1. Expert opinion method
2. Sales force opinion method
3. Consumer jury method

Statistical Methods

1. Percentage of sales method
2. Least square method
3. Correlation and regression method
4. Time series analysis
5. Exponential smoothing method
6. Moving averages method

Capacity Planning

Capacity is number of units which can be produced over a period of time. It is basically a factory's maximum productive capability usually expressed as volume of output per period of time. Plant capacity is decided on the basis of demand.

Production Planning and Control

Production planning is that function of management which decides about the resources the firm will require for its manufacturing operations and of allocating these resources to produce the desired output in required amount at least cost. Production control is a mechanism which constantly keeps a watchful eye on the production flow and use of resources to find out the deviations, if any, and then taking corrective actions so that everything takes place according to plans. These are done to plan and control the production process. Firstly raw material needs are identified, secondly flow of semi-finished goods are decided, thirdly machines and their operators are decided, fourthly production is compared with the targets, lastly corrective actions are taken if any deviation comes to light.

Quality Control

A separate quality control section is established to control the quality and to ensure that the goods are being manufactured according to specifications. The quality certifications like FPO and ISO must be obtained by the department.

Work Study

It is conducted in organisations to increase the efficiency and productivity. It has two components, namely Time and Motion study. Time study is conducted to reduce the time taken in completing various activities. Motion study is conducted to eliminate the unnecessary activities.

Value Engineering

It is used to reduce the cost of product. In this techniques the product features and components are analysed and then an effort is made to eliminate unnecessary features. Cheaper substitute of components used are also found out.

Maintenance

A separate section is created to provide maintenance services. An effort is done to reduce the number of breakdowns. The concept of preventive maintenance must be used by this department.

Inventory Management

Inventory is stock of goods in godowns of the company. Inventory levels must be controlled. If inventory level is high then funds will be tied up in finished inventories but if the levels are low then smooth supply can't be made.

Waste Management

It is necessary in modern day times as environment protection is the need of the hour. Waste management is concerned with affluent treatments so that environment is protected. Recycling is also used to reduce the waste.

Marketing

It is finding out the needs and wants of the people and then offering products and services prior to competitors to satisfy those needs and wants. Basically it is disposing of products at a profit. It calls for decisions on product features, styles, colours, packaging, labeling, pricing, advertising, distribution channels etc.

Finance

It is concerned with procurement of funds from various sources at cheaper rates and then utilising those funds in profitable ventures. It involves decisions on stock and shares, dividends, accounting, taxation, auditing etc.

Production

Production is an intentional act of producing something in an organised manner. It is the fabrication of a physical object through the use of men, material and equipment. It is an act of transformation, *i.e.;* inputs are processed and transformed into some output.

Human Resource Development

It is concerned with recruitment, selection, training, development, compensation, motivation, and evaluation of all employees. This department is also concerned with employee welfare and social security.

Research and Development

It is the most important function for any organisation. This is needed for innovation and upgradation of products and procedures.

REFERENCES

Kotler, Philip. (2001. *Marketing Management*. Millennium edition, Prentice Hall of India Pvt. Ltd., New Delhi.

Ramaswamy, V.S., and Namakumari, V. (1999). *Marketing Management: Planning, Implementation, and Control: The Indian Context"*, 2nd edition, Macmillan India Ltd., New Delhi.

Stevenson, W.J. *Production/Operations Management*. Irwin/Toppan Publication, Tokyo, Japan.

Acharya, S.S. and Agarwal, N.L. (1999). *Agricultural Marketing in India*, 3rd edition. Oxford and IBH Publishing Co. Pvt. Ltd., New Delhi.

50

Marketing of Processed Food Products: Indian Scenario

Atul Dhingra

Department of Business Management, Chaudhary Charan Singh Haryana Agricultural University, Hisar – 125 004

India has tremendous potential for the development of food processing industry considering the fact that it has emerged as the largest producer of milk, the second largest producer of fruits and vegetables, the largest producer of spices, groundnut, rapeseed and musterd, and the fourth largest producer of wheat. India is among the top ranking nations in terms of total production of various raw materials required for development of food processing industry in general and fruit and vegetable processing industry in particular. Despite major advantage in terms of raw material availability, India is nowhere in the world map in terms of processing when compared with other countries like Brazil, Philippines, Thailand, and Malaysia which are processing 70 per cent, 78 per cent, 80 per cent, and 83 per cent respectively, of their total horticulture produce. The corresponding figure for India is only about 1 per cent.

The inadequate development of food processing sector in the country is due to the poor domestic demand. Traditionally, Indians prefer home cooked food. They always have doubts about the quality of agricultural raw material that go into the production of processed foods and also find such foods very costly. The cost of packaging, advertising and development of brands is also very high in India. Moreover, 70 per cent of people live in rural areas and they do not consume most of the processed food items to any significant extent hence most of the demand comes from urban areas only and that too mainly from metropolitan towns.

The scenario, however, is in for a change due to many factors. Firstly, the population in cities is rising due to migration of people from rural to urban areas. Secondly, females have taken to jobs, which has enhanced their dependence on ready to cook and ready to eat food. Thirdly, packaging innovations have also helped in the preservation of freshness and quality and in increasing the shelf life of processed foods. Finally, the government has reduced the duty on various processed food items, which has helped in reducing their prices.

The processing of agricultural commodities can take any of the following forms:

Food Processing

1. Milling of grains: Wheat flour milling, rice milling, and pulse milling.
2. Manufacture of edible oils and oilcakes from oilseeds.
3. Manufacture of hydrogenated and vanaspati oils.
4. Manufacture of sugar, gur, and khandsari from sugarcane.
5. Manufacture of dairy products like ghee, butter, khoa, milk powder etc.
6. Canning and preservation of fruits and vegetables.
7. Manufacture of juices and other products from fruits and vegetables.
8. Slaughtering of animals and preparation of meat.
9. Canning and preservation of fish.
10. Manufacture of bakery and confectionery products.
11. Tea processing.

12. Coffee curing and roasting.
13. Cashew nut processing.
14. Manufacture of starch.
15. Manufacture of prepared animal feed.

Agro-related Processing (other than food processing)

1. Processing of tobacco leaves and manufacture of tobacco products.
2. Brewing and manufacture of beverages.
3. Manufacture of yarn, cloth and made ups (cotton and jute).
4. Leather processing and manufacture of leather and fur products.
5. Manufacture of wood products, furniture and fixtures.
6. Manufacture of paper and paper products.

Present Status of Indian Food Processing Industry

The food processing industry contributes about 18 per cent to the industrial gross domestic product and employs around 1.5 million persons. It is primarily functioning in informal and unorganised sectors, which accounts for 87 per cent of the processed food items, although bigger players like Hindustan Lever Ltd., Nestle India Ltd., Godrej Foods Ltd., Pepsi Foods Ltd., Hamdard Laboratories, HPMC, NDDB, Markfed, and Pachranga Foods etc. are expanding their shares day by day.

Processing of Wheat

In India, about 90 per cent of the wheat is consumed in the form of chapatis and 10 per cent as bread, biscuits, buns and other bakery products. The first roller flour mill in India was set up in 1880. The number increased to 40 in 1940, 206 in 1970, 232 in 1980, 464 in 1986, and 812 in 1996. At present, around 10.5 million tonnes of wheat is converted to various wheat products by these roller flour mills. In India till 1991, the roller flour mills were regulated by the government through licensing. However, since July 1991, licensing has been abolished for setting up of new flour mills or for expansion in the capacity of existing units. No license is required for the manufacture of wheat products. There are no controls on price and distribution of wheat products either. The mills are free to obtain their requirement of wheat from any source.

Bread and biscuit manufacturing in India is reserved for small scale sector. The production of bread and biscuits, both in organised and unorganised sectors, is estimated to be 14 lakh tonnes and 10 lakh tonnes respectively out of the total biscuits manufactured in India nearly one third is in the organised sector and remaining two thirds in the small scale and unorganised sector.

Processing of Paddy

Paddy rice milling is one of India's largest industries, for the output of this industry exceeds the total of a11 the other food grain processing industries. In 1969, there were 52192 mechanized rice mills of all types in India. This number increased to 123761 in 1988 and further to 138867 in 1996.

Table 50.1: Number of Rice Mills and Roller Flour Mills in India

Year	*Rice Mills*	*Wheat Roller Flour Mills*
1969	52192	202
1970	59022	206
1975	91333	232
1980	92725	232
1986	115641	464
1988	123761	NA
1996	138867	812

Source: Ministry of Food Processing.

Processing of Oilseeds

Oilseeds are a group of farm products that are consumed only after they have been processed. Most of the oil mills in India have capacity of 1 to 5 tonnes per day. Nearly 10 per cent of the oil mills in the country have the capacity of more than 25 tonnes per day. Number of oil mills with a capacity of more than 50 tonnes per day is only around 1 per cent of the total mills.

Table 50.2: Number and Capacity of Oilseed Processing Units in India

Type of Processing Unit	*Number of Units*	*Installed Capacity (million tonnes)*
Ghanies	250000	10.0
Oil mills (Expeller)	15000	25.0
Solvent extraction plants	360	13.0
Refineries	100	0.8
Vanaspati plants	100	1.6

Fruits and Vegetables Processing

The main fruits and vegetables being processed are mango, pineapple, guava, banana, papaya, apple, lemon, orange, and peach in the fruits group and beans, potato, tomato, chili, garlic, and onions in vegetables group. This sector is also attracting lot of foreign investments and till June 1997, 4,098 Industrial Entrepreneur Memoranda envisaging an investment of Rs. 47,146 crore had been received.

Though only an estimated 1.3 per cent of fruits and vegetables are processed in the country but this is not due to lack of installed capacity. There is surplus capacity available but only 35 per cent of this is utilised because of various reasons ranging from narrow product base to the technological obsolescence. The installed capacity in 1980 was 0.27 million tonnes but it quadrupled in 1992. The installed capacity is still rising at a fast rate and till 1996 it has reached approximately 2 million tonnes.

Table 50.3: Installed Capacity of Fruit and Vegetable Processing Industry in India

Year	*Installed Capacity (Metric Tonne/year)*	*No. of Units*	*Production (M.T./Year)*
1981	2,70,000	N.A.	N.A.
1985	4,00,000	N.A.	N.A.
1989	7,08,000	3629	2,40,000
1992	11,08,000	4057	4,69,000
1993	12,60,000	4127	5,59,000
1994	14,02,000	4270	6,80,000

Source: Ministry of Food Processing.

Table 50.4 shows that fruit pulp and fruit juices followed by ready to serve beverages and pickles constitute the major share of fruit and vegetable processing industry.

Table 50.4: Break-up of India's Processed Fruit and Vegetable Products

Products	*% Share*
Fruit juice and Fruit pulp	27
Ready to serve beverages	13
Pickles	12
Jams and Jellies	10
Synthetic syrups	8
Squashes	4
Tomato ketchup and puree	4
Canned vegetables	4
Others	18

Source: Agricultural and Processed Food Products Export Development Authority.

Processed fruits and vegetables also have great potential for exports as India has the advantage of producing vegetables and fruits round the year due to the climatic variations in different regions of the country. There seems to be keen interest among NRIs and other foreign investors in this sector due to ever expanding domestic and international markets. The rising trend of exports in fruits and vegetables is shown in Table 50.5 below:

Table 50.5: Export of Fruits and Vegetables (Rs. Crore)

Items	*1993–94*	*1994–95*	*1995–96**	*1996–97**
1. Fruits and vegetables	414	437	528	579
2. Processed fruits/vegetables	156	249	347	326

*Provisional.

Source: Economic Survey (1997–98), p. 117.

SWOT Analysis of Food Processing Industry

SWOT analysis highlights the strengths, weaknesses, opportunities, and threats for the food processing industry.

Strengths

The major strengths for the industry are availability of adequate raw material and cheap manpower. Food grains, cereals, and fruits and vegetables are available throughout the year in India because of variations in climatic conditions of different regions. India is the second largest producer of fruits and vegetables in the world. There is also no shortage of manpower and the processing industry employs 1.5 million people. Another favourable thing is the acceptance of Indian processed foods specially fruits and vegetables among foreign consumers.

Weaknesses

A large number of constraints are hindering the growth of food processing industry. The main problems of this industry are, low productivity of perennial fruits and plantation crops due to small size of holdings, poor crop management, shortage of planting seeds of improved varieties, post harvest handling which account for 37 per cent of the losses at different stages of storage, grading, and packaging, poor infrastructure, inadequate research and development support, and poor port and transportation facilities. There is also a problem of finance as the banks have been rather slow to enter these areas which are comparatively high risk prone. Land ceiling laws have also acted as hindrance for the industry and taxation in Indian food processing industry is highest in the world. This is proving to be a disadvantage to the exporters who have to compete with international prices. High taxes are hurting the small-scale sector on the one hand and consumers on the other hand as these result in higher final prices.

Processing units agree that uniform quality of processed foods become virtually impossible because of the large variety of products grown by a number of small farmers. Marketing is also not well organised as growers realise only 30 to 40 per cent of the price ultimately paid by the consumers.

Opportunities

The major opportunity is that processing sector is attracting lot of foreign investments and till June 1997, 4,098 Industrial Entrepreneur Memoranda envisaging an investment of Rs. 47,146 crore had been received. HACCP certification scheme has also been launched by the bureau of Indian Standards, which will enhance the marketability of Indian food products in the global market. Rising income levels, increase in nuclear families, and increase in the number of working wives is also increasing demand of processed foods.

Threats

The main threat is that majority of people in India still prefer fresh food grains, fruits and vegetables and are not accepting processed food items. This may result in stagnating sales for processing units. Also, taxes on processed foods are highest in India, which may discourage new entrants and result in low consumption as taxes increase the final price of the product. Lack of skilled manpower and managerial expertise and low capacity utilisation of units are also major areas of worry.

The major cause of concern for the small-scale units in the industry may be the entry of multinational companies which on one hand will increase exports but on the other hand pose threat due to their financial and managerial power.

Strategies for Success

The Government, being well aware of the prospects for development of food processing sector, has taken various initiatives for its promotion. These include setting up of a separate Ministry for Food Processing and promotional bodies such as Marine Products Export Development Authority (MPEDA), Agricultural and Processed Food Products Export Development Authority (APEDA) and National Horticulture Board (NHB).

Beside all these, certain other measures are needed for the success in food processing area. They are, development of infrastructure for post harvest handling and, marketing, product diversification, area expansion and use of latest technology, export enhancement by participation in trade fairs, reducing taxes and providing incentives, intensive training programmes for manpower employed in industry and for educating the farmers, facilities of cold storage at major air and sea ports, and improvement in food and land ceiling laws. Wide publicity, for creating awareness regarding processed foods among Indians and building 'Made in India' image among foreign consumers, is also needed. Creation of National Processing Grid similar to NDDB for eliminating, multiplicity of investment of resources will also help the industry.

References

Acharya, S.S. and Agarwal, N.L. (1999). *Agricultural Marketing in India*, 3rd edition. Oxford and IBH Publishing Co. Pvt. Ltd., New Delhi, p. 121-133.

Bana, S. (1998). Cultivating global markets. *Business India*, p. 63-67.

Bansal, R.K. (1998). Fruits and vegetables processing industry in India. *Indian Food Packer*, p. 33-36.

Bhatti, S. and Verma, U. (1995). *Food and Vegetables Processing Organisation and Institutions*. CBS Publishers and Distributors, New Delhi.

Government of India (1997). Thematic Issues and Sectoral Programmes, 9th Five year plan (1997–2002), Vol. II, Planning Commission, New Delhi, p. 437-651.

Government of India (1998). Economic Survey, (1997–98), Economic Division, Ministry of Finance, New Delhi, p. 113-123.

Subramanyam, K.V. (1998). Horticulture in India: organisation, production, marketing, and processing. *Indian J. Agril. Eco.*, p. 21-23.

Yadav, S.S. (1995). Problems and prospects of export of fruit and vegetables. *Indian J. Agril. Market.*, 9(2); 127-136.

51

Nutraceuticals and Its Implications on Human Health

Bhupendar Singh Khatkar
Department of Food Technology, Guru Jambheshwar University, Hisar – 125 001, Haryana

Introduction

It is well established that every cell in the human body originates from food that we eat. The omnivorous diet of humans provides the diverse blends of nutrients needed for growth, maintenance and overall health. Thus, lack of foods or of essential nutrients in food can cause disorder or illness in our body. In the recent past, research has shown specific evidence to support the belief that certain foods provide medical and/or health benefits including the prevention and treatment of diet related diseases. Such food products are known as nutraceuticals. It has its origin in two words, namely nutrients and pharmaceuticals *i.e.,* nutraceuticals refer to those food products or food ingredients, which in addition to satisfying sensory and basic nutritional needs also provide health benefits including prevention and treatment of diet related disorders, such as colon cancer, heart disease, diabetes, obesity, hypertension, osteoporosis, arthritis, macular degeneration, cataracts, menopausal symptoms, diminished memory, digestive upsets and constipation, headaches, etc. This category of foods is called Foods for Specific Health Use (FOSHU) in Japan. Such food products may range from isolated nutrients, dietary supplement, herbal products and processed foods, *e.g.*, supplemented bakery products, breakfast cereals, soups and beverages. In India, diet related diseases or disorders are common either due to lack of knowledge or due to inability to purchase adequate amount of variety of foods for a balanced diet.

Single and multiple deficiencies of essential dietary elements (*e.g.* essential amino acids, vitamins, minerals, fibre and overall low doses of food intakes in diet) have been found to exert direct bearing on the malfunction of specific organ systems, particularly within the poorer economic group. High income group people in our country, on the other hand, suffer quite often from disorders, such as diabetes, cardiovascular diseases, gastrointestinal disease, obesity and colon cancer due to the habitual over intake of fibre free energy rich modem foods. Appropriate nutraceutical food products can often remedy, such chronic, subacute and acute disorder. It appears therefore logical to develop suitable technologies for nutraceutical food products to meet the dietary need of different groups of people.

Categories of Nutraceuticals

The nutraceutical products are classified as potential and established nutraceuticals. A potential nutraceutical is one that holds promise of a health and for medical benefit. It becomes an established nutraceutical only after sufficient clinical data demonstrate such benefit. A stream of clinical studies continues to define the potential or established clinical benefits of a rapidly growing range of nutraceuticals on an expanding array of specific disease processes and altered health states. Examples include the use of niacin to reduce plasma lipids, folic acid to prevent neural tube defects in babies, fish oil to treat a specific type of nephropathy, magnesium to reduce insulin resistance, and vitamin A to treat measles.

Thus, business opportunities exist for the food industries to manufacture specially designed, value-added products to meet the special nutritional and therapeutic needs of young children, adolescents, women of childbearing age, sports enthusiasts, the middle-aged and elderly. Simply adding the appropriate ingredients, however, cannot develop nutraceuticals. Efforts of processing have to be considered, and the effect of these ingredients on the taste and quality of the final product is also a very important consideration. Regardless of the category of the product, safety must be the first and foremost criteria in today's market. As shown in Table 51.1, nutraceuticals currently fall into four major categories.

Table 51.1: Categories of Nutraceuticals with Specific Health Benefits

Category	*Example of Food Source*	*Health Benefit*
Basic foods	Carrots	Neutralize free radicals which may cause damage to cells
Basic foods with enhanced functional components (Through genetic engineering)	Tomatoes with higher levels of lycopene (carotenoid)	Reduce the risk of prostate cancer
Processed foods	Whole grain bread/high fiber bread	Reduce risk of heart disease
Processed foods with added functional ingredients	Folic acid enriched biscuits	Prevent neural tube defects in babies

Health Benefits of Nutracuticals

As interest in healthy eating has developed in recent years, one of the main responses was the removal of ingredients perceived as unhealthy, such as fat, sugar and salt. Consumers are now more aware of how lifestyle and eating habits influence their health. Because of the negative image of drug use and uncertainty about supplements, nutraceuticals are becoming more popular particularly in economically developed countries. As the incidences of many diseases, including heart diseases, diabetes, arthritis and cancer continue to increase, consumer interest in the therapeutic foods will gain momentum in years ahead. After years of media hype, the public now starting to realize that health can be improved significantly by avoiding junk foods and eating foods that offer a special health benefits. They sought to keep themselves healthy through "positive eating".

Although the connection between diet and health has long been recognized, however, our understanding of the roles that nutrients play at the cellular and subcellular levels continues to grow. It is clear that drugs and nutrients share similarities in biological effects, because they either participate or intervene in selected biochemical and cellular processes. The difference between foods and drugs are important and go beyond their intended use to their biological purpose. Food provides nutrients to fuel all body cell reactions, and nutrients must be continuously replenished. Drugs, taken to cure or treat a disorder, are intended to alter a biological process that has gone awry. Once, a drug has done its job, the body detoxifies it and disposes of it.

Recent studies have indicated that a numerous type foods, such as cereal grains, legumes, oilseeds, fruits and vegetables contain wide variety of phytochemicals that are known to deliver health benefits as outlined below to human beings. Technically, the term phytochemical refers to every naturally chemical substance present in plants especially those that are biologically active. Major phytochemicals include phenolic acids, flavonoids (such as flavones, isoflavones and flavonols) coumarin derivatives and polyphenols. These phytochemicals are used in preparation of nutraceutical products in the form of isolated nutrients, dietary supplement and food additives.

Anti-carcinogenic Effect

Cancer is still the number one health concern of majority of people worldwide. Breast cancer is quite common among women. On the other hand, prostate cancer is the most prevalent form of cancer among men. It is well known that Asian women, who typically eat a soy-based diet, have a much lower incidence of breast cancer than Western women. Daidzein and genistein are the two primary isoflavones found in soybeans. These compounds may reduce the risk of a number of cancers, including those of the breast, prostate, lung, colon, rectum and stomach. While it appears that daidzein is more bioavailable than genistein. However, most of the research interest in the anticancer effects of soybean isoflavones has centered on genistein. Genistein is thought to act against cancer in several ways:

1. By interfering with cancer promoting enzymes. Protease inhibitors found in soybean cause an irreversible suppression of the carcinogenic process and have been shown to inhibit carcinogenic-induced protease activity.
2. By blocking the activity of hormones in the body. Isoflavones exert anti tumor effect by blocking estrogen receptors.
3. By interfering with the process by which tumors receive nutrients and oxygen. Genistein is found in soy, clove and a few other green plants. However, soy-based foods represent the only practical way the food processor can incorporate genistein in the processed food. Any type of soyfood including soymilk, tofu, tempeh, textured vegetable protein or whole soybean may offer cancer-prevention benefits.

Cereal dietary fiber particularly wheat bran inhibit carcinogenesis by binding bile acids, which are thought to be cancer promoting agents. The other mechanisms by which cereal fibers play anti-carcinogenic role include:

1. Reduction of transit time in colon. Cereal brans, especially those that have a higher lignin fraction, such as wheat bran, are well known to contribute to faster stool passage and stool bulking effects in humans.
2. Cereal fibers serve as a substrate for bacterial fermentation, which contributes to increase stool bulking and the production of volatile fatty acids (VFAs). VFAs such as butyrate have demonstrated anticarcinogenic effects in vitro and may be a preferred fuel source for colon epithelial cells. An indirect effect of VFAs is to lower gut -pH, which should slow the conversion of primary bile acids to secondary bile acids via pH -dependent bacterial enzymes.

References

Chauvan, J.K. and Kadam, S.S. (1993). Nutritional enrichment of bakery products by supplementation with nonwheat flours. *Critical Reviews in Food Science and Nutrition*, 33(3): 189-226.

Stephen, L. and Defelice, M.D. (1998). Where is the first nutraceutical cereal company? *Cereal Foods World*, 43(5): 351-354.

Tungland, B. (2002). Examining the role of pre- and probiotics in digestive health. *Nutraceuticals World*, pp. 50.

Wilson, D. (1998). Nutraceuticals: A natural choice for grain based products. *Cereal Foods World*, 43(9): 718-719.

52

Role of Dietary Fibre in Human Health

Bhupendar Singh Khatkar
Department of Food Technology, Guru Jambheshwar University, Hisar – 125 001, Haryana

A complex series of changes in the dietary patterns have been witnessed during urbanization in our country. Recently, intake of fibre free energy-concentrated modern foods such as refined sugar, sweets, ice-cream, soft drinks, jams, cakes, white breads, meat, fish, poultry and fried foods has been increased. The consumption of high fibre starchy foods, mainly whole wheat flour and brown rice, has gone down considerably and have been replaced by low fibre refined wheat flour and polished rice. Depletion of fibre in our present day diet has resulted in emergence of several diseases including diabetes, obesity, cardiovascular disease, gastrointestinal diseases, appendicitis, cancer of colon and constipation. All these diseases are more commonly noticed in urban population than their counterpart in rural masses (Kumar and Jha, 2002; Gorden, 1989; Kelsey, 1978; Shukla, 1998; Schneeman, 1986).

There is no single definition of dietary fibre. However, most researchers agree that total dietary fibre includes the portion of plant products that cannot be digested in the human digestive system. There are two components of total dietary fibre; water-insoluble and water-soluble. Insoluble fibre includes cellulose, hemicellulose and lignin, which are found mostly in cereals. Soluble fibres, although not fibrous in nature including pectin, gums and mucilage are found mostly in fruits, vegetables and beans (Shukla, 1998). Fibre has two major functional properties–its absorption capacity (chelating ability) and its water-binding properties. The cardiovascular diseases relate to the ability of fibre to absorb or bind materials, which reduce blood cholesterol. The gastrointestinal diseases appear to be related to fibre's water-binding properties. Obesity is indirectly related to both properties, but mostly to factors in the diet itself (Scala, 1976).

Fibre and Obesity

The interaction of dietary fibre and obesity has received increasing attention during the past decade. There is some indication that fibre-rich diet forces one to eat more slowly than would be the case with a fibre depleted diet. This is because such fibre-rich diets are calorically more dilute, thus a diet rich in fibre could be an obstacle to overeating and obesity (Vanl, 1978). Therefore, the arguments for using fibre to combat obesity centres on the bulking properties of fibre, which produces chewy and low-density foods with greater satiety value (Scala, 1976).

Fibre and Cardiovascular Diseases

Atherosclerosis, the principal cause of the most cardiovascular diseases, is modern man's greatest scare. Fifty per cent of deaths past the age of 45 can be directly attributed to it. It is characterized by material deposited on and within the arterial walls called plaque, which consists of fat, protein and cholesterol (its major component): Cholesterol is not only ingested with foods but is synthesised in large quantities by our body tissues, especially liver. Cholesterol is converted to bile acids that are secreted into the small intestine where they help in the digestion and absorption of fat and dietary cholesterol. The mechanism by which fibre exerts this influence involves the bile acids. People on a high fibre diet excretes more bile acids, more sterols and fat. This implies that fibre binds the bile acids and sterols, thereby preventing bile acids' reabsorption. Since bile acids excretion is the main elimination pathway of internally produced cholesterol, the increase in excretion reduces the cholesterol pool. This reduction is followed by a lowering of blood cholesterol and hence increases in life expectancy (Anderson, 1977; Leveille, 1976; Roger and Duensing, 1984). Research studies show that life is expected more in population, which eat a diet rich in starchy foods such as cereals, pulses, roots, fruits and vegetables than in those that eat a western-type diet (Vorster, Ubbink, Lotter, and Steyn, 1987).

With increasing levels of carbohydrates in the diet, there is a proportionate decrease in fat. The high-carbohydrates diet, therefore, seems to provide an additional benefit for the diabetic. Low-fat intakes in association with a reduced level of blood lipids lead to lesser incidence of cardiovascular disease. Such a high-carbohydrate, high-fibre, low fat diet favourably influences the reduction of risk factors for the development of atherosclerosis (Bierman, 1985).

Fibre and Diverticular Disease

The disease is considered a deficiency disease of urban civilization caused by the refining of foods, which entails the removal of plant fibre from the diet (Smith, Drummand and Eastwood, 1981; Wrick, Robertson, Vansoert, Lewis, Rivers, Roe, and Hackler, 1983; Moore, 1978). The disease involves ballooning that develops in weak areas in the bowel wall. When they are numerous and become inflammed, diverticulitis develops. This condition is often accompanied by pain on the lower left side, altering diarrhoea and constipation with flatulencel (Kumar and Jha, 2002). Statistical correlation indicates that people who have diverticular disease are likely to have appendicitis, hemorrhoids and varicose veins. These associations indicate the hereditary role, but fibre apparently also plays a significant role. Research reports recommend dietary fibre, more specifically cereal fibre for curing diverticulitis (Leveille, 1976).

Fibre and Cancer of Colon

Intestinal cancer is the second ranked killer among adults (lung cancer being first). The urban population is more likely to be affected by intestinal cancer (Scala, 1976). Some health professionals believe that dietary fibre may provide protection from cancer of the colon and the rectum. Intake of about 20–36 g of dietary fibre per day would be beneficial. This can be obtained by eating fresh fruits, vegetables and cereals. The hypothesis relating dietary fibre to colon cancer presumes that the slow movements of the faeces which occurs with a low-fibre diet allows more time for any carcinogens presents in the colon to initiate cancer. In addition, according to this hypothesis, the extra water, bile acid, salts and fat bound to dietary fibre act as solvent to remove a wide variety of chemical factors, which might be carcinogenic. Also a high fibre diet may alter type and number of microorganisms in the colon, and may possibly inhibit production of potential carcinogens (Leveille, 1976).

Fibre and Constipation

Fibre's value for relieving constipation lies in its ability to increase water content of the faeces producing bulkier but soft and well formed stools (Kelsey, 1978). It has been proposed that the increased volume and softness of the stools, by reducing straining during defecation is a factor in preventing hemorrhoids and varicose veins. Available evidence suggests that the fibre from starchy foods and from cereals in particular, relative to quantity eaten, is more efficacious in restoring normal bowel movement than is the fibre from fruit and vegetables (Kelsey, 1981). It has repeatedly been shown that the addition of two grams of crude fibre to the diet enormously improves gastrointestinal function (Kelsey, Jacob, and Prather, 1979; McCounell, Eastwood, and Mitchell, 1974).

Fibre and Diabetes

Diabetes is the body's inability to utilize sugar in a normal way. It is a common disease among Indian population and primarily related to hereditary and diets (Anderson, 1981). People who eat modem diet consume over doses of sugar through tea, coffee, soft drinks, sweets and ice cream. Refined form of sugar in these items is absorbed rapidly and the blood sugar level rise is faster which require more insulin secretion from the pancreas. This puts extra load on pancreas, whereas in high fibre diet, sugar is absorbed slowly and requires less insulin for digestion and therefore, puts lesser load on pancreas. It may play an important role in precipitation of diabetes (Crapo, Kolterman, Waldeck, Reaven, and Olefsky, 1980). High carbohydrate, high-dietary fibre intake seems to benefit the diabetic patient by reducing the serum glucose and insulin response. However, researchers are unsure of mechanism by which such benefits are obtained (Hollenbeek, Leklem, Riddle, and Connor, 1983; Jankin, Wolever, Bacon, Ninchem, Lees, Rowden, Love and Hockaday, 1980).

Fibre and Dermatological Disorders

Frequent consumption of sweets, ice creams and other sugar rich diets causes pimples on face. Researchers believe that excess consumption of sugar has strong correlation with aging and development of wrinkles on face as well as in the neck region. Skin diseases are more common in people who eat non-vegetarian and refined carbohydrate foods (Scala, 1976). The vegetarian diet has been found to maintain a skin condition better than non-vegetarian and high sugar diet. For healthy and beautiful skin on must eat frequently fresh fruits, vegetables, high fibre cereals and avoid fat rich products (Leveille, 1976).

Food Source of Dietary Fibre

Whole cereal grains, brown rice (unpolished rice), whole-wheat flour, whole pulses, fresh fruits and vegetables are the potential food sources of dietary fibre. In both fresh fruits and vegetables, the total dietary fibre appears to be relatively low, because of their high water content. Potatoes and starchy vegetables supply significant amounts of fibre (Kelsey, Jacob, and Prather, 1979). Substitution of whole wheat flour for white flour provides 3-fold increase in total dietary fibre. Consumer interest in this subject has promoted a number of food industries to develop fibre-rich foods such as popped, puffed and flaked cereals, high-fibre extruded products and whole wheat flour bread (Collins, Kalantari, and Post, 1982; Cadden, Sosulsky and Olson, 1983; Dreher and Padmanaban, 1983). Fibre from wheat bran, oat bran, rice bran and isabgol are successfully tried in biscuits, cookies, yogurt, meat products, chocolate and soup recipes (Plizzoto, Tinsley, Weber, and Berry, 1983; Toma, Orr, D'Applonia, Dintzis, and Tabekhia, 1979).

Recommendations

It will be right to remind the reader that many nutrients including vitamins, minerals and proteins have been found to be harmful in large dosages while being essential to health. The consumption of nutrients must be rational and based on reasonable scientific evidence (Toma, and Curtis, 1986; Riaz, 1999). In the absence of such evidence, moderation should be adopted. A daily intake of 25–30 g fibre can be recommended half from fruits and vegetables and remaining half from grains (Scala, 1976; Leveille, 1976). A variety of whole grain products, fruits and vegetables will ensure a good mixture of fibre constituents and make a positive combination for the overall nutritional value of our diet, resulting in significant health benefits.

References

Anderson, J. (1981). Can dietary fibre playa role in the treatment of diabetes. *Therapacia*, 4: 44.

Anderson, J.W. (1977). High polysaccharides diet studies in patients with diabetes and vascular disease. *Cereal Foods World* 22: 12.

Bierman, E.L. (1985). Diet and diabetes. *Am. J Clin. Nutri.*, 41: 1113.

Cadden, A.M., Sosulsky, F.W. and Olson, J.P. (1983). Physiological response of rats to high-fibre bread diets containing several sources of hulls or bran. *J. Food Sci.*, 48: 1151.

Collins, J.L., Kalantari, S.M. and Post, A.R. (1982). Peanut hull flour as dietary fibre in wheat bread. *J. Food Sci.*, 47: 1899.

Crapo, P.A., Kolterman, O.G., Waldeck, W., Reaven, G.M. and Olefsky, J.M. (1980). Postprandial hormonal responses to different types of complex carbohydrates in individuals with impaired glucose intolerance. *Am. J. Clin. Nutri.*, 33: 1723.

Dreher, M.L. and Padmanaban, G. (1983). Sunflower hull flour as a potential dietary fibre supplement. *J. Food Sci.*, 48: 1151.

Gorden, D.T. (1989). Functional properties vs physiological action of total dietary fibre. *Cereal Foods World*, 34(7 -12): 517.

Hollenbeek, C.B., Leklem, J.E., Riddle, M.C. and Connor, W.E. (1983). The composition and nutritional adequacy of subject selected high carbohydrates, low fat diet in insulin dependent diabetes mellitus, *Am. J. Clin. Nutri.*, 38: 41.

Jankin, D.J.A., Wolever, T.M.S., Bacon, S., Ninchem, R., Lees, R., Rowden, R., Love, M. and Hockaday, T.D.R. (1980). Diabetic diets: High carbohydrate combined with high-fibre. *Am. J. Clin. Nutri.*, 33: 1729.

Kelsey, J.L. (1978). A review of research on effects of fibre intake on man. *Am. J. Clin. Nutri.*, 31: 142.

Kelsey, J.L., Jacob, R.A. and Prather, E.S. (1979). Effects of fibre from fruits and vegetables on metabolic response of human subjects. *Am. J. Clin. Nutri.*, 32: 2307.

Kelsey, T.L. (1981). Effect of diet fibre level on bowel function and trace mineral balances in human subjects. *Cereal Chem.*, 58: 2.

Kumar, S. and Jha, Y.K. (2002). Nutraceuticals and its implication in human health. *Bev. and Food World*, pp. 34.

Leveille, G.A. (1976). Dietary fibre. *Cereal Foods World*, 21(1-12): 255.

McCounell, A.A., Eastwood, M.A. and Mitchell, W.D. (1974). Physical characteristics of vegetable foodstuffs that could influence bowel function. *J. Sci. Food Agri.*, 25: 1457.

Moore, W.E.C. (1978). Workshop II-Fibre and bacteria in the gut: Summary and recommendation. *Am. J. Clin. Nutri.*, 31: 5111.

Plizzoto, L.M., Tinsley, A.M., Weber, C.W. and Berry, J.W. (1983). Dietary fibres in Muffins, *J. Food Sci.*, 48: 111.

Riaz, M.N. (1999). Soybeans as functional foods. *Cereal Foods World*, 44: 88.

Roger, M.B. and Duensing, W.J. (1984). Processing and dietary fibre ingredient applications of corn bran. *Cereal Foods World*, 34(7-12): 535.

Scala, J. (1976). Has fibre a role to play in preventive medicine. *Cereal Foods World*, 21(1-12): 356.

Schneeman, E.O. (1986). Dietary fibre: Physical and chemical properties, method of analysis and physiological effects. *Food Technol.*, 40(1-6): 104.

Shukla, T.P. (1998). Nutraceutical novelties: trends in proprietary technology. *Cereal Foods World*, 43: 388-389.

Smith, A.N., Drummand, E. and Eastwood, M.A. (1981). The effect of coarse and fine Canadian red spring wheat and French soft wheat bran on colonic motility in patient in the diverticular disease. *Am. J Clin. Nutri.*, 34: 2460.

Toma, R.B. and Curtis, D.J. (1986). Dietary fiber: Effect on mineral bioavailability. *Food Technol.*, 40(1-6): 111.

Toma, R.B., Orr, P.H., D'Applonia, B., Dintzis, F.R. and Tabekhia, M.M. (1979). Physical and chemical properties of potato peel as a source of dietary fibre in bread. *J. Food Sci.*, 44: 805.

Vanl, T.B. (1978). Dietary fibre and obesity. *Am. J Clin. Nutrition*, 31: 543.

Vorster, H.H., Ubbink, J.B., Lotter, A.P. and Steyn, H.S. (1987). Effects of two dietary fibre supplements in healthy volunteers on a western diet. *Nutrition Reports Int.*, 36(5): 971.

Wrick, K.L., Robertson, J.B., Vansoert, P.J., Lewis, B.A., Rivers, J.M., Roe, D.A. and Hackler, L.R. (1983). The influence of dietary fibre source on human intestinal transit and stool output. *J. Nutri.*, 113: 1464.

Index